Optical Nanoscopy and Novel Microscopy Techniques

Optical Nanoscopy and Novel Microscopy Techniques

Edited by Peng Xi

CRC Press is an imprint of the
Taylor & Francis Group, an **informa** business

First published in paperback 2024

First published 2015
by CRC Press
2385 NW Executive Center Drive, Suite 320, Boca Raton FL 33431

and by CRC Press
4 Park Square, Milton Park, Abingdon, Oxon, OX14 4RN

CRC Press is an imprint of Taylor & Francis Group, LLC

Publisher's Note
The publisher has gone to great lengths to ensure the quality of this reprint but points out that some imperfections in the original copies may be apparent.

ISBN: 978-1-4665-8629-1 (hbk)
ISBN: 978-1-03-292467-0 (pbk)
ISBN: 978-0-429-16878-9 (ebk)

DOI: 10.1201/b17421

Visit the Taylor & Francis Web site at
http://www.taylorandfrancis.com

and the CRC Press Web site at
http://www.crcpress.com

Contents

Preface

Microscopy is at the forefront of multidisciplinary research. It was developed by physicists, made specific by chemists, and applied by biologists and doctors to better understand the function of the subunits of the human body. For this very reason, the field has been revolutionized constantly in past decades.

The objective of this book is to choose some of those revolutionary ideas and help a general audience from broad disciplines to achieve a fundamental understanding of these technologies and to better apply them in their daily research. It is organized as follows: Chapters 1–3 are on super-resolution optical microscopy, in which the targeted modulation, such as STED and SIM, or the localization methods, such as PALM, are discussed; Chapters 4–7 are on the novel development of fluorescent probes, such as organic small-molecule probes, fluorescent proteins, and inorganic labels such as quantum dots; Chapters 8–10 are about advanced optical microscopy, such as fluorescence lifetime imaging, fiber optic microscopy, scanning ion conductance microscopy, and the joining of optics and acoustics—photo-acoustic microscopy. Following each chapter is a detailed list of references so that the readers can use this book as a review of the relevant fields. In addition, to confirm understanding of the materials presented, we provide several problems after each chapter.

I express my deepest appreciation to the contributors of each chapter. Usually readers only remember the name of the editor, but the contributors are the real teachers and where the credit is due. I thank Li-Ming Leoung for her invitation to edit the book and for continuous support during the manuscript's preparation. I thank Amit Lal for proofreading the manuscript. I would like to thank all the CRC editorial team for their professional service, which is essential for the book to reach readers. Also, my sincere gratitude to my wife, Guanqin Chang, and our daughter, Yueming Xi, for their understanding and for sacrificing their quality time with me.

And thank you, my dear reader, for choosing this book. I guarantee that it will be a delightful and resourceful journey.

Peng Xi
Zhong Guan Yuan, Peking University

Contributors

Hao Chang
Institute of Biophysics
Chinese Academy of Sciences
Beijing, China

Peng R. Chen
College of Chemistry and
Molecular Engineering
Peking University
Peking, China

Dan Dan
Xi'an Institute of Optics and
Precision Mechanics
Chinese Academy of Sciences
Xi'an, China

Yichen Ding
Peking University
Peking, China

Zhicong Fei
IMM
Peking University
Peking, China

Jianhong Ge
State Key Laboratory of Modern
Optical Instrumentation
Zhejiang University
Hangzhou, China

Yuchun Gu
IMM
Peking University
Peking, China

Jin U. Kang
Department of Electrical and
Computer Engineering
Johns Hopkins University
Baltimore, Maryland

Fu-Jen Kao
Institute of Biophotonics
National Yang-Ming University
Taipei, Taiwan

Cuifang Kuang
State Key Laboratory of Modern
Optical Instrumentation
Zhejiang University
Zhejiang, China

Ming Lei
Xi'an Institute of Optics and
Precision Mechanics
Chinese Academy of Sciences
Xi'an, China

Changhui Li
Department of Biomedical
Engineering
Peking University
Peking, China

Jie Li
College of Chemistry and
Molecular Engineering
Peking University
Peking, China

Po-Yen Lin
Institute of Physics
Academia Sinica
Taipei, Taiwan

Xuan Liu
Biomedical Engineering
Michigan Technological University
Houghton, Michigan

Yujia Liu
Faculty of Science
Macquarie University
Sydney, Australia

Qiushi Ren
Department of Biomedical
Engineering
Peking University
Beijing, China

Hui Shi
IMM
Peking University
Peking, China

Andrew M. Smith
Department of Bioengineering
University of Illinois at
Urbana-Champaign
Champaign, Illinois

Yujie Sun
College of Chemistry and
Molecular Engineering
Peking University
Peking, China

Jie Wang
College of Chemistry and
Molecular Engineering
Peking University
Peking, China

Peng Xi
Department of Biomedical
Engineering
Peking University
Peking, China

Hao Xie
Department of Biomedical
Engineering
Peking University
Peking, China

Pingyong Xu
Institute of Biophysics
Chinese Academy of Sciences
Beijing, China

Maiyun Yang
College of Chemistry and
Molecular Engineering
Peking University
Peking, China

Yi Yang
College of Chemistry and
Molecular Engineering
Peking University
Peking, China

Baoli Yao
Xi'an Institute of Optics and
Precision Mechanics
Chinese Academy of Sciences
Beijing, China

Mohammad U. Zahid
University of Illinois at
Urbana-Champaign
Champaign, Illinois

Mingshu Zhang
Institute of Biophysics
Chinese Academy of Sciences
Beijing, China

Yanjun Zhang
Tianjin Medical University
Tianjin, China

chapter one

Optical nanoscopy with stimulated emission depletion

Peng Xi, Hao Xie, Yujia Liu, and Yichen Ding

Contents

1.1 Introduction

Since its invention in the seventeenth century, optical microscopy has been applied to biological study. For the past 400 years it has been at the forefront of biological study, from the macroscopic level to the microscopic, cellular level. For "seeing is believing," optical microscopy has been widely extended to almost all disciplines of scientific research. Yet, due to the fact that optics has a wave nature, the diffraction limit becomes the resolution barrier. In the past decade, numerous super-resolution techniques have been created. Stimulated emission depletion (STED) is one of the techniques that employs the feature of fluorescence to generate a smaller effective point spread function (PSF), or PSF engineering.

In this chapter, we first discuss the diffraction limit and its appearance as a PSF; then, the STED process in a single molecule energy level, as well as its behavior in a fluorescent medium. The combination of other techniques with STED, such as time-gating, iso-STED, STED and 4Pi, two-photon excitation, and self-aligned STED techniques are discussed. Next, the technical notes on the experimental instrumentation of STED are presented. Finally, the applications of STED to a variety of biological specimens are listed.

1.2 *Point spread function and diffraction limit*

1.2.1 *Diffraction limit criteria*

Before breaking it, we need to investigate how the barrier of optical diffraction limit is formed. We need to take a history tour, to a beautiful small town in Germany called Jena. We understand that the development of optics requires three key elements: theory, engineer, and material. The time was back in the 1870s, when three giants met magically here. Ernst Karl Abbe, an optical theory master; Carl Zeiss, a great founder of optical engineering; and Otto Schott, a pioneer in optical materials. They have made concerted efforts to lead optics to a higher level for the whole world. Nowadays, the enterprises founded by Schott and Zeiss, respectively, are still renowned around the world.

When light propagates through a barrier such as the aperture of objective, diffraction happens. With wave optics, one can describe the optical distribution of the focal plane as

$$I(0,\upsilon)=\left[\frac{2J_1(\upsilon)}{\upsilon}\right]^2 I_0, \tag{1.1}$$

where

$$\upsilon=\frac{2\pi}{\lambda}NA\cdot r,$$

$NA = n\sin\alpha$ is the numerical aperture of the objective, and r is the radius of the point of interest. I_0 and λ are the intensity and wavelength of the incident beam, respectively. This gives us a Gaussian-like function as illustrated in Figure 1.1, called the point spread function (PSF), that is, the image (spread function) of a perfect point (or equivalently, a planar wave). It is known as Airy pattern, or Airy disk in the two-dimensional form.

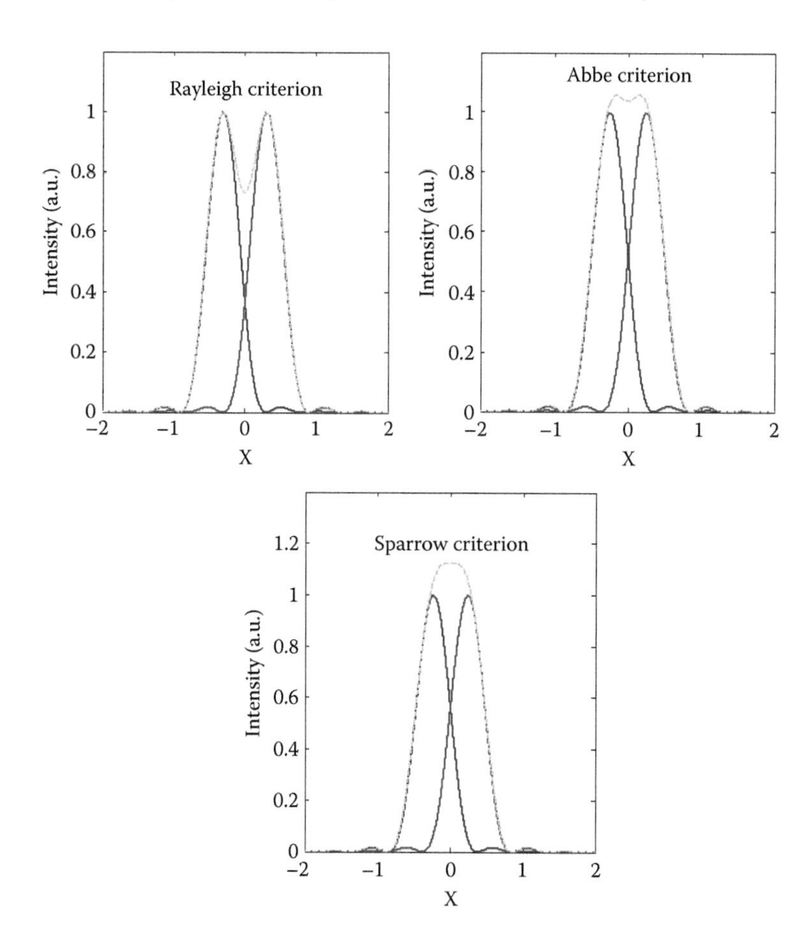

Figure 1.1 **(See color insert.)** Comparison of the various diffraction limit criteria.

As one can see immediately from the plot, the closer two PSFs are, the more difficult it is to separate them. Lord Rayleigh formulated the Rayleigh criterion (Lord Rayleigh 1896):

$$d_{\text{Rayleigh}} = 0.61\frac{\lambda}{NA} \tag{1.2}$$

This criterion is measured from the maximum of the PSF to the first minimum. More technically, the full width at half maximum (FWHM) is often used, which can be expressed as:

$$d_{\text{FWHM}} = 0.51\frac{\lambda}{NA} \tag{1.3}$$

By putting two PSFs at the distance of FWHM, one can notice that, there is a dip of ~3%. A more elegant expression is given by Abbe (Abbe 1873):

$$d_{\text{Abbe}} = \frac{\lambda}{2NA} \tag{1.4}$$

Clearly, to distinguish the two points, they cannot get too close to each other. However, there is still ambiguity as to how well these combined PSFs can be resolved. Alternative evaluations are like Sparrow's resolution limit (Sparrow 1919), which states that the resolution limit is where the dip disappears (becomes flat, or the second-order derivative equals zero).

$$d_{\text{Sparrow}} = 0.47\frac{\lambda}{NA} \tag{1.5}$$

In a confocal setup, a pinhole is placed before a point-detector (such as a photomultiplier tube, PMT, or an avalanche photodiode, APD). Under ideal conditions (the pinhole is infinitely small), the FWHM of such a system can be described as:

$$d_{\text{Confocal}} = \frac{0.51\lambda}{\sqrt{2}NA} \approx \frac{0.36\lambda}{NA} \tag{1.6}$$

Note that this indicates the detector cannot get any photon since its size equals to zero. Increasing the pinhole will get more signal, with a scarification of resolution.

1.2.2 *PSF Measurement*

The PSF of a confocal system is the excitation PSF modulated by the pinhole's modulation spread function (MSF):

$$\text{PSF}_{\text{Confocal}} = \text{PSF}_{\text{Excitation}} \cdot \text{MSF} \tag{1.7}$$

When the pinhole size is a point (delta function), the MSF equals the PSF at the fluorescent wavelength. Since practically the size of the confocal pinhole is not infinitely small, the MSF can be calculated with the two-dimensional convolution of the PSF, and the size of the pinhole is given by:

$$\text{MSF} = \text{PSF}_{\text{Detection}} \otimes \text{Pinhole} \tag{1.8}$$

where ⊗ denotes the convolution process.

The measurement of the PSF is usually much more difficult than imaging a specimen. The measured PSF is actually the convolution of the PSF of the imaging system with the object:

$$PSF_{Measured} = PSF_{System} \otimes Obj \tag{1.9}$$

Therefore, it is only when the object equals to a delta function that the measured PSF equals the PSF of the system. However, practically in STED, in order to overlap the excitation PSF with the depletion doughnut PSF, 80-nm gold nanoparticles are often used. Since only the tip of the nanosphere contributes the reflection signal, it works effectively as a delta function in measuring the PSFs for overlapping purpose.

On the contrary, when benchmarking the resolution, 20-nm or 40-nm fluorescent nanobeads are employed. In this situation, the excitation will ignite the entire bead; therefore, the size of the bead needs to be considered. In order to get accurate measurement of the PSF, the beads need to be very sparse to avoid self-organization/accumulation. Consequently, the signal can only be obtained from a single bead, in which 20-nm beads have eight times fewer fluorescent molecules than that of the 40-nm counterparts. Moreover, when measuring the PSF, the field of view is restricted to a region not much larger than a focal size (three to four orders of magnitude smaller than we normally collect an image), making the photobleaching severe. As STED induces more photobleaching because of switching on-off operation, it is very crucial to control the excitation to its minimal level of detection for smaller beads, when STED resolution needs to be measured. The same requirement holds true to the imaging of fine subcellular organelles; this is discussed in Section 3.5.

1.3 Stimulated Emission Depletion (STED)

To evaluate the best resolution that an optical microscope can get, Ernst Abbe developed the famous Abbe criteria. This is a double-sided sword: it gives the microscopic engineer a clear goal to reach, yet also set a clear barrier.

The Abbe diffraction limit states basically that by focusing ONE incident beam, we cannot get a focal spot smaller than Equation (1.4).

Further, in confocal, we have seen that through setting a pinhole to block the peripheral contribution, better resolution can be achieved. However, because the pinhole is also diffraction modulated, its effect on shrinking the resolution is limited.

So here is the question: Is it possible for us to use TWO beams, with one generating the excitation PSF, and the other functioning as the pinhole of confocal to block (switch off) the peripheral contribution? Further, is it possible to apply a very "hard" block on the focal spot, so that the PSF that results can be infinitely improved?

This question was answered in 1994 by Stefan W. Hell (Hell and Wichmann 1994). Of the many optical nanoscopy techniques, STED was the first to be invented. It was successfully demonstrated experimentally in 1999 with a spatial resolution of 106 ± 8 nm (Klar and Hell 1999).

Back to our question:

1. Can we switch off or modulate the fluorescent light emission, by using another light?
2. Can we generate a certain (circular for example) distribution around the periphery of the PSF, to apply the mechanism of Question 1?

1.3.1 The STED Process

As fluorescence can be used to tag the subcellular organelles, it plays a critical role in biological microscopy. Fluorescence is convenient also because it absorbs high-energy photons and emits lower-energy photons; therefore, even if the organelle can only absorb very little fluorescent dye, it can still be visualized. This is far superior to the chromatic dyes, with which sufficient concentration is necessary for visualization.

Because of its efficient photon conversion capability, fluorescence is also the fundament of laser material: before lasing, the laser has to fluoresce first. As fluorescence is a spontaneous emission, it has a broad emission spectrum. Then, through continual popping the electron to upper state, one can maintain a situation that there are always electrons at higher energy level, which is called population inversion. At this point, if a photon with energy $h\nu = E_1 - E_0$ is presented, then stimulated emission can happen, which generates an identical photon. This means that its energy (spectrum), polarization, propagation direction, and phase, are all identical.

Now, with a simple dichroic filter, we can BLOCK the stimulated emission from spontaneous emission. Or, we can switch off the fluorescence with stimulated emission. Hence the name STED.

If fluorescent dyes were once the leading material for lasers, then all sorts of laser medium can function as efficient dyes for STED nanoscopy, provided that they can be fabricated small enough. STED has achieved a lateral resolution of 5.8 ± 0.8 nm for NV-center nanodiamonds (Rittweger et al. 2009).

For an organic fluorophore with energy level illustrated in Figure 1.2, the interaction between absorption, thermal quenching, vibrational relaxation, and spontaneous and stimulated emission can be described as (Schrader et al. 1995):

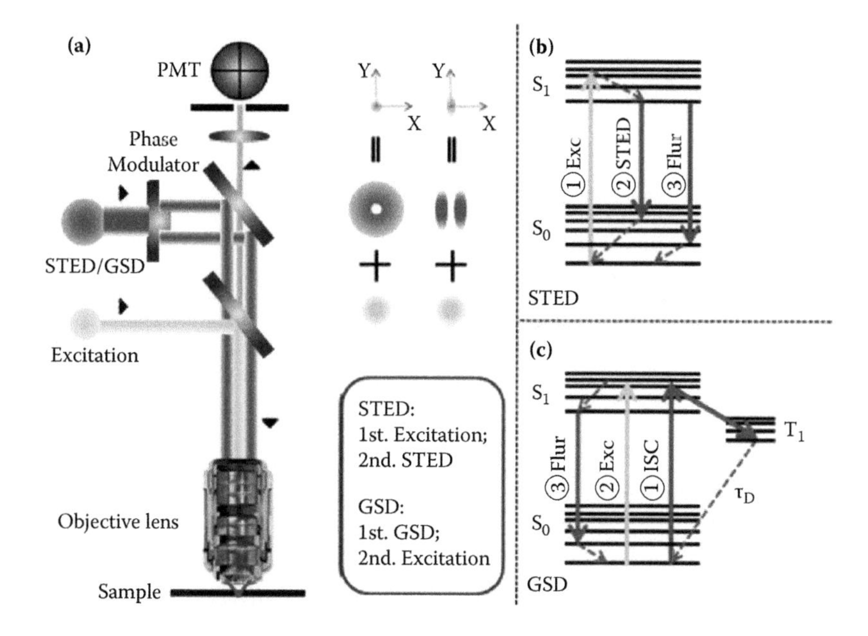

Figure 1.2 (a) The diagram of STED and GSD. (b) Process of STED. (c) Process of GSD.

$$\frac{dn_0}{dt} = h_{\text{exc}}\sigma_{\text{exc}}\left(n_1^{\text{vib}} - n_0\right) + k_{\text{vib}}n_0^{\text{vib}} \tag{1.10a}$$

$$\frac{dn_0^{\text{vib}}}{dt} = h_{\text{STED}}\sigma_{\text{STED}}\left(n_1 - n_0^{\text{vib}}\right) + \left(k_{fl} + k_Q\right)n_1 - k_{\text{vib}}n_0^{\text{vib}} \tag{1.10b}$$

$$\frac{dn_1^{\text{vib}}}{dt} = k_{\text{vib}}n_1^{\text{vib}} - h_{\text{STED}}\sigma_{\text{STED}}\left(n_1 - n_0^{\text{vib}}\right) - \left(k_{fl} + k_Q\right)n_1 \tag{1.10c}$$

$$\frac{dn_1^{\text{vib}}}{dt} = h_{\text{exc}}\sigma_{\text{exc}}\left(n_0 - n_1^{\text{vib}}\right) - k_{\text{vib}}n_1^{\text{vib}} \tag{1.10d}$$

Therefore, the stimulated emission dominates the emission process only when

$$h_{\text{STED}}\sigma_{\text{STED}}n_1 \gg \left(k_{fl} + k_Q\right)n_1, h_{\text{STED}}\sigma_{\text{STED}}n_0^{\text{vib}}, k_{\text{vib}}n_1^{\text{vib}} \tag{1.11}$$

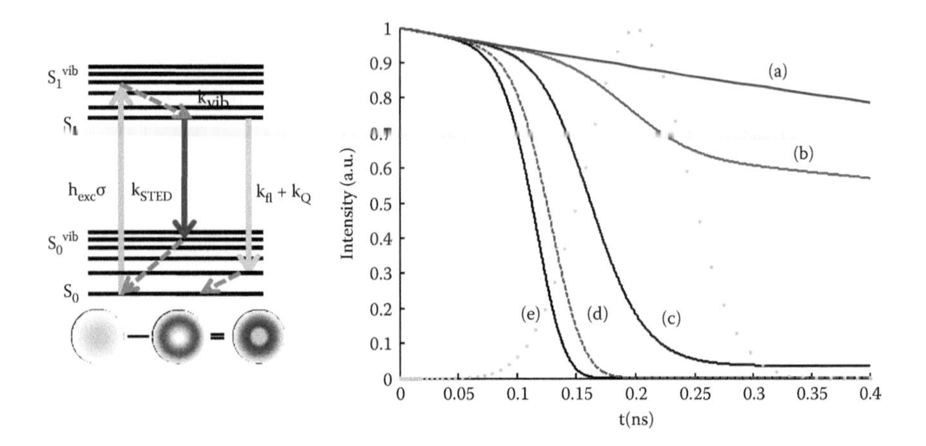

Figure 1.3 Jablonski diagram of an organic fluorophore, and the population of the first excited state as a function of time when exposed to a STED beam of 100 ps with peak intensities of (a) 0; (b) 10, (c) 100, (d) 500, and (e)1000 MW/cm².

where $(k_{fl} + k_Q)n_1$ is the possibility of spontaneous emission and quenching, and $h_{STED}\sigma_{STED}n_0^{vib}$ is the possibility of reverse excitation $S_0^{vib} \rightarrow S_1$, and $k_{vib}n_1^{vib}$ is the vibrational relaxation possibility $S_1^{vib} \rightarrow S_1$.

We can evaluate the depletion process by putting typical physical parameters:

The cross-section of stimulated emission process $\sigma_{STED} \approx 3 \times 10^{-17}$ cm², $k_{fl} = 1/\tau_{fl} = 3 \times 10^8$ s^{-1}, $k_{vib} = 1/\tau_{vib} = 1 \times 10^{12}$ s^{-1}, the photon numbers on the S1 can be plotted in Figure 1.3(b).

1.3.2 *Resolution Scaling with STED*

The principle of this method is shown in Figure 1.3: S_0 represents the ground or dark state; S_1 represents the excited or bright state. Until a group of ultra-short laser pulses illuminate the sample, the molecules of the sample do not change from S_0 to S_1. From the perspective of the relaxation time and lifetime of electrons, the STED beam is an ultra-short, red-shifted pulse laser with a spot-shape modulated to be like a donut with zero center intensity that illuminates the same focal point of a sample several picoseconds after the illumination pulse. The molecules covered by the second laser leap from S_1 to S_0 while its center still spontaneously radiates fluorescence. Therefore, the PSF is ideally confined as long as the center of the ring is small enough after passing through a filter that is applied in order to eliminate other fluorescence wavelengths and illumination that are not useful for detection. Notably, neither the illumination nor STED beam is beyond the diffraction limit barrier.

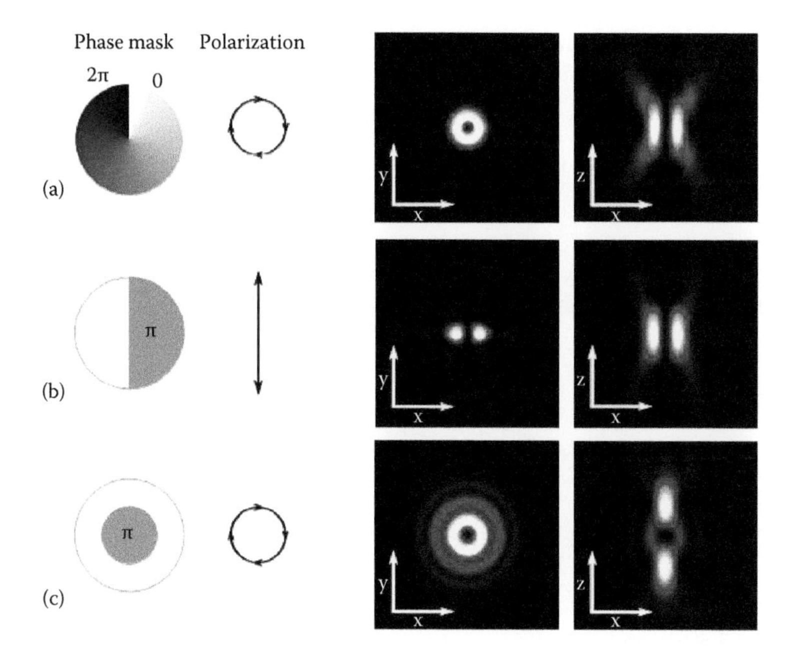

Figure 1.4 Relationship of phase mask, polarization, and PSF distribution. (From R.R. Kellner. 2007. STED microscopy with Q-switched microchip lasers. Faculties for the Natural Sciences and for Mathematics, University of Heidelberg. With permission.)

It is therefore critical to evaluate the generation of the doughnut structure. There are several phase mask and polarization combinations to generate such doughnut PSF, as shown in Figure 1.4 (Kellner 2007):

It should be noted that, in Figure 1.4(a), the direction of the circular polarization has to be the same as the direction of the vortex phase plate, or the central intensity is nonzero (Hao et al. 2010). As a rule of thumb, the experimental central intensity should not exceed 5% of the maximum or the fluorescent intensity will drop significantly, resulting in low signal to noise ratio (SNR) of the STED image (Willig et al. 2006a). This dramatically challenges the application of STED in a scattering condition.

Once the distribution of the doughnut PSF and the excitation PSF are given, the resolution of STED can be readily estimated with the efficiency of stimulated emission. Let the modulation intensity of the STED beam be I_{mod}, and the saturation intensity of a certain molecule be I_{sat}, and when $I_{mod} = I_{sa}$ the possibility of spontaneous emission and stimulated emission is equal (the fluorescent intensity dropped to its half). As generally the intensity is measured with the power of the laser, the focal spot size should be considered.

To obtain the resolution of STED microscopy, we can first take the PSF of confocal as a Gaussian distribution:

$$I_{\text{confocal}}(x) = \exp\left[-\frac{1}{8\ln 2}\left(\frac{x}{d}\right)^2\right] \tag{1.12}$$

where d is the FWHM of the confocal PSF. The STED distribution can be expressed with a parabolic function:

$$I_{\text{sted}}(x) = 4I_{\text{sted}}a^2x^2 \tag{1.13}$$

Considering the depletion efficiency

$$\eta(x) = \exp\left(-\ln 2 \cdot \frac{I_{\text{sted}}(x)}{I_{\text{sat}}}\right) \tag{1.14}$$

The resulting STED PSF can be written as

$$\begin{aligned} I_{\text{STED}}(x) &= I_{\text{confocal}}(x)\eta(x) \\ &= \exp\left\{-\frac{1}{8\ln 2}\frac{1}{d^2}\left[1+8(\ln 2)^2 \cdot 4a^2(I_{\text{sted}}/I_{\text{sat}})d^2\right]x^2\right\} \end{aligned} \tag{1.15}$$

Therefore, we can have the resolution of STED microscopy (Westphal and Hell 2005):

$$d = \frac{\lambda}{2NA\sqrt{1+\alpha I_{\text{sted}}/I_{\text{sat}}}} \tag{1.16}$$

where $\alpha = 8(\ln 2)^2 \cdot 4a^2$. Figure 1.5 gives a simulation result between STED beam intensity and the resolution. Principally, there is no limitation to resolution scaling as long as I_{mod} increases (Hell and Wichmann 1994; Hell et al. 2006).

To reach a confined PSF, three parameters of the STED beam should be carefully chosen: wavelength, intensity, and cross-section:

1. The wavelength of the STED beam should be longer than that of the illumination beam if it is to achieve an effective transition $S_0 \rightarrow S_1$ instead of anti-Stokes excitation (note that this is also related to the absorption spectrum of the fluorescent molecule). According

to Einstein's theory of stimulated emission, the STED beam has an equal probability of a stimulated absorption and emission effect on the molecules. However, because the lifetime of the vibrational level of the ground state S_0^{vib} is much shorter than the first excited state S1, the S_0^{vib} state molecules are rapidly exhausted and are thus highly advantageous for S1 state molecules returning to S0 on a large scale. Thus, those molecules in the S_0^{vib} state escape being stimulated to the S_1 once more.

2. The nonlinear depletion effect is the core of STED. When $I \gg I_s$, and a large number of molecules are depleted from S_1 to S_0 following the stimulated emission process. With an increasing I/Is, the number of S_1 molecules is continuously eliminated and FWHM of PSF is substantially decreased consequently, so that resolution diffraction limit can be broken. However, with an increase of the depletion beam, photobleaching and photodamage will cause rapid deterioration of the tissue and image quality will decrease due to the effect of the signal-to-noise ratio on the resolving power of the system. Therefore, the resolution of the system is limited by the stability of the molecules.
3. The smaller the STED zero beam area (compared with the saturation intensity of the fluorescent dye), the higher the achievable resolution. Therefore, choosing a dye with lower saturation intensity (Liu et al. 2012), or a wavelength closer to the emission maximum can help improve the resolution (Vicidomini et al. 2012).

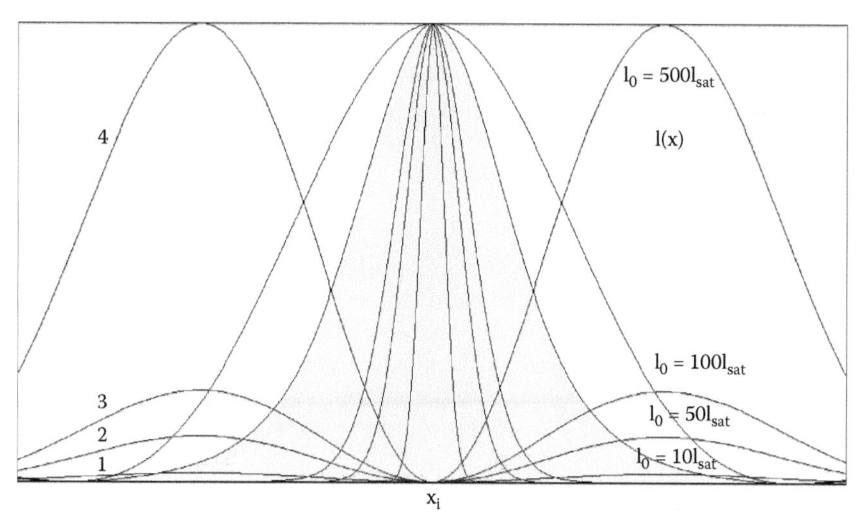

Figure 1.5 The relationship between STED beam intensity and the residual PSF. With an increasing ζ, the number of S1 molecules can be severely constrained and FWHM of PSF is substantially decreased, resulting in improvement of the resolution.

The ring-shaped spot can deactivate fluorescence along all directions perpendicular to the propagation of light, but it is not the only choice. To meet different needs, modulating the phase of the STED beam may produce distinct cross sections of the STED spot, so constraining FWHM in diverse directions (see Figure 1.2a), for example, the cross-sectional shrinkage can either be along one axis or all directions in the focal plane (Bretschneider, Eggeling, and Hell 2007). The use of azimuthal polarization can also be effective in generation of the ring-shaped distribution (Xue et al. 2012).

1.3.3 STED and Other Imaging Modalities

Generally, STED is used for the improvement of lateral resolution. Through the combination with other imaging modalities, new applications can be benefited.

1.3.3.1 Time-gated STED

Continuous wave (CW) STED plays an important role in STED, as it requires no temporal sequencing of the excitation and the depletion beams (Willig et al. 2007). However, as it is constantly excited and depleted, the fluorescence endures more cycles of switching than that of pulsed STED (p-STED). Moreover, since the depletion efficiency cannot be 100%, there is always a background in CW STED image if comparing with the p-STED. If noticing an important fact that, at the STED process, the molecules at the center follow more spontaneous emission, yet the peripheral molecules have a high ratio of stimulated emission whose lifetime is one order of magnitude shorter than that of spontaneous emission, one can then apply a temporal gate to further discriminate the emissions from different locations, at a whole new temporal domain.

1.3.3.2 STED + 4Pi

The PSF formed by the objective is generally elongated in the axial direction, which is due to the insufficient collecting angle in the z direction. By applying an opposing lens to increase the collecting angle, 4Pi can improve the axial resolution to its 1/3 (Schrader and Hell 1996). Therefore, combining the STED depletion with coherent 4Pi excitation interference can improve both lateral and axial resolution (Dyba, Keller, and Hell 2005).

1.3.3.3 ISO-STED

In Section 2.2 we have discussed that a variety of phase modulations can be used in conjunction with polarization states for enhancing both lateral and axial resolution. It should be noted that the destructive interference of the depletion beam can be used to improve the axial resolution as well, and more efficiently with an opposing lens. The combination of STED in lateral with STED in axial can yield an isotropic size PSF (Schmidt et al. 2008, 2009).

1.3.3.4 Multiphoton-STED

Multiphoton microscopy has the advantage of less scattering. Through applying an ultrashort laser source as two-photon excitation in place of the single-photon excitation in STED, the 2PE-STED has been demonstrated (Figure 1.6; Li, Wu, and Chou 2009; Moneron and Hell 2009). Moreover, as STED is a single-photon process with duration of ~250 ps, one can stretch the ultrashort pulse, so that 2PE STED can even be done with one single wavelength (Bianchini et al. 2012).

1.3.3.5 Automatic Spatially Aligned STED

To decrease the difficulty of temporal overlapping in pulsed STED, a few techniques have been presented to generate the excitation and the STED laser pulse with one single laser (Rankin, Kellner, and Hell 2008; Wildanger et al. 2008), or with CW laser (Willig et al. 2007; Vicidomini et al. 2011; Liu et al. 2012). However, the most challenging is still the spatial overlap of the PSFs.

For materials with a different refractive index, to the excitation wavelength the refractive index is identical, whereas to the depletion wavelength there is a difference (Figure 1.7). Therefore, the combination phase

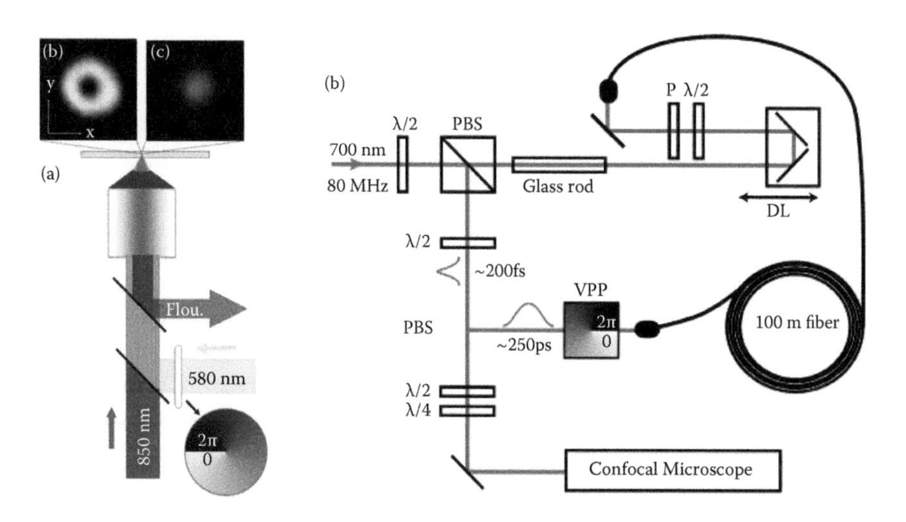

***Figure 1.6* (See color insert.)** (a) Two-photon excitation STED; (b) single-wavelength 2PE STED. (Panel A From Q. Li, S.S.H. Wu, and K.C. Chou. 2009. *Biophysical Journal* 97 (12):3224–3228. With permission. Panel B from P. Bianchini et al. 2012. *Proceedings of the National Academy of Sciences* 109 (17):6390–6393. With permission.)

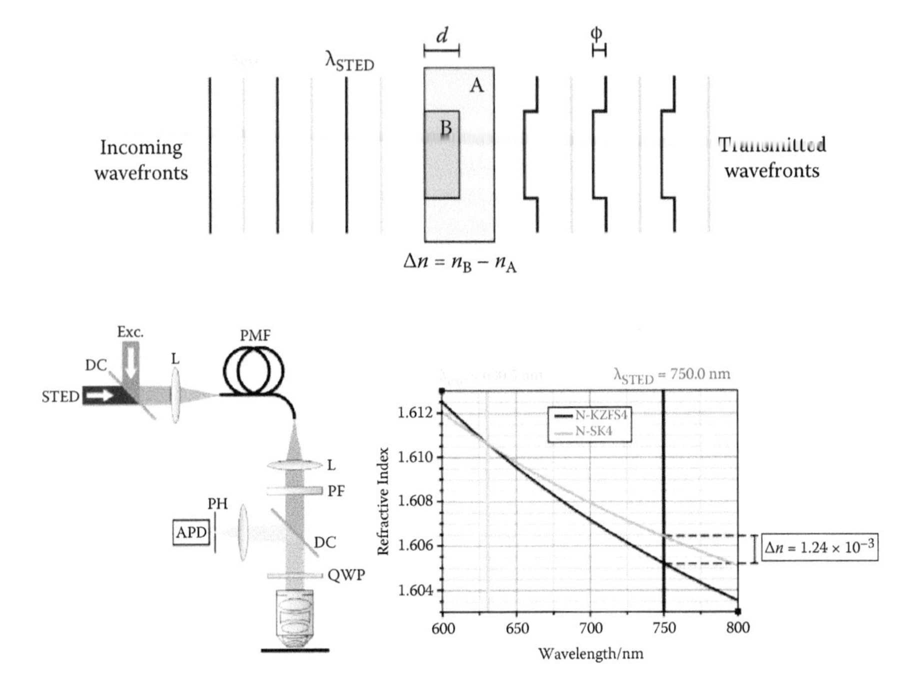

Figure 1.7 **(See color insert.)** The principle of phase modulation with two optical materials for automatic alignment of the STED setup. (From Wildanger, D. et al. A STED microscope aligned by design. *Optics Express* 17:16100–16110 (2009). With permission.)

plate is "transparent" to the excitation beam, but the depletion beam will be modulated (Wildanger et al. 2009a).

An alternative means to this approach, is to apply a specifically designed birefringent retarder before the objective, as illustrated in Figure 1.8 (Reuss, Engelhardt, and Hell 2010). Since the birefringent retardation is sensitive to the wavelength, the overall polarization after the retarder for excitation is linear, yet for the STED beam it can be radial or azimuthal. Consequently, the PSF is automatically overlapped, and a confocal setup can be easily modified to be STED nanoscope with this add-on.

1.4 Technical Notes on STED Optical Nanoscopy

1.4.1 Generation and Measurement of the PSFs

In the STED experiment, there are a few PSFs that need to be measured and carefully overlapped: (1) excitation Gaussian PSF, (2) depletion beam Gaussian PSF for the measurement of the saturation intensity; and (3) depletion beam ring PSF after phase modulation. Before the measurement

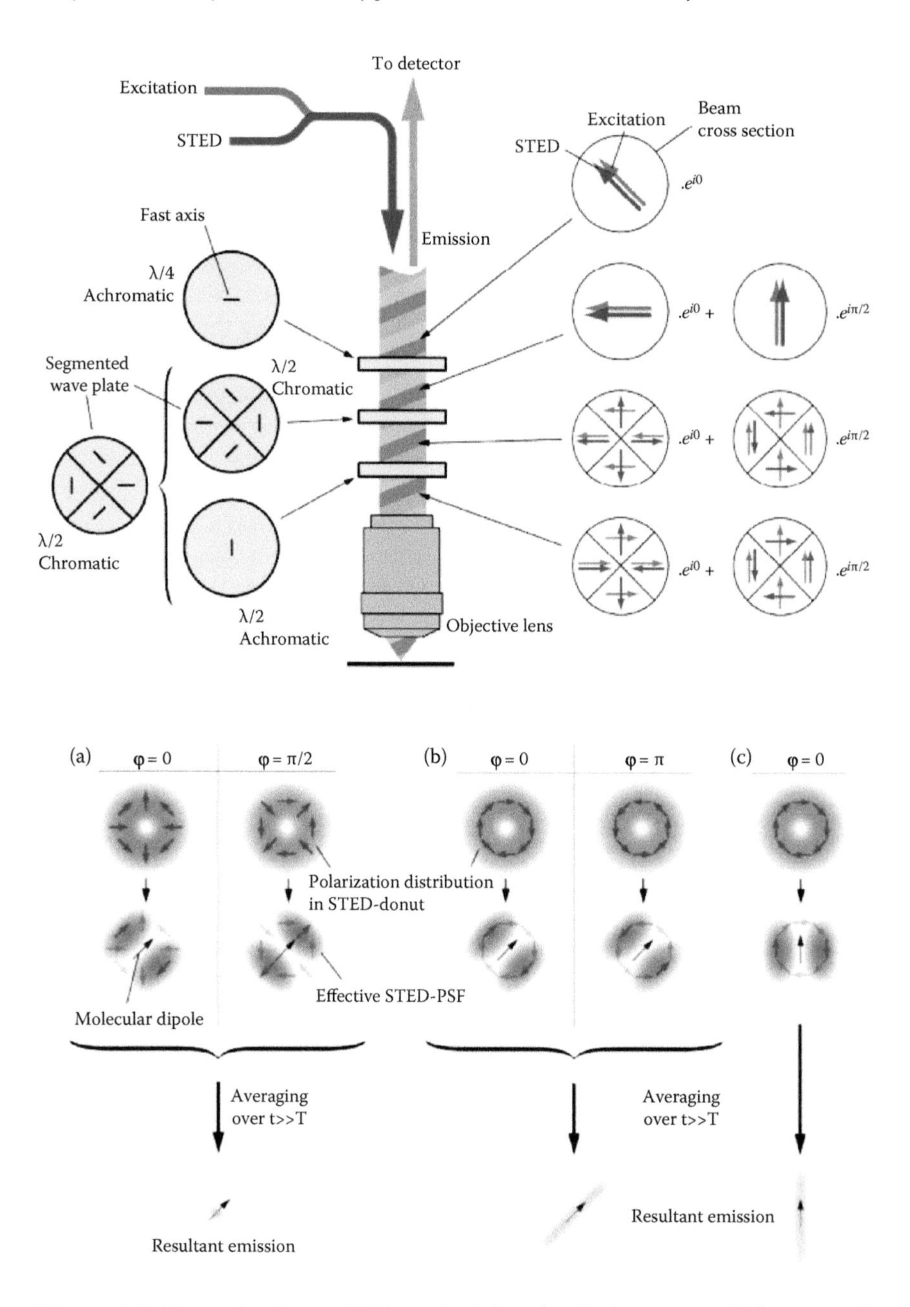

Figure 1.8 **(See color insert.)** The principle of polarization modulation with birefringent retarders sensitive to wavelength for automatic alignment of the STED setup. (From Reuss, M. et al., Birefringent device converts a standard scanning microscope into a STED microscope that also maps molecular orientation. *Optics Express* 18:1049–1058 (2010).

of PSF, a gold nanoparticle specimen should be prepared. The particle should have a diameter <200 nm, and reflection signal is detected. Since the tip of the nanoparticle serves as the source, it can only be regarded as a point when the radius is small (generally smaller than the theoretical FWHM of the PSF). To collect the reflection, a pellicle can be used.

The generation of the depletion ring (doughnut) PSF is the most critical part of the STED experiment. The circular polarization has to be perfect, or the center of the ring PSF will not be <5%. Practically, it can be measured with rotating a linear polarizer and detecting the output signal intensity. This works for both the Gaussian and the ring PSFs. The PSF is best measured right before the objective, therefore the beam is parallel and all the reflection mirrors are present. It should be noted that, due to different reflectivity of TE and TM, the polarization can be altered with mirrors. Even with good circular polarization, the measured PSF may not be zero as the objective can still change the polarization a bit. So, double check the polarization with the PSF generated is the last step before a good STED image can be collected. A rule of thumb is, the center of the PSF should not exceed 5%, or the fluorescent signal is greatly reduced.

1.4.2 Measurement of the Saturation Intensity

Generally, the saturation intensity I_s is measured with the corresponding fluorescent solutions. Here the phase modulation should not present, and the two PSFs of excitation and depletion beams should be perfectly overlapped. With increasing STED intensity, the fluorescent intensity decays accordingly. As discussed before, the point at which $I_{sted} = I_s$ is where the depletion power of the fluorescence is halved. However, the measurement of the fluorescence contribution is taken at an integral (usually with fluorescent dye solution). As $I_{sted} = P_{sted}/\sigma$, where σ is the focal FWHM size, considering the distribution of the PSF is Gaussian in three dimensions, the measurement of P_{sted} (as an integral) can be experimentally taken at where the fluorescence dropped to its 1/4 with an assumption of uniform distribution of the PSF. This should not be confused with the expansion of the doughnut shape PSF versus the Gaussian PSF, which is also fourfold.

1.4.3 Sample Preparation

It should be kept in mind that a smaller STED PSF, or a better resolution, comes at a price of fewer photons. An increase of resolution of 10× will reduce the photons at least 100×. If taking into account the central STED ring PSF is never zero, such loss is much more. Next, the STED image has a better resolution because the fluorophores are forced to stimulate emission. This process will also introduce photobleaching to the sample.

So, the fluorophore for STED should be photobleaching robust, with large cross-section of absorption and quantum yield (which means bright), and longer lifetime. For antibody binding to specific cellular organelles, a multiple clone is always superior than a single clone as more fluorophores are presented. A detailed description of the sample preparation for STED nanoscopy can be found in Wurm et al. (2010).

1.4.4 *Resolution Measurement*

The resolution of the STED system is usually measured with fluorescent nanobeads. Unlike the measurement of PSFs with gold nanoparticles, the entire nanobead contributes fluorescence during the measurement, and consequently the final result is a two-dimensional convolution of the size of the bead and the STED effective PSF. Due to the error of nanobead fabrication, a smaller bead is preferable when plotting and calculating the final PSF (Willig et al. 2007).

The resolution can also be measured with the small feature of the specimen, for example, the cell skeleton has a diameter of 5 to 25 nm so it is often used as a test bench for resolution demonstration (Liu et al. 2012). It should be noted that STED resolution is sensitive to refractive index mismatch, therefore the optimal resolution can be obtained at the distance right after the cover glass for high NA, oil-immersion objective. As the feature of the cell is generally not at this layer, the resolution on a biological sample can be lower than that of bead measurement, even with the same fluorophore and STED power.

1.5 *Conclusion*

In this chapter, we have provided an overview of the origin of diffraction limits, and how STED has utilized the nature of fluorescence to break the diffraction barrier. The theoretical effect of the STED process and the resolution of a STED system have been analyzed. With the combination of different techniques, STED has demonstrated its further improved resolution with time-gating, isotropic resolution with multiple STED patterns, single wavelength STED with multiphoton excitation, or self-aligned STED with novel wavelength-sensitive phase or polarization modulation techniques. A technical note on the STED instrumentation was given in Section 1.4.

So far, STED has been applied on a series of subcellular organelles to study the morphology and cellular interaction with better resolution. Table 1.1 is a list of some applications.

In the future, optimization of the STED technique relies on advances in the following aspects: (1) the calibration and compensation of the spatial astigmastism, so that STED can demonstrate its wide application in three-dimensional super-resolution imaging; (2) the development of

Table 1.1 Some Biological Applications of STED Optical Nanoscopy

	Cellular organelle	Uniqueness of the technique	References
Fixed cells and tissues	Mitochondria	Iso-STED	Schmidt et al. 2008, 2009
		Two-color STED	Pellett et al. 2011; Neumann et al. 2010; Donnert et al. 2007
	Membrane	Two-color iso-STED	Schmidt et al. 2008
		Two-color STED	Pellett et al. 2011
		Fast tracking	Sahl et al. 2010
	Syntaxin clusters	CW STED	Willig et al. 2007
	Actin filament	CW STED	Liu et al. 2012
	Microtubules	CW STED	Liu et al. 2012
		Fast CW STED	Moneron et al. 2010
	Intermediate filaments	CW STED	Liu et al. 2012
		g-STED	Vicidomini et al. 2011
		SNAP-tag fusion proteins	Hein et al. 2010
		Fast CW STED	Moneron et al. 2010
	Neural filaments	CW STED	Willig et al. 2007
		Triplet relaxation	Donnert et al. 2006
		Single-source STED	Wildanger et al. 2008
		easySTED	Reuss, Engelhardt, and Hell 2010
		Compact STED source	Wildanger et al. 2009b
	Endoplasmic reticulum	GFP labeling	Willig et al. 2006a
	Nuclei	Triplet relaxation	Donnert et al. 2006
	Nuclear lamina	CW STED	Willig et al. 2007
	Endosomes	Triplet relaxation	Donnert et al. 2006
	RNA	CW STED	Liu et al. 2012
	Synaptic vesicle		Willig et al. 2006b
	Nicotinic acetylcholine receptors		Kellner et al. 2007
	Olfactory sensory neurons	Two-color STED	Lin et al. 2007

Table 1.1 (continued) Some Biological Applications of STED Optical Nanoscopy

	Cellular organelle	Uniqueness of the technique	References
Living cells and tissue cultures	Neural synaptic vesicle	Video-rate STED	Lauterbach et al. 2010; Westphal et al. 2008
		Two-color STED	Opazo et al. 2010
	Endoplasmic reticulum	FP label	Hein, Willig, and Hell 2008
		Fast CW STED	Moneron et al. 2010
	Membrane protein cluster		Sieber et al. 2007
	Membrane lipids	STED FCS	Eggeling et al. 2008
	Dendritic spines		Nägerl et al. 2008
Living animal	Neuron	Living mouse brain	Berning et al. 2012

robust, non-photobleaching, bright yet less toxic fluorescent dyes; and (3) the combination of STED techniques with other techniques for the study of the cellular dynamics of living cells.

The past 300 years have pushed research in biology to the limit of optical microscopy. Today, optical nanoscopy has enabled the study of more fundamental questions which were beyond the resolution limit. With more and more researchers becoming aware of and applying these revolutionary optical nanoscopic techniques, we believe that super-resolution microscopy will set the new standard in the next decade's biological research.

References

Abbe, E. 1873. Beiträge zur Theorie des Mikroskops und der mikroskopischen Wahrnehmung. *Arch. f. Mikroskop. Anatomie* 9 (1):413–418.

Berning, S., K.I. Willig, H. Steffens, P. Dibaj, and S.W. Hell. 2012. Nanoscopy in a living mouse brain. *Science* 335 (6068):551.

Bianchini, P., B. Harke, S. Galiani, G. Vicidomini, and A. Diaspro. 2012. Single-wavelength two-photon excitation–stimulated emission depletion (SW2PE-STED) superresolution imaging. *Proceedings of the National Academy of Sciences* 109 (17):6390–6393.

Bretschneider, S., C. Eggeling, and S.W. Hell. 2007. Breaking the diffraction barrier in fluorescence microscopy by optical shelving. *Phys Rev Lett* 98 (21):218103.

Donnert, G., J. Keller, R. Medda, M.A. Andrei, S.O. Rizzoli, R. Lührmann, R. Jahn, C. Eggeling, and S.W. Hell. 2006. Macromolecular-scale resolution in biological fluorescence microscopy. *Proceedings of the National Academy of Sciences* 103 (31):11440–11445.

Donnert, G., J. Keller, C.A. Wurm, S.O. Rizzoli, V. Westphal, A. Schönle, R. Jahn, S. Jakobs, C. Eggeling, and S.W. Hell. 2007. Two-color far-field fluorescence nanoscopy. *Biophysical Journal* 92 (8):L67–L69.

Dyba, M., J. Keller, and S.W. Hell. 2005. Phase filter enhanced STED-4Pi fluorescence microscopy: Theory and experiment. *New Journal of Physics* 7:134.

Eggeling, C., C. Ringemann, R. Medda, G. Schwarzmann, K. Sandhoff, S. Polyakova, V.N. Belov, B. Hein, C. Von Middendorff, and A. Schönle. 2008. Direct observation of the nanoscale dynamics of membrane lipids in a living cell. *Nature* 457 (7233):1159–1162.

Hao, X., C. Kuang, T. Wang, and X. Liu. 2010. Effects of polarization on the de-excitation dark focal spot in STED microscopy. *Journal of Optics* 12:115707.

Hein, B., K.I. Willig, C.A. Wurm, V. Westphal, S. Jakobs, and S.W. Hell. 2010. Stimulated emission depletion nanoscopy of living cells using SNAP-tag fusion proteins. *Biophysical Journal* 98 (1):158–163.

Hein, B., K.I. Willig, and S.W. Hell. 2008. Stimulated emission depletion (STED) nanoscopy of a fluorescent protein-labeled organelle inside a living cell. *Proceedings of the National Academy of Sciences* 105 (38):14271–14276.

Hell, S.W., and J. Wichmann. 1994. Breaking the diffraction resolution limit by stimulated emission: Stimulated-emission-depletion fluorescence microscopy. *Opt Lett* 19 (11):780–782.

Hell, S.W., K. Willig, M. Dyba, S. Jakobs, L. Kastrup, and V. Westphal. 2006. Nanoscale resolution with focused light: STED and other RESOLFT microscopy concepts. In *Handbook of Biological Confocal Microscopy*, edited by J.B. Pawley. New York: Springer-Verlag.

Kellner, R.R. 2007. STED microscopy with Q-switched microchip lasers. Faculties for the Natural Sciences and for Mathematics, University of Heidelberg.

Kellner, R.R., C.J. Baier, K.I. Willig, S.W. Hell, and F.J. Barrantes. 2007. Nanoscale organization of nicotinic acetylcholine receptors revealed by stimulated emission depletion microscopy. *Neuroscience* 144 (1):135–143.

Klar, T.A, and S.W. Hell. 1999. Subdiffraction resolution in far-field fluorescence microscopy. *Optics Letters* 24 (14):954–956.

Lauterbach, M.A., J. Keller, A. Schönle, D. Kamin, V. Westphal, S.O. Rizzoli, and S.W. Hell. 2010. Comparing video-rate STED nanoscopy and confocal microscopy of living neurons. *Journal of Biophotonics* 3 (7):417–424.

Li, Q., S.S.H. Wu, and K.C. Chou. 2009. Subdiffraction-limit two-photon fluorescence microscopy for GFP-tagged cell imaging. *Biophysical Journal* 97 (12):3224–3228.

Lin, W., R. Margolskee, G. Donnert, S.W. Hell, and D. Restrepo. 2007. Olfactory neurons expressing transient receptor potential channel M5 (TRPM5) are involved in sensing semiochemicals. *Proceedings of the National Academy of Sciences* 104 (7):2471.

Liu, Y., Y. Ding, E. Alonas, W. Zhao, P.J. Santangelo, D. Jin, J.A. Piper, J. Teng, Q. Ren, and P. Xi. 2012. Achieving $\lambda/10$ resolution CW STED nanoscopy with a Ti:sapphire oscillator. *PLoS One* 7 (6):e40003.

Lord Rayleigh. 1896. On the theory of optical images with special reference to the microscope. *Philos Mag* 42:167–195.

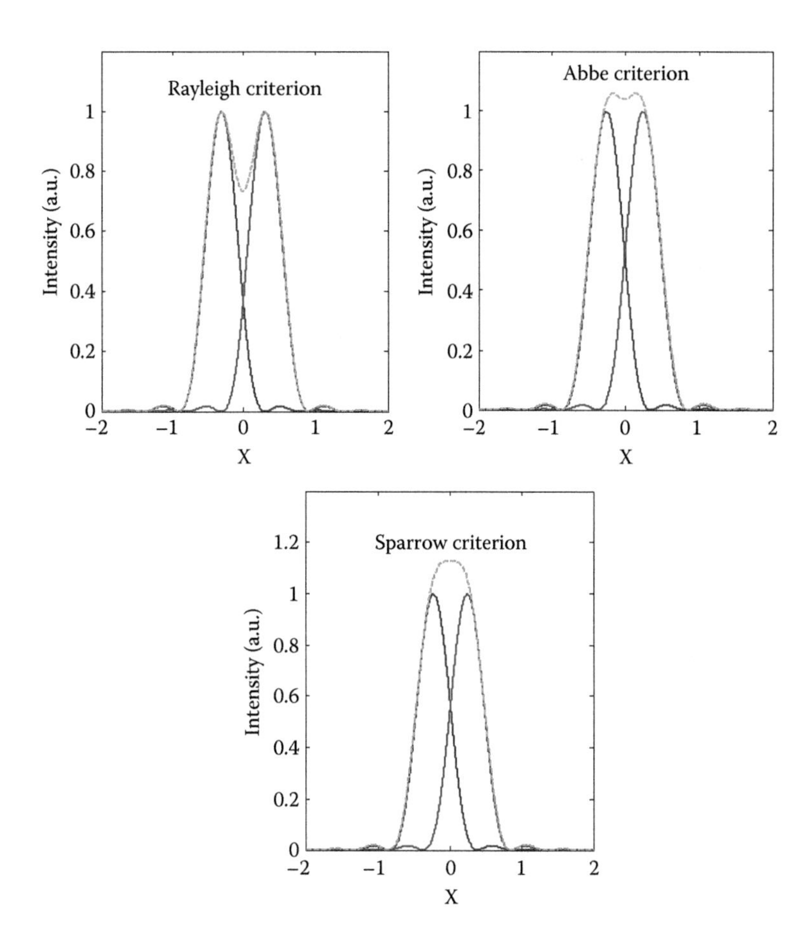

Figure 1.1 Comparison of the various diffraction limit criteria.

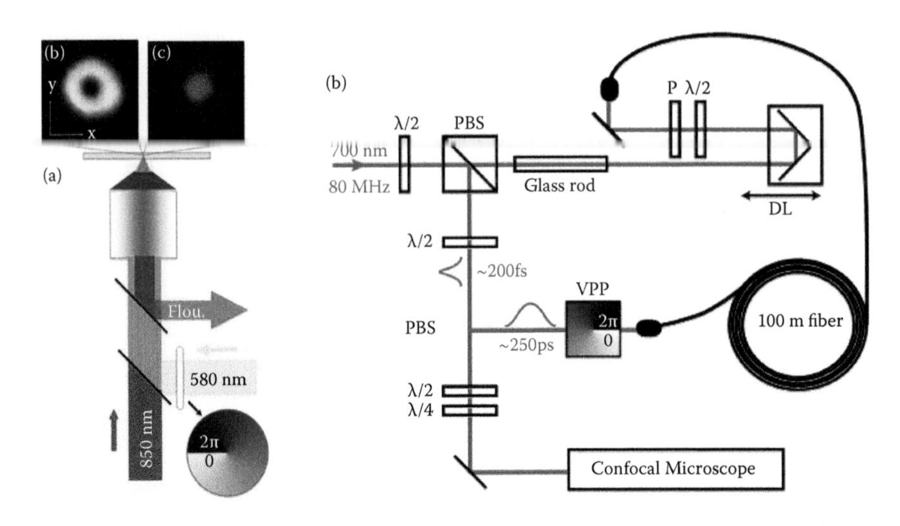

Figure 1.6 (a) Two-photon excitation STED; (b) single-wavelength 2PE STED. (Panel A From Q. Li, S.S.H. Wu, and K.C. Chou. 2009. *Biophysical Journal* 97 (12):3224–3228. With permission. Panel B from P. Bianchini et al. 2012. *Proceedings of the National Academy of Sciences* 109 (17):6390–6393. With permission.)

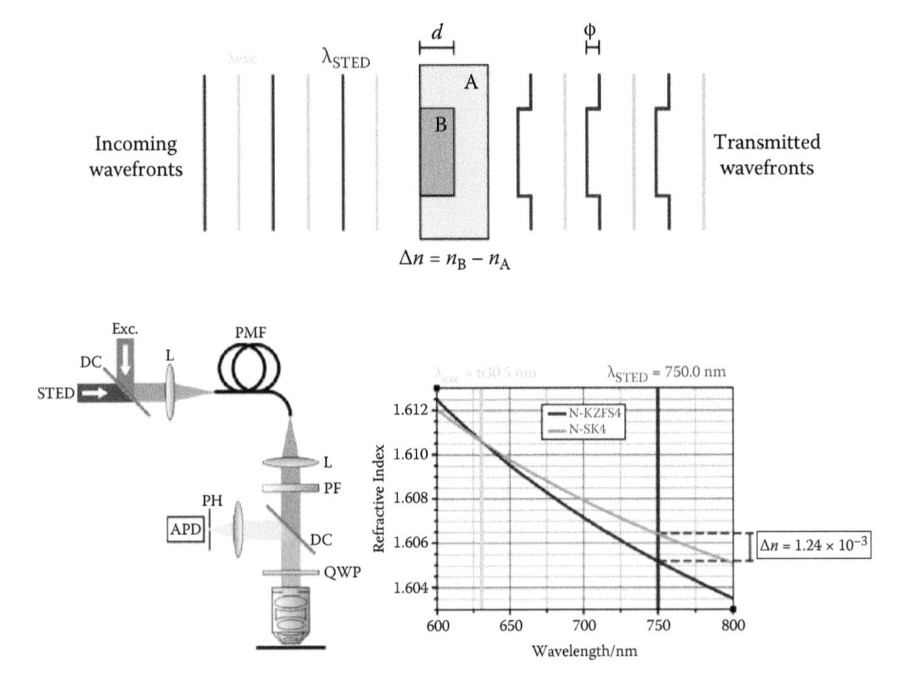

Figure 1.7 The principle of phase modulation with two optical materials for automatic alignment of the STED setup.

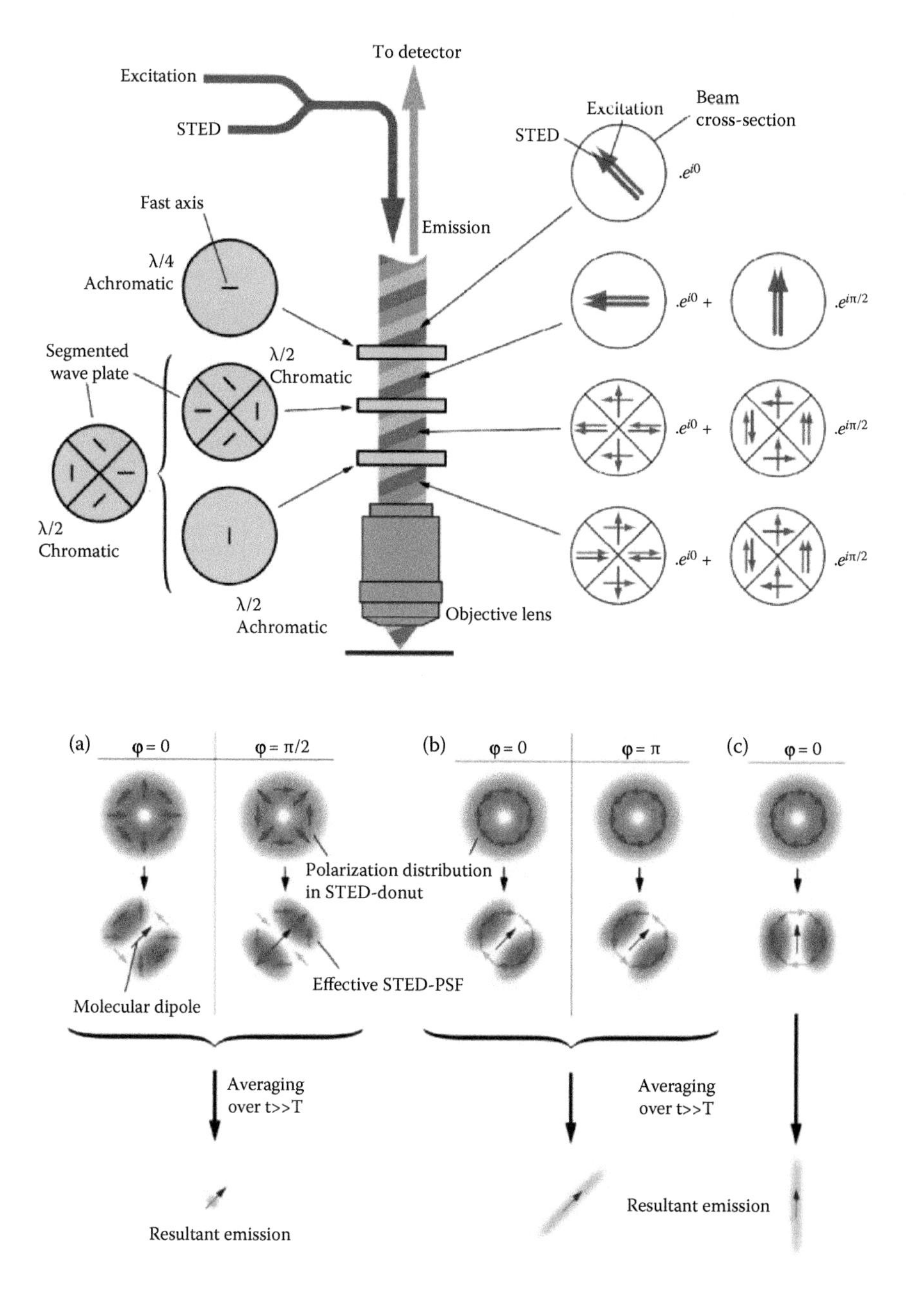

Figure 1.8 The principle of polarization modulation with birefringent retarders sensitive to wavelength for automatic alignment of the STED setup.

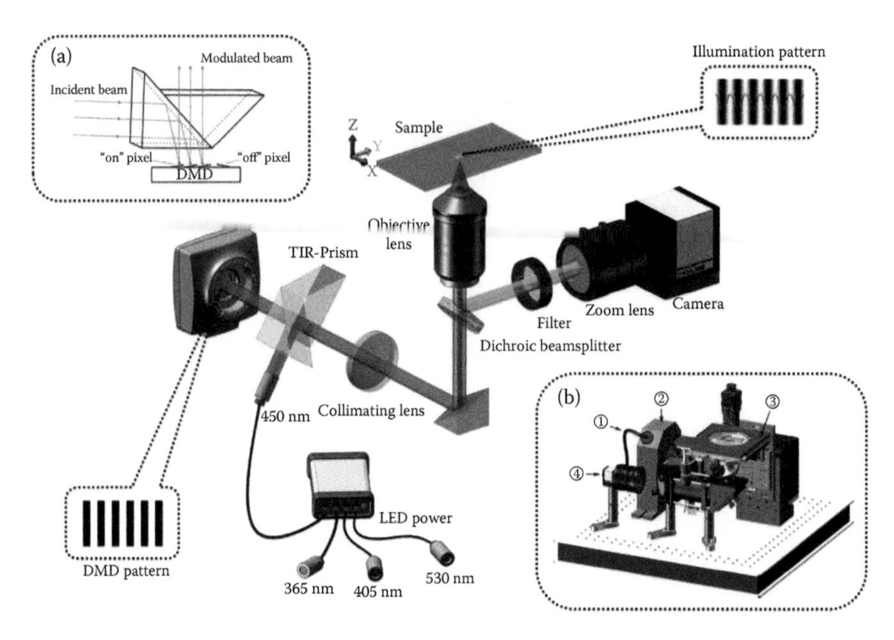

Figure 2.8 Scheme of the DMD-based LED-illumination SIM microscope. A low-coherence LED light is introduced to the DMD via a TIR prism. The binary fringe pattern on the DMD is demagnified and projected onto the specimen through a collimating lens and a microscope objective lens. Higher orders of spatial frequencies of the binary fringe are naturally blocked by the optics, leading to a sinusoidal fringe illumination in the sample plane. Fluorescence or scattered light from the specimen is directly imaged onto the CCD camera. Inset (a) shows the principle of the TIR prism. The illumination light enters the prism and reflects through total internal reflection (TIR) within the prism and illuminates the DMD. Each micromirror of the DMD has an "on" position of +12° and an "off" position of –12° from the normal of the DMD. As a result, the illumination light is introduced at 24° from the normal so that the mirrors reflect the light at 0° when working at the "on" state. Inset (b) gives the configuration of the system, including the LED light guide ϕ, the DMD unit κ, the sample stage λ, and the CCD camera μ.

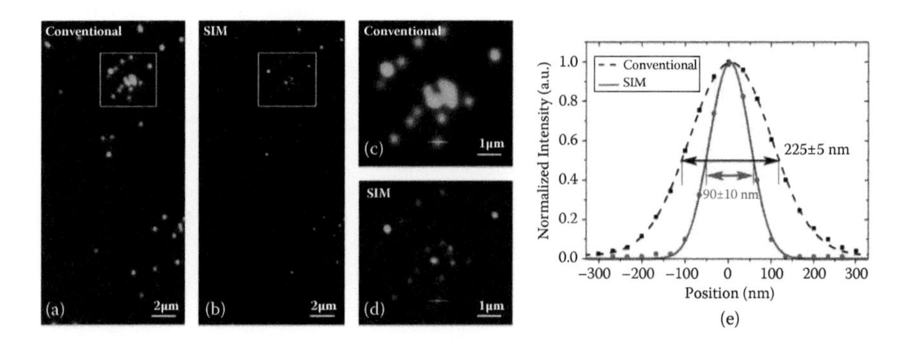

Figure 2.10 Determination of the lateral SIM resolution using 80-nm-diameter gold nanoparticles. As a comparison, image (a) is captured in the conventional illumination mode, and (b) is the corresponding SIM image. Zooms are shown in (c) and (d). The line-scans of a selected nanoparticle in (c) and (d) are plotted in (e). The experiment was performed with the 100× objective of NA = 1.49 with LED wavelength of 450 nm.

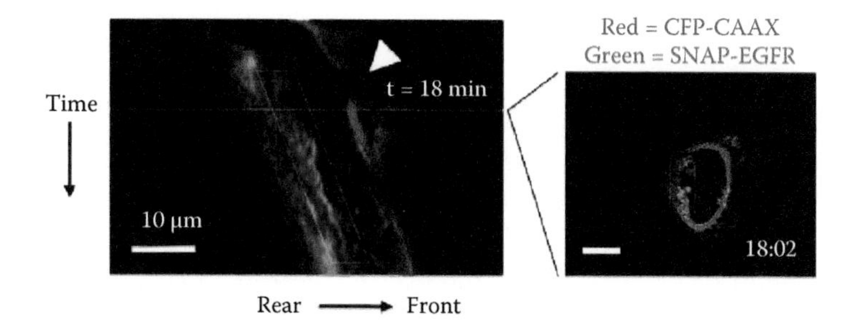

Figure 4.15 Difference in internalization of SNAP-EGFR from rear to front during cell migration in epidermal growth factor-stimulated cells. (From T. Komatsu et al. (2011). *J. Am. Chem. Soc.* **133**(17): 6745–6751. Copyright American Chemical Society. Reprinted with permission.)

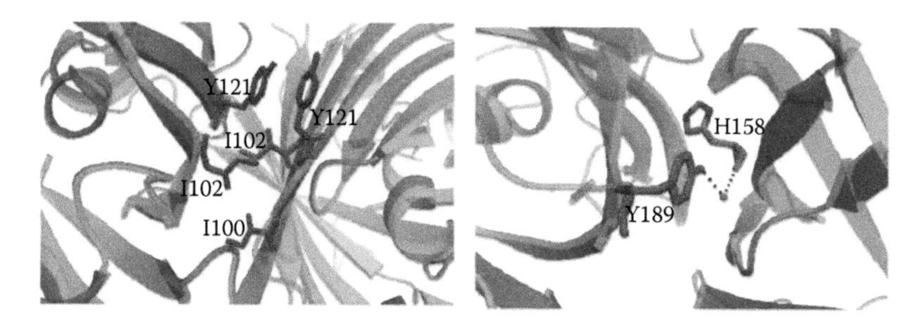

Figure 5.2 Interfaces in tetrameric mEos2.

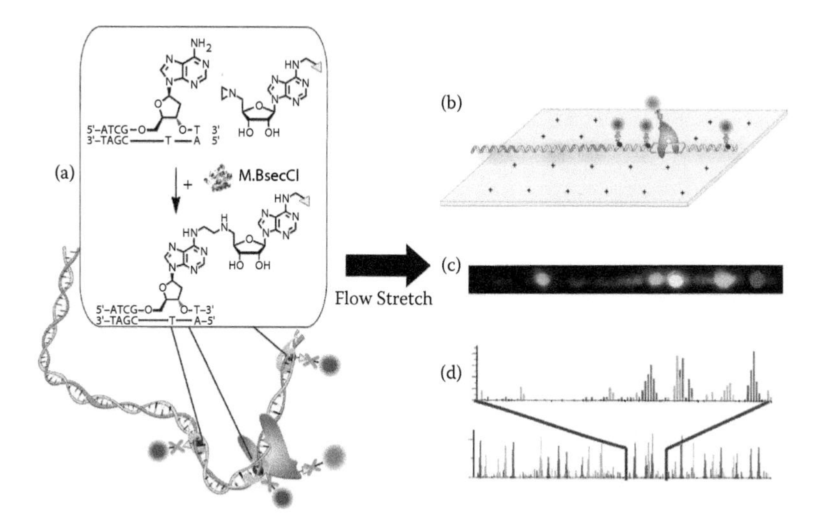

Figure 6.7 (a) Schematic representation of RNA polymerase labeled with green QDs and bound to T7 bacteriophage DNA. The DNA has been enzymatically labeled with biotin (yellow triangle) at rare sequences (5′-ATCGAT-3′), which can be labeled with streptavidin-QDs to serve as reference points on the genome. (b) RNA polymerase-bound T7 bacteriophage genome stretched over a polylysine surface. (c) Image of flow-stretched, dye-stained T7 bacteriophage DNA (white) with QD-labeled RNA polymerase (green) and sequence-specific labels with red QDs as sequence references. Overlapping red and green signals are shown in yellow. (d) Conceptual representation of a genome-wide map of promoters (green) and reference sites (red). (From S. Kim et al. 2012. *Angewandte Chemie* 124 (15): 3638–3641. With permission.)

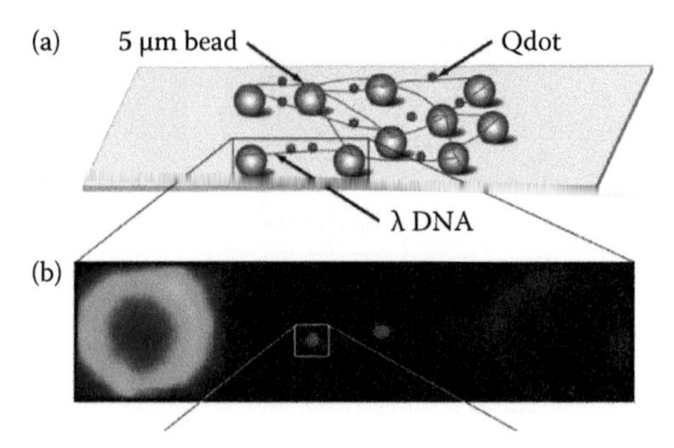

Figure 6.8 Single-molecule N-glycosylase imaging. (a) DNA "tightropes" extended between stationary 5-μm silica beads suspend DNA off of the substrate. (b) Image of faint DNA strung between beads (green) with bound QD-labeled N-glycosylase (red). (From A. R. Dunn et al. 2011. *Nucleic Acids Research* 39 (17): 7487–7498. With permission.)

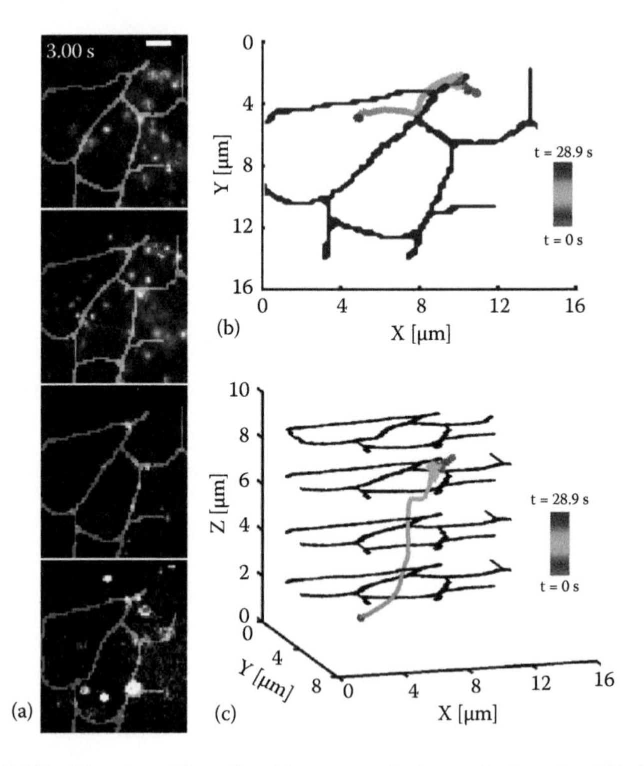

Figure 6.12 Multifocal imaging of live cells taking up transferrin proteins bound to QDs. In panel *A*, four focal planes are shown for a single time point in which cell plasma membranes are marked green and QDs are shown in gray scale. Scale bar = 5 μm. Panels *B* and *C* show the X-Y projection and the full three-dimensional trajectory, respectively, of the transferrin-QD molecule highlighted with a red arrow in panel *A*. The trajectories are color-coded to indicate time. The Transferrin-QD molecule is initially seen inside one cell and then moves toward the lateral plasma membrane and undergoes exocytosis. The molecule is then immediately endocytosed by the adjacent cell. (From S. Ram et al. 2012. *Biophysical Journal* 103 (7): 1594–1603. With permission.)

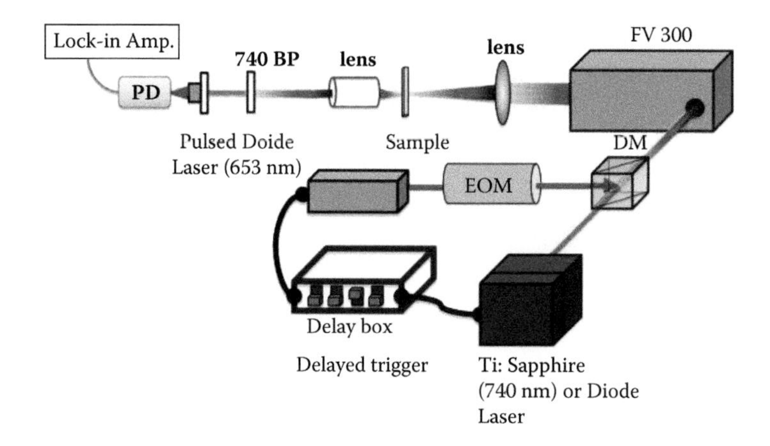

Figure 7.3 Setup for low-NA stimulated emission imaging. EOM, electro-optical modulator; BP, band pass filter; DM, dichroic mirror; PD, photodiode.

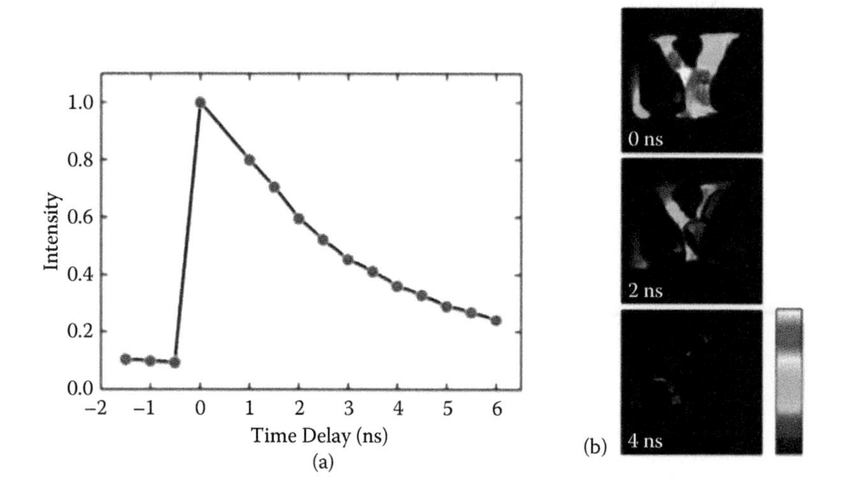

Figure 7.5 Time-resolved fluorescence images. (a) Stimulated signal versus time delay between excitation and stimulation pulses. (b) Stimulated emission images of ATTO 647 N sample injected into Y-shape microfluidic sample with various interpulse delays. Scanned size: 600 μm × 600 μm.

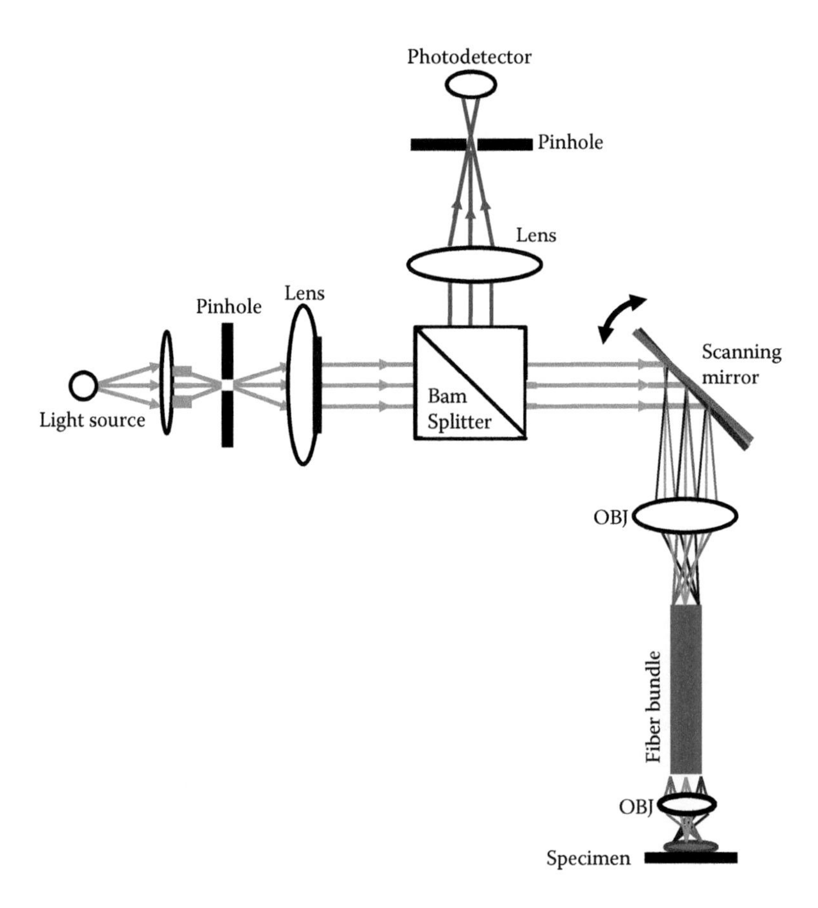

Figure 8.8 Laser-scanned fiber bundle microscopy.

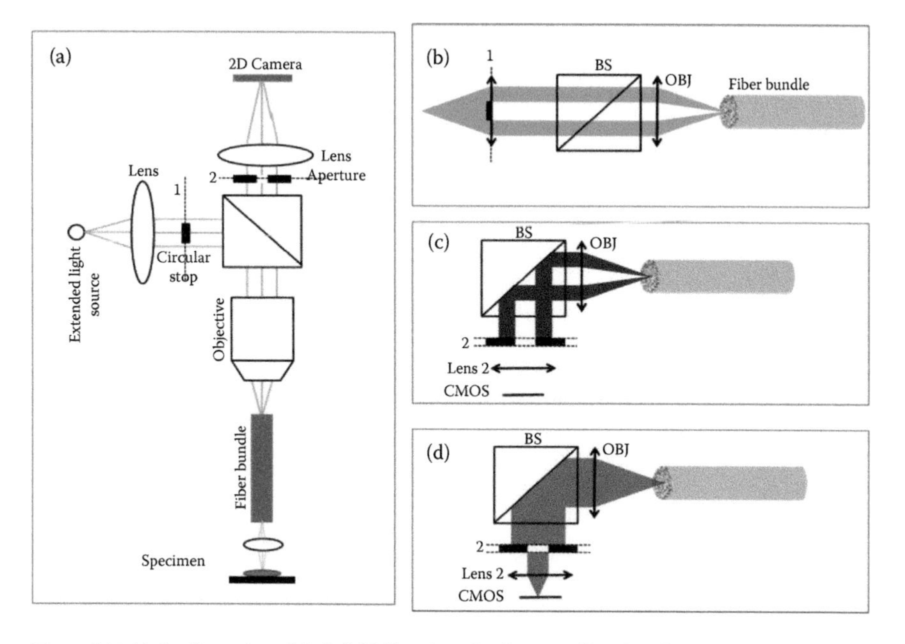

Figure 8.14 (a) Configuration of dark field illuminated reflectance fiber bundle microscope; (b) optical path of illumination light; (c) optical path of specular reflection from the fiber bundle end; (d) optical path of signal light back scattered by the specimen and guided by the fiber bundle.

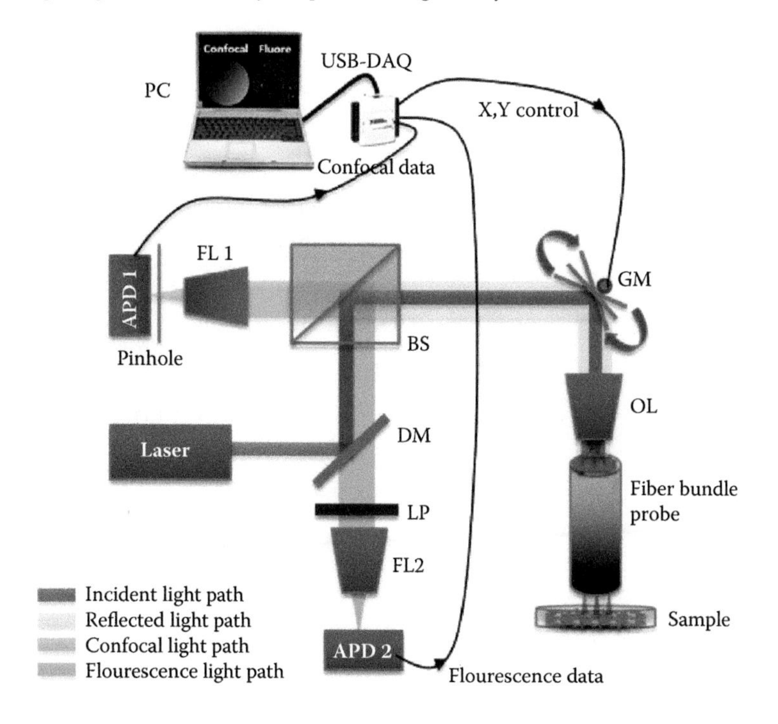

Figure 8.17 System configuration (DM, dichroic mirror; BS, 50:50 beam splitter; GM, galvo mirror; OL, objective lens; FL 1&2, focusing lens; LP, longpass filter; APD 1&2, avalanche photodetector; DAQ, digital-to-analog and analog-to-digital).

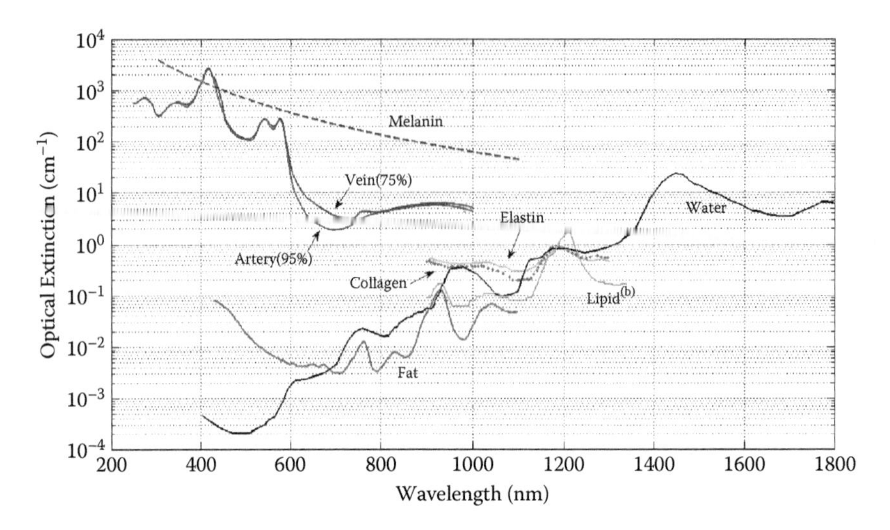

Figure 10.1 Absorption coefficients of variable components in tissue at different wavelengths. Data for artery (HbO_2 95%, 150 gl^{-1}, red) and vein (HbO_2 75%, 150 gl^{-1}, blue) from http://omlc.ogi.edu/spectra/hemoglobin/summary.html; water (black) from Hale and Querry 1973; melanin (green dashed) from http://oomlc.ogi.edu/spetra/melanin/mua.html; lipid$^{(b)}$ (purple), elastin (yellow), and collagen (blue dashed) from Tsai et al. 2001); and fat (orange) from http://omlc.ogi.edu/spectra/fat/fat.txt.

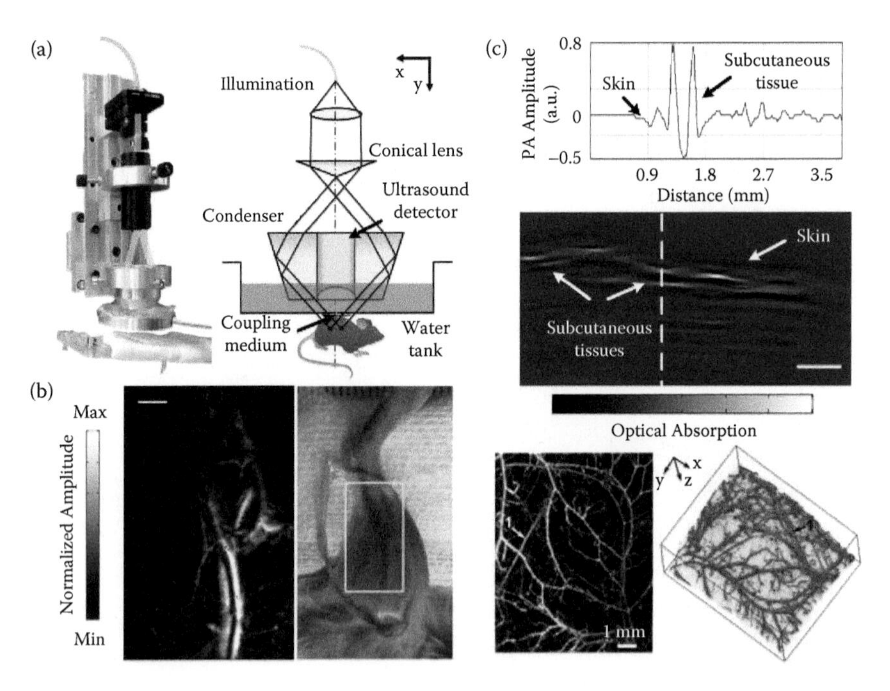

Figure 10.2 AR-PAM system and imaging results. (a) System setup. (b) The result of subcutaneous vasculature of the mouse hind limb with 20 MHz central frequency, scale bar: 1 mm.(c) The A-scan, B-scan, maximum amplitude projection, and three-dimensional vasculature results of mouse, scale bar: 1 mm. (From H.F. Zhang et al. 2006 (panel a); H.F. Zhang et al. 2006 (panel b); and Ye et al. 2012 (panel c). Reproduced with permission.)

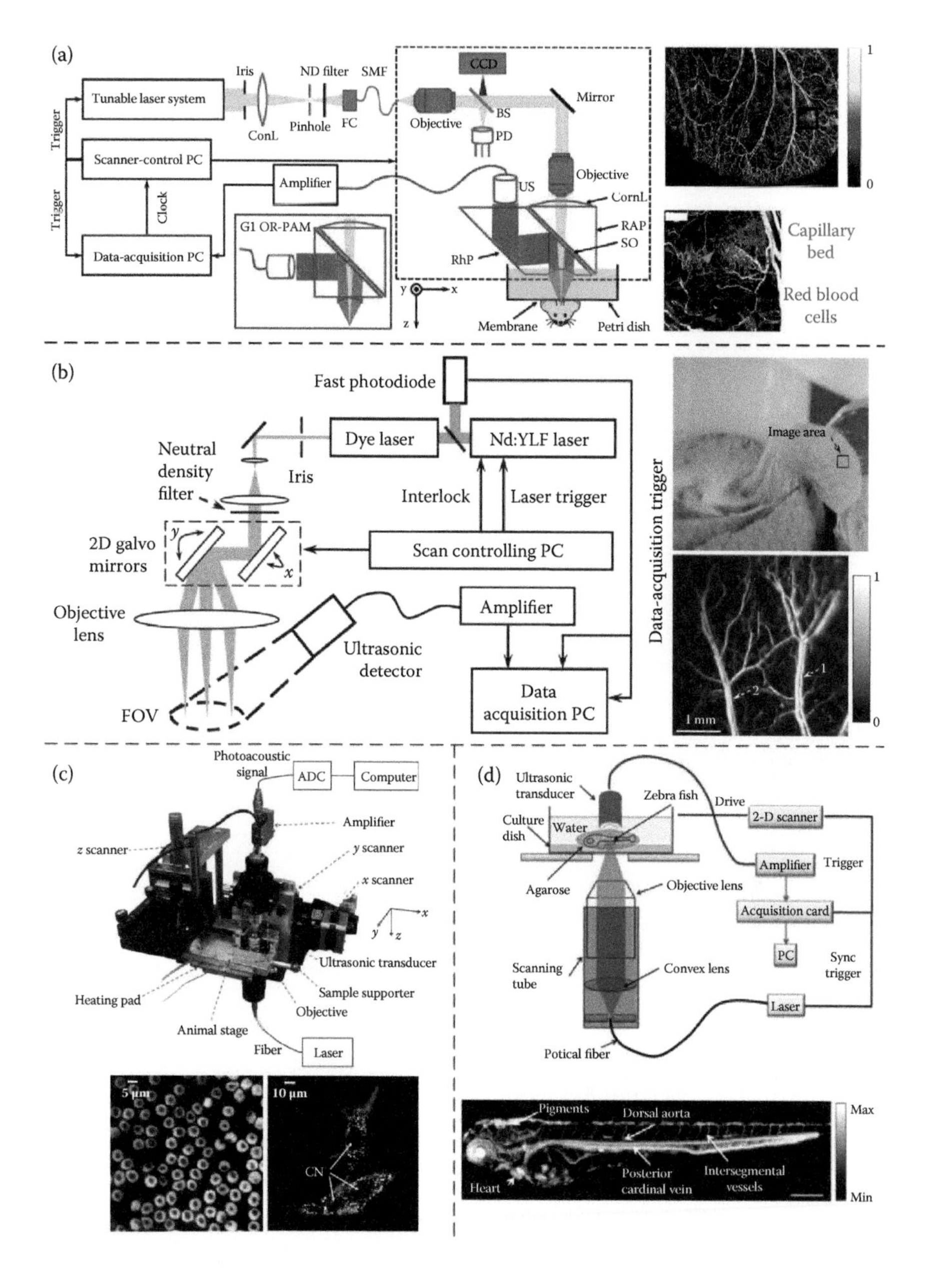

Figure 10.3 OR-PAM implementation and imaging results. (a) One type of reflection-mode system and total hemoglobin concentration of the living mouse ear, with 50 MHz central frequency. Scale bar: 150 μm. (b) Laser scanning reflection-mode system and vasculature in a mouse ear in vivo, with 50 MHz central frequency. Scale bar: 1 mm. (c) Transmission-mode OR-PAM and ex vivo images of red blood cells (left, scale bar: 5 μm) and melanoma cells (right, scale bar: 10 μm), with 40 MHz central frequency. (d) Transmission-mode system and in vivo image of zebra fish larva, with 40 MHz central frequency. Scale bar: 250 μm. (From (Xie et al. 2009 (panel a); Zhang et al. 2010 (panel b); Hu et al. 2011 (panel c); Ye et al. 2012b (panel d). Reproduced with permission.)

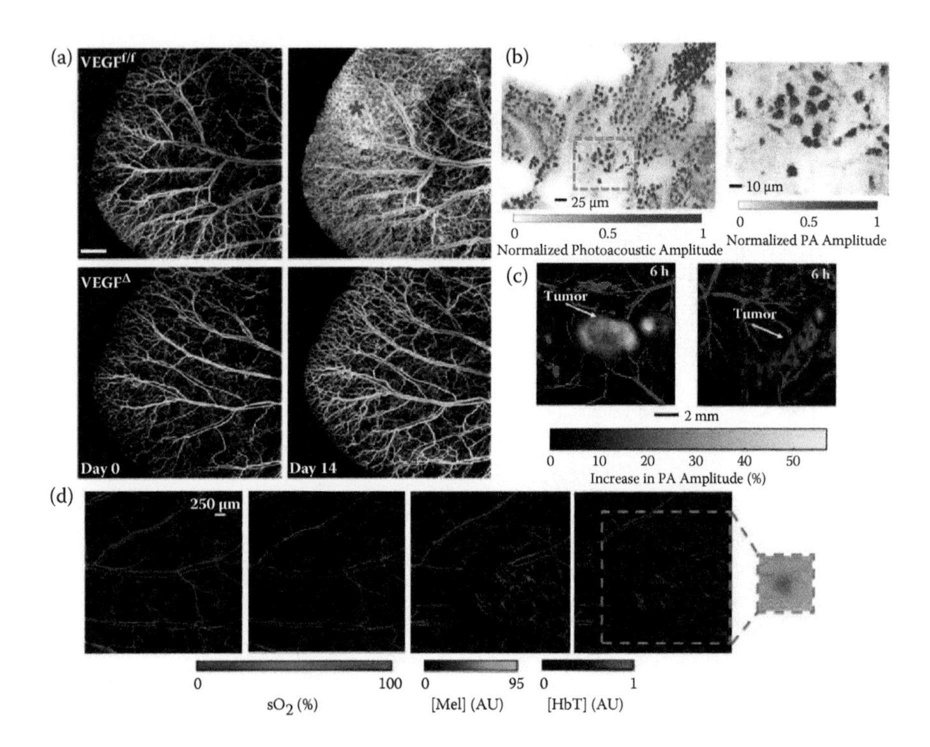

Figure 10.4 (a) In vivo label-free study of neovasculature, scale bar: 500 μm. (b) Label-free cell nuclei images. (c) Noninvasive photoacoustic results of melanomas labeled by gold nanocages. (d) Tyrosinase reporter gene in vivo PAM study. (From C. Kim et al. 2010. *ACS Nano* **4**(8): 4559–4564 (panel a); D.K. Yao et al. 2010. *Opt. Lett.* **35**(24): 4139–4141 (panel b); A. Krumholz et al. 2011. *J. Biomed. Opt.* **16**(8): 080503 (panel c); S. Oladipupo et al. 2011. *Proc. Natl. Acad. Sci. USA* **108**(32): 13264–13269 (panel d). Reproduced with permission.)

Moneron, G., and S.W. Hell. 2009. Two-photon excitation STED microscopy. *Optics Express* 17 (17):14569–14573.

Moneron, G., R. Medda, B. Hein, A. Giske, V. Westphal, and S.W. Hell. 2010. Fast STED microscopy with continuous wave fiber lasers. *Optics Express* 18 (2):1302–1309.

Nägerl, U.V., K.I. Willig, B. Hein, S.W. Hell, and T. Bonhoeffer. 2008. Live-cell imaging of dendritic spines by STED microscopy. *Proceedings of the National Academy of Sciences* 105 (48):18982.

Neumann, D., J. Bückers, L. Kastrup, S.W. Hell, and S. Jakobs. 2010. Two-color STED microscopy reveals different degrees of colocalization between hexokinase-I and the three human VDAC isoforms. *BMC Biophysics* 3 (1):4.

Opazo, F., A. Punge, J. Bückers, P. Hoopmann, L. Kastrup, S.W. Hell, and S.O. Rizzoli. 2010. Limited intermixing of synaptic vesicle components upon vesicle recycling. *Traffic* 11 (6):800–812.

Pellett, P.A., X. Sun, T.J. Gould, J.E. Rothman, M.Q. Xu, I.R. Corrêa, and J. Bewersdorf. 2011. Two-color STED microscopy in living cells. *Biomedical Optics Express* 2 (8):2364–2371.

Rankin, B.R., R.R. Kellner, and S.W. Hell. 2008. Stimulated-emission-depletion microscopy with a multicolor stimulated-Raman-scattering light source. *Optics Letters* 33 (21):2491–2493.

Reuss, M., J. Engelhardt, and S.W. Hell. 2010. Birefringent device converts a standard scanning microscope into a STED microscope that also maps molecular orientation. *Optics Express* 18:1049–1058.

Rittweger, E., K.Y. Han, S.E. Irvine, C. Eggeling, and S.W. Hell. 2009. STED microscopy reveals crystal colour centres with nanometric resolution. *Nature Photonics* 3 (3):144–147.

Sahl, S.J., M. Leutenegger, M. Hilbert, S.W. Hell, and C. Eggeling. 2010. Fast molecular tracking maps nanoscale dynamics of plasma membrane lipids. *Proceedings of the National Academy of Sciences* 107 (15):6829.

Schmidt, R., C.A. Wurm, S. Jakobs, J. Engelhardt, A. Egner, and S.W. Hell. 2008. Spherical nanosized focal spot unravels the interior of cells. *Nature Methods* 5 (6):539–544.

Schmidt, R., C.A. Wurm, A. Punge, A. Egner, S. Jakobs, and S.W. Hell. 2009. Mitochondrial cristae revealed with focused light. *Nano Letters* 9 (6):2508–2510.

Schrader, M., and S.W. Hell. 1996. 4Pi-confocal images with axial superresolution. *Journal of Microscopy* 183 (2):110–115.

Schrader, M., F. Meinecke, K. Bahlmann, M. Kroug, C. Cremer, E. Soini, and S.W. Hell. 1995. Monitoring the excited state of a fluorophore in a microscope by stimulated emission. *Bioimaging* 3 (4):147–153.

Sieber, J.J., K.I. Willig, C. Kutzner, C. Gerding-Reimers, B. Harke, G. Donnert, B. Rammner, C. Eggeling, S.W. Hell, and H. Grubmuller. 2007. Anatomy and dynamics of a supramolecular membrane protein cluster. *Science* 317 (5841):1072.

Sparrow, C.M. 1919. Theory of imperfect gratings. *Astrophysical Journal* 49 (2):65–95.

Vicidomini, G., G. Moneron, C. Eggeling, E. Rittweger, and S.W. Hell. 2012. STED with wavelengths closer to the emission maximum. *Optics Express* 20 (5):5225–5236.

Vicidomini, G., G. Moneron, K.Y. Han, V. Westphal, H. Ta, M. Reuss, J. Engelhardt, C. Eggeling, and S.W. Hell. 2011. Sharper low-power STED nanoscopy by time gating. *Nature Methods* 8:571–573.

Westphal, V, SO Rizzoli, MA Lauterbach, D Kamin, R Jahn, and SW Hell. 2008. Video-rate far-field optical nanoscopy dissects synaptic vesicle movement. *Science* 320 (5873):246.

Westphal, V., and S.W. Hell. 2005. Nanoscale resolution in the focal plane of an optical microscope. *Physical Review Letters* 94 (14):143903.

Wildanger, D., J. Bickers, V. Westphal, S.W. Hell, and L. Kastrup. 2009a. A STED microscope aligned by design. *Optics Express* 17:16100–16110.

Wildanger, D., R. Medda, L. Kastrup, and S.W. Hell. 2009b. A compact STED microscope providing 3D nanoscale resolution. *Journal of Microscopy* 236 (1):35–43.

Wildanger, D., E. Rittweger, L. Kastrup, and S.W. Hell. 2008. STED microscopy with a supercontinuum laser source. *Optics Express* 16 (13):9614–9621.

Willig, K.I., R.R. Kellner, R. Medda, B. Hein, S. Jakobs, and S.W. Hell. 2006a. Nanoscale resolution in GFP-based microscopy. *Nature Methods* 3 (9):721–723.

Willig, K.I., S.O. Rizzoli, V. Westphal, R. Jahn, and S.W. Hell. 2006b. STED microscopy reveals that synaptotagmin remains clustered after synaptic vesicle exocytosis. *Nature* 440 (7086):935–939.

Willig, K.I., B. Harke, R. Medda, and S.W. Hell. 2007. STED microscopy with continuous wave beams. *Nature Methods* 4 (11):915–918.

Wurm, C.A., D. Neumann, R. Schmidt, A. Egner, and S. Jakobs. 2010. Sample preparation for STED microscopy. *Methods in Molecular Biology* 591:185–199.

Xue, Y., C. Kuang, S. Li, Z. Gu, and X. Liu. 2012. Sharper fluorescent super-resolution spot generated by azimuthally polarized beam in STED microscopy. *Optics Express* 20 (16):17653–17666.

Problems

1. Simulate the inverse image of the confocal pinhole under diffraction of the microscopic objective, and the effect of the final PSF. Compare its FWHM with that of wide-field counterparts.
2. Duplicate Figure 1.1 and extend the diffraction limit criteria into two dimensions.
3. With the models of (1) $\sin^2(u)$, (2) u^2, (3)$J_1^2(u)$, simulate the STED resolution versus different STED power. Calculate the integration of the fluorescence at confocal and STED situations. How does the total fluorescent intensity change with the resolution?
4. Add a Gaussian function on top of your simulation of Problem 3 to simulate an imperfect doughnut, and see how the noise may affect the final fluorescent intensity.
5. Numerically solve the differential equation of Equation (1.8), and plot Figure 1.4.
6. Aside from stimulated emission, what other mechanisms can you think of to effectively inhibit the fluorescence emission? How can these mechanisms be applied in super-resolution microscopy?

chapter two

Structured illumination microscopy

Dan Dan, Baoli Yao, and Ming Lei

Contents

2.1 *Introduction*

2.1.1 *Illumination for microscopy*

Illumination of the specimen is the most important variable in achieving high-quality images in microscopy. Köhler illumination was first introduced in 1893 by August Köhler as a method of providing the optimum specimen illumination (Köhler 1893). Köhler illumination has since become the predominant technique for sample illumination in modern scientific light microscopy. It acts to generate an extremely uniform illumination of the sample and ensures that an image of the illumination source (for example, a halogen lamp filament) is not visible in the resulting image, thus allowing the user to realize the microscope's full potential. With the improvement and accuracy of lens manufacture, some kinds of aberration and blur can be corrected to get a better image quality. However, even if all the optics elements are made perfectly, a resolution limit of about 200 nm still exists in microscopy. This is the so-called diffraction barrier or Abbe diffraction limit (Abbe 1873). Its value is equal to $\sim 0.5\lambda/NA$, determined only by the wavelength of source light and the numerical aperture (NA) of objective lens. Moreover, the wide-field geometry nature of light microscopy leads to a certain depth-of-view on the focal plane. When a recording camera is put on the conjugated position of that focal plane, the in-focus and out-of-focus information are fused into a two-dimensional image. The targeted in-focus information cannot be separated from this fusion two-dimensional image directly. That is the reason why the conventional light microscopy cannot obtain optical sectioning, and the captured image is usually low both in contrast and signal-to-noise ratio (SNR).

2.1.2 *Super-resolution fluorescence microscopy*

The invention of green fluorescence protein (GFP) opened the era of fluorescence microscopy (Patterson et al. 1997). Browse through any journal on cell biology and the impact of fluorescence microscopy will be obvious, with >80% of the images of cells in books usually being acquired with a fluorescence microscope (Huang et al. 2010). Many new fluorescent labels such as fluorescence proteins, chemicals (such as rhodamine, quantum dots, and so on) (Fernández-Suárez and Ting 2008), and various labeling techniques have emerged to specify the interested part of the specimen (Gonçalves 2008). Thanks to these fluorescence labeling methods, the contrast and SNR of images are largely improved. Based on fluorescence microscopy, a number of methods have been proposed to deal with the weaknesses of resolution and optical sectioning in light microscopy, some aimed at one of them, some at both. For example, in stimulated emission depletion (STED) microscopy (Hell and Wichmann 1994, Klar and Hell 1999), a normal laser light is focused into a diffraction limited spot to excite the fluorophores to the higher energy state (i.e., the excited state) on the focal plane. At the same time, a doughnut light called the STED light is also focused and superposed on the former spot. In the region of overlap, the excited fluorophores are brought down to the lowest energy state (i.e., the ground state) by STED, leaving only the smaller center area to emit fluorescence signals. This reduces the size of the region of molecules that fluoresce, as if the "focal spot" of the microscope is sharpened. Scanning this sharpened spot across a sample then allows a super-resolution image to be recorded. Thus, this approach elegantly generates an image without the need for any postacquisition processing.

Another category of super-resolution method such as photoactivated localization microscopy (PALM) (Betzig et al. 2006, Hess et al. 2006) and stochastic optical reconstruction microscopy (STORM) (Rust et al. 2006; Bates et al. 2007; Huang et al. 2008) utilizes the "on-off" switchable characters of specific fluorescence molecules. In practice, each time, sparse fluorophores are excited to the "on" state to fluoresce. Because the positions of these fluorophores are far enough away from each other, each fluorophore can be precisely located to several nanometers through a localization algorithm. Then, those fluorophores are brought down to the "off" state. Finally, after hundreds and thousands of repeats, a super-resolution image is reconstructed by using all the localized fluorescent points.

2.1.3 *Optical sectioning microscopy*

The confocal microscope has gained popularity in the science and industry communities for optical sectioning (Wilson 1990). By using point illumination and a spatial pinhole to eliminate out-of-focus light in specimens, optical resolution and contrast can be increased by this technology. After point scanning through the whole specimen both in lateral and axial, the confocal microscope enables the reconstruction of three-dimensional structures from the acquired signals. As an alternative to the confocal microscope, the two- or multi-photon excitation fluorescence microscopy (So et al. 2000; So 2004) allows for imaging of tissues with higher depth due to its deeper penetration and efficient light detection. Being a special variant of the multi-photon fluorescence microscope, the two-photon microscope uses red-shifted excitation light (normally infrared) to excite fluorescent dyes, during which two photons of the infrared light are absorbed simultaneously to the excited state. Due to the nonlinear threshold effect of the multi-photon absorption, the background signal is strongly suppressed. Using infrared light minimizes scattering in the tissue. Both effects lead to an increased penetration depth for the multi-photon excitation fluorescence microscopy.

Selective/single plane illumination microscopy (SPIM) (Huisken et al. 2004) or light sheet microscopy (LSM) (Keller et al. 2008) is another fluorescence microscopic technique with an intermediate optical resolution, but good sectioning capabilities. In contrast to epi-fluorescence microscopy, only a thin slice (usually a few hundred nanometers to a few micrometers) of the sample is illuminated perpendicularly to the direction of observation. As only the actually observed section is illuminated, this method reduces the phototoxicity induced on a living sample dramatically, and also the background signal is greatly reduced, resulting in an image with higher contrast, comparable to confocal microscopy.

2.1.4 *Emergence of structured illumination microscopy*

The techniques described above, whether for super-resolution or for optical sectioning, are time-consuming and complex in configuration. STED, confocal, and two-photon microscopy require a point scanning process to work. For a point detector (such as APD or PMT), exposure time should be long enough (usually >10 μs) to maintain an accepted SNR of imaging (Křížek and Hagen 2012). Meanwhile, the scanning steps between two adjacent points should be short (usually less than the half size of resolution) enough to follow the Nyquist sampling theorem. These two factors make the scanning process low speed. An additional configuration is needed to trigger feedback for synchronization during the entire scanning procedure. It inevitably leads to a complex configuration with low stability. The pointillism methods of PALM and STORM require hundreds and thousands of raw images to accomplish a super-resolution image. Therefore, time and cost are also their drawbacks. Moreover, accuracy and speed must be trade-off due to the localization algorithm of fluorescence molecules. In order to overcome these shortcomings of the above techniques, some scientists put insights into the illumination way in microscopy. In 1997, Neil et al. used a structured illumination to obtain optical sectioning on a conventional microscope. In 2000, Gustafsson et al. proposed a structured illumination microscopy to break the diffraction limit by a factor of two. These two structured illumination microscopies (SIMs) use similar methods of illumination (sinusoidal fringe illumination) and a phase-shifting method, but with different data processing algorithms. Over ten years of development, structured illumination is no longer limited to a sinusoidal fringe pattern. It may be sparse speckles (Waterman-Storer et al. 1998; Walker 2001; Ventalon and Mertz 2005; Heintzmann and Benedetti 2006), chess-like (Frohn et al. 2000; Fedosseev et al. 2005), or even random unknown patterns (Mudry et al. 2012), dependent on different purposes such as super-resolution, optical sectioning, or range measurement. As an alternative to Köhler illumination, SIM is essentially based on wide-field geometry; therefore, it requires minimal modification on a conventional microscope with a lower cost. It also has all the advantages of conventional microscopy: (1) less complex optical configuration not requiring point-wise scanning modules (2) wide-field full-frame imaging with high recording speed, and (3) low light dose, avoiding harming live specimens.

2.1.5 *Development of structured illumination microscopy*

In this section, we concentrate on the structured illumination microscopy for super-resolution and optical sectioning. As far as we know, structured illumination has also wide applications in metrology. So, a brief description about this is also mentioned in the end.

2.1.5.1 SIM-SR

In SIM for super-resolution (SIM-SR), a sinusoidal fringe patterned light is used to illuminate a sample (Gustafsson 2000). When the sinusoidal fringe overlaps with the high spatial frequency component of the sample, a new sparse pattern appears due to the well-known "Moiré pattern" phenomenon. The spatial frequency of Moiré fringe is lower than both those of the structured illumination and the high frequency components of the sample. The Moiré pattern encodes the unresolved structure of the sample (i.e., the high frequency components) into the "observable region" of the limited bandwidth of microscopic objective. Decoding the high frequency information requires a three-step phase-shifting operation of the illumination pattern. In order to obtain a nearly isotropic frequency spectrum along all directions in lateral, multiple orientation structured illumination (at least in two or three different orientations) is necessary. Finally, all the decoded high frequency components are restored to their proper positions in the frequency spectrum, and then an enhanced resolution image beyond the diffraction limit can be obtained by making an inversed Fourier Transform transform for the enlarged frequency spectrum. Two similar techniques called laterally modulated excitation microscopy (LMEM) (Heintzmann and Cremer 1999) and harmonic excitation light microscopy (HELM) (Frohn et al. 2000; Fedosseev et al. 2005) were proposed by Heintzmann and coworkers in 1999 and Frohn J.T. et al. in 2000, respectively. It should be noticed that the recently proposed image scanning microscopy (ISM) (Heintzmann and Cremer 1999; Muller C.B. and J. Enderlein 2010; York et al. 2012) and the structured illumination microscopy with unknown patterns (Mudry et al. 2012) are based on different super-resolved principles from the SIM-SR described here.

In practice, the generation of structured illumination and data post-processing are the two essential issues in SIM. The increased resolution attained by SIM is based on the degree to which high spatial frequencies can be down converted into the passband of the imaging system. To do this effectively, a high-contrast, high-frequency illumination pattern is required. With a laser source, interference by two or multiple beams is a common means to get the required structured patterns (Ryu et al. 2006; Littleton et al. 2007). A grating is often used to diffract incident light into +1st order and −1st order beams, which will interfere with each other to form the sinusoidal fringe pattern (Wang et al. 2011). Considering the speed and accuracy in phase-shifting, gratings are replaced by SLM (Chang et al. 2009; Hirvonen et al. 2009; Kner et al. 2009; Shao et al. 2011; Fiolka et al. 2012), either LC-SLM or Ferro-SLM, DOE (Gardeazábal Rodríguez et al. 2008) or hole-arrays (Docter et al. 2007) with no moving parts in optical setup, to maintain the stability of the system and improve the recording speed. In use of incoherent light source (e.g., LED), pattern projection with a structured mask or a modulated DMD chip through objective lens allows a compact system without coherent speckle noise (Dan et al. 2013). In the post-processing field, much research is focused on optical transfer function (OTF) compensation (Shroff et al. 2008; Somekh et al. 2008, 2009, 2011), estimation of phase-shifting (Shroff et al. 2007, 2010; Wicker et al. 2013), algorithm efficiency (Ryu et al. 2003; Orieux et al. 2012), and aberration correction (Ryu et al. 2003; Beversluis et al. 2008) to enhance the image quality (SNR, contrast, and resolution). OTF compensation extends the concept of the conventional transfer function by incorporating noise statistics, thus giving a measure of the SNR at each spatial frequency. Since the knowledge of phase shift in the structured illumination is critical for the data post-processing, any inaccuracy in this knowledge will lead to artifacts in the reconstruction. Estimation of phase shifting is necessary to ensure the accuracy. New algorithms (Ryu et al. 2003; Orieux et al. 2012) and general programming based on a parallel GPU framework are proposed to reduce the data-processing time (Lefman et al. 2011). Adaptive optics models are used to deal with the aberrations leading by the high NA objective lens (Ryu et al. 2003; Beversluis et al. 2008).

Based on the SIM-SR principle, spatial resolution enhancement in all three dimensions (3D) can be carried out by a three-dimensional structured illumination (3D-SIM) (Frohn et al. 2001; Gustafsson et al. 2008). A resolution of 100 nm has been reported with 3D-SIM both in lateral and axial directions (Gustafsson 2000; Schermelleh, Carlton et al. 2008; Kner et al. 2009). Another choice to get such 3D super-resolution is to incorporate an axial super-resolution scheme (for example TIRF) with the lateral SIM (Fiolka et al. 2008; Kner et al. 2009). Since the spatial frequency of Moiré fringe is also limited by the passband of the optical system, the maximum resolution extension of SIM is only twofold over conventional microscopy. Further improvement of resolution can be achieved by exploiting the nonlinear response of fluorescence molecules with saturated excitation, which is called saturated structured illumination microscopy, SSIM) (Heintzmann 2003; Gustafsson 2005; Zhang et al. 2011). The practical resolving power is determined by the SNR, which in turn is limited by photobleaching (Gurn et al. 2011). A 2D point resolution of 50 nm is possible on sufficiently bright and photostable samples (Gustafsson 2005). Recently, cellular structure at resolution of 50 nm is revealed with a photoswitchable protein (Hirvonen et al. 2008; Rego et al. 2012).

SIM-SR is compatible with various microscopies, such as light sheet microscopy (Hirvonen et al. 2008; Planchon et al. 2011; Gao et al. 2012), imaging interferometric microscopy (Neumann et al. 2008; Shao et al. 2012), line-scanning microscopy (Mandula et al. 2012), 4Pi microscopy (Shao et al. 2008), differential interference contrast (DIC) (Chen et al. 2013) and so forth to extend their functionalities, and thus has broad applications in biology (Schermelleh et al. 2008; Cogger et al. 2010; Fitzgibbon et al. 2010; Best et al. 2011; Shao et al. 2011; Markaki et al. 2012; Sonnen et al. 2012; Strauss et al. 2012), medicine (Bullen 2008), fluidics (Lu et al. 2013), nanomorphology (Hao et al. 2013), and nanoparticles (Chang et al. 2011). Now, there are a number of commercial SIM-SR instruments available on the market, such as DeltaVision from GE Healthcare, N-SIM from Nikon Inc., and ELYRA S.1 from Carl Zeiss Inc.

2.1.5.2 SIM-OS

The basic idea of SIM for optical sectioning (SIM-OS) is to extract the in-focus information and reject the background of an out-of-focus portion of the image by structured illumination. It is mainly classified into two categories: speckle illumination and grid patterns illumination. Both kinds of microscopy are based on the geometry of wide-field and aimed at imaging thick samples with high speed, large field-of-view.

Speckle-illuminated fluorescence microscopy, proposed by Walker and coworkers (Gustafsson 2000; Jiang and Walker 2004, 2005, 2009), is a nonscanning imaging technique that uses a random time-varying laser speckle pattern to illuminate the specimen, and then record a large number of wide-field fluorescence frames and a sequence of corresponding illumination speckle patterns, or reference speckle patterns, with a CCD camera. Then, the recorded frames are processed by an averaging formula over the set of recorded frames. Mertz and coworkers (Ventalon and Mertz 2005, 2006; Ventalon et al. 2007; Lim et al. 2008; Choi et al. 2011; Lim et al. 2011) have demonstrated a technique called dynamic speckle illumination (DSI) microscopy that produces fluorescence sectioning with illumination patterns that are neither well defined nor controlled. The idea of this technique is to illuminate a fluorescent sample with random speckle patterns obtained from a laser. Speckle patterns are granular intensity patterns that exhibit inherently high contrast. Fluorescence images obtained with speckle illumination are therefore also granular; however, the contrast of the observed granularity provides a measure of how in focus the sample is: high observed contrast indicates that the sample is dominantly in focus, whereas low observed contrast indicates it is dominantly out of focus. The observed speckle contrast thus serves as a weighting function indicating the in-focus to out-of-focus ratio in a fluorescence image. While the exact intensity pattern incident on a sample is not known, the statistics of the intensity distribution are well known to obey a negative exponential probability distribution. According to this distribution, the contrast of a speckle pattern scales with average illumination intensity. Thus, weighting a fluorescence image by the observed speckle contrast is equivalent to weighting it by the average illumination intensity (as in standard imaging); however, with the benefit that the weighting preferentially extracts only in-focus signals.

Generally, speckle illumination microscopy needs tens or hundreds of speckle-illuminated raw images varied in space and time to obtain a sectioning image by algorithms in statistics. The drawback is obvious in time and cost. In fact, as early as 1997, Neil and coworkers (1997, 1998) proposed a grid pattern illumination microscopy with only three phase-shifted raw images to reduce the recording time on a conventional microscope. This method is based on the fact that under a grid pattern illumination with high spatial frequency, only the in-focus part of the image is modulated while the out-of-focus region is out of modulation. This is because the high spatial frequency component attenuates very fast with defocus. The out-of-focus signal cannot carry on the high frequency illumination patterns. The following step is to eliminate the residual patterns in a modulated in-focus signal, usually adopting a three-step phase-shifting method. Since Neil's approach has the same structured illumination pattern as SIM-SR, it is also referred to as SIM in many journals. To avoid confusion, we refer to this approach as Neil-SIM in the following sections.

The attractive feature of Neil-SIM is its simplicity in system configuration and a data post-processing algorithm. A grating is employed as a plug-in module to a conventional microscope to produce fringe pattern illumination. The algorithm is based on simple elementary operations to three raw images. The structured illumination pattern must be known in advance and the movement (phase-shifting) of the illumination pattern should be accurate. It is a challenge for the grating approach due to the physical movement in precision with stability. Fortunately, the grating can be replaced by various programmable spatial light modulators (SLM) with no moving parts to ensure the phase-shifting both in accuracy and speed (as described above in SIM-SR) (Fukano and Miyawaki 2003; Monneret et al. 2006; Wong et al. 2006; Delica and Blanca 2007; Mazhar et al. 2010; Křížek et al. 2012). Moreover, the phase-shifting process can be avoided by color-coded or polarization-coded structured illumination microscopy, which allows for providing optical sectioning in a single shot (Krzewina and Kim 2006; Wicker and Heintzmann 2010). In theoretical scope, studies are mostly concentrated on the aspects of image formation (Karadaglić

2008; Karadaglić and Wilson 2008), quantitative analysis of sectioning and noise (Barlow and Guerin 2007; Hagen et al. 2012), optimization and characterization for microscopy (Chasles et al. 2007; Hagen et al. 2012).

Combining Neil-SIM with other imaging techniques is of interest in a wide range of research fields, such as fluorescence lifetime imaging microscopy (FLIM) (Cole et al. 2000, 2001; Siegel et al. 2001; Webb et al. 2002), image mapping spectrometer (IMS) (Gao et al. 2011), optical tomography (Lukic et al. 2009, Bélanger et al. 2010; Ducros et al. 2013), plane illumination microscopy (light-sheet microscopy) (Edouard and Mattias 2008; Kristensson et al. 2008a; Schröter et al. 2012), two-photon excitation microscopy (Dal Maschio et al. 2010), spectral imaging (Siegel et al. 2001), and holography (Ansari et al. 2002). So far, Neil-SIM has found a variety of applications in endomicroscopy (Karadaglić et al. 2002; Wong et al. 2006; Bozinovic et al. 2008), multiple-scattering suppression (Edouard and Mattias 2008), visualization of gas flow (Kristensson et al. 2008b), cell biology (Langhorst et al. 2009), hydrocarbon-bearing fluid inclusions (Blamey et al. 2008), retinal imaging (Gruppetta and Chetty 2011), and so on. Nowadays, there are several types of commercial Neil-SIM apparatus available on the market, such as ApoTome (Carl Zeiss Inc.), Optigrid M (Olympus Inc.), and OptiGrid module (which can be integrated into specific microscopes from Leica Inc.)

2.1.5.3 SI-SP

Besides its utilization in microscopy, the structured illumination strategy is also popular in the community of 3D surface measurement for relatively large objects (Su and Zhang 2010). Projecting a narrow band of light onto a three-dimensional surface produces a line of illumination that appears distorted from other perspectives than that of the projector, and can be used for an exact geometric reconstruction of the surface shape. The structured illumination for surface profilometry (SI-SP) is of great attraction in the domain of range measurement (Caspi et al. 1998), 3D imaging (Rocchini et al. 2001), and pattern recognition (Caspi et al. 1998). It has found widespread use in product inspection, driver assistance systems in cars, stereo imaging, automation, and so on.

2.2 Theoretical basis of SIM

2.2.1 Linear SIM for super-resolution

The microscope, as an imaging system, can be treated and analyzed by the theory of signals and systems which is currently widely used in the field of electronics. In most cases, the microscope is a time-invariant and linear system. According to the principle of linear signal systems, the characteristic of the whole system is determined by the impulse response, that is, once

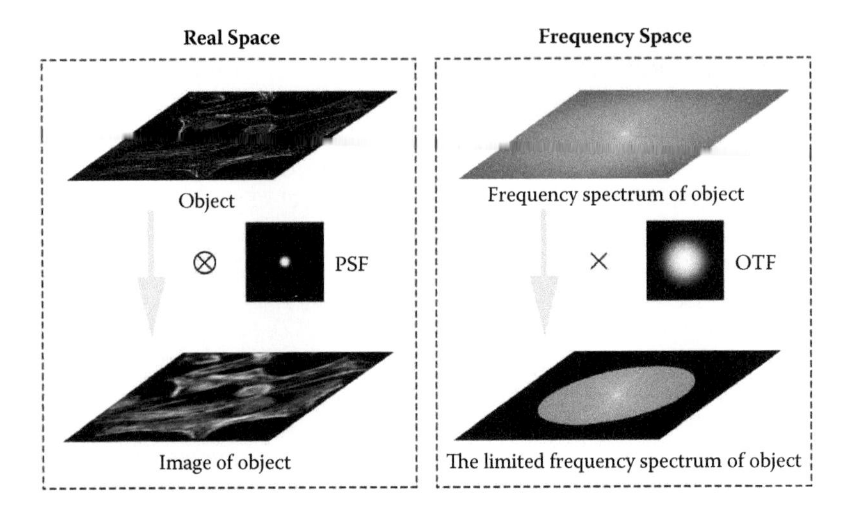

Figure 2.1 Schematic of image formation in real space and in spatial frequency domain.

the impulse response function is given, the output of the system can be calculated by the convolution of the input and the impulse response function. In a microscope, the point spread function (PSF) describes the impulse response function to an ideal point source or a point object. It maps the emission distribution of the sample to the intensity distribution on the detector in fluorescence microscopy. Thus, an image of any objects is a convolution of the emission intensity and the PSF. Meanwhile, in the spatial frequency domain, the optical transform function (OTF), which is the Fourier transform of PSF, indicates the allowed passband and the weighted modulation of each frequency of the emission intensity distribution. This basic knowledge on the image formation in a microscope is illustrated in Figure 2.1.

In mathematics, the image formation is expressed as:

$$D(x) = \left[E_m(x) \otimes PSF(x)\right] \tag{2.1}$$

where $D(x)$ is the distribution of photons on the detector, $E_m(x)$ is the emission distribution of the sample, $PSF(x)$ is the point spread function of the microscope, x indicates the space coordinate component, and $\otimes$ denotes a convolution operation. Making a Fourier transform to Equation (2.1) and using the convolution theorem, an expression is obtained to describe the spatial frequencies accessible to the microscope:

$$\tilde{D}(k) = \left[\tilde{E}_m(k) \cdot OTF(k)\right] \tag{2.2}$$

where the tilde denotes the function after Fourier transform, OTF is the Fourier transform of the PSF, and k is the corresponding spatial frequency component. In a conventional fluorescence microscope, the fluorescence emission intensity $E_m(x)$ is linearly dependent on the illumination intensity $I(x)$ and the fluorescence yield rate of the dye-labeled sample. Actually, the distribution of fluorescence yield rate across the sample indicates the structure of the sample $S(x)$:

$$E_m(x) = I(x) \cdot S(x) \tag{2.3}$$

or expressed in the frequency domain by:

$$\tilde{E}_m(k) = \tilde{I}(k) \otimes \tilde{S}(k) \tag{2.4}$$

Substituting Equation (2.4) into Equation (2.2), the frequency spectrum of the observed image is presented in the form:

$$\tilde{D}(k) = \left[\tilde{I}(k) \otimes \tilde{S}(k)\right] \cdot OTF(k) \tag{2.5}$$

In the case of uniform illumination, which is the situation of conventional fluorescence microscopy, $I(x)$ = constant, we have:

$$\tilde{D}(k) = \tilde{S}(k) \cdot OTF(k) \tag{2.6}$$

This means all spatial frequencies of the sample are confined by the *OTF*. If the illumination is not uniform, for example, being one-dimensional sinusoidal structured, $I(x)$ is expressed in

$$I(x) = I_0\left[1 + m \cdot \cos(2\pi k_0 \cdot x + \varphi)\right] \tag{2.7}$$

where k_0 and φ denote the spatial frequency and the initial phase of the sinusoidal fringe illumination pattern, respectively. I_0 and m are constants called mean intensity and modulation depth, respectively. By making the Fourier transform, Equation (2.7) can be expressed in frequency space:

$$\tilde{I}(k) = I_0\left[\delta(k) + \frac{m}{2} \cdot e^{i\varphi}\delta(k - k_0) + \frac{m}{2} \cdot e^{-i\varphi}\delta(k + k_0)\right] \tag{2.8}$$

Substituting Equation (2.8) into Equation (2.5), the frequency spectrum of the observed image is presented by:

$$\tilde{D}(k) = I_0\left[\tilde{S}(k) + \frac{m}{2}\mathrm{e}^{i\varphi}\,\tilde{S}(k - k_0) + \frac{m}{2}\mathrm{e}^{-i\varphi}\,\tilde{S}(k + k_0)\right] \cdot OTF(k) \tag{2.9}$$

The first term in Equation (2.9) represents the normal frequency spectrum observed by the conventional microscope, in which the value scope of k is $|k| \leq f_c$, where $f_c = 2NA/\lambda$ is the cut-off frequency of OTF, NA is the numerical aperture of the objective lens, and λ is the wavelength. The second and third terms in Equation (2.9) contribute additional information. They have the same form as the first term, but the central frequencies are shifted by $+k_0$ and $-k_0$, respectively, and k satisfies $|k - k_0| \leq f_c$ and $|k + k_0| \leq f_c$, respectively. Thus, the value scope of k is extended to $|k| \leq |f_c + k_0|$. Because the maximum value of k_0 can reach to f_c, the maximum spectral bandwidth for linear SIM will be doubled over the conventional microscope. This is the reason why the resolution of the linear SIM-SR is able to be enhanced theoretically by a factor of two.

To obtain the three terms of frequency spectra in Equation (2.9), three different phases φ_1, φ_2, φ_3 for the illumination pattern can be set to get the following three independent linear equations, provided that m and $OTF(k)$ are known beforehand:

$$\begin{pmatrix} \tilde{D}_1(k) \\ \tilde{D}_2(k) \\ \tilde{D}_3(k) \end{pmatrix} = I_0 \begin{pmatrix} 1 & \frac{m}{2}e^{i\varphi_1} & \frac{m}{2}e^{-i\varphi_1} \\ 1 & \frac{m}{2}e^{i\varphi_2} & \frac{m}{2}e^{-i\varphi_2} \\ 1 & \frac{m}{2}e^{i\varphi_3} & \frac{m}{2}e^{-i\varphi_3} \end{pmatrix} \begin{pmatrix} \tilde{S}(k) \\ \tilde{S}(k-k_0) \\ \tilde{S}(k+k_0) \end{pmatrix} \cdot OTF(k) \tag{2.10}$$

The resultant spectral terms $\tilde{S}(k)$, $\tilde{S}(k-k_0)$ and $\tilde{S}(k+k_0)$ solved from Equation (2.10) consist of the separated frequency spectra of the sample, as illustrated in Figure 2.2(a). Taking the overlapped frequency region and the central frequencies shifted by $+k_0$ and $-k_0$ into account, we can assemble these three parts together to form an extended frequency spectrum of the sample, as shown in Figure 2.2(b). To obtain nearly isotropic resolution enhancement, the single-orientation sinusoidal fringe pattern should be rotated in at least two orthogonal orientations.

A simulation of SIM-SR is conducted in the following procedure, illustrated in Figure 2.3. Two sets of sinusoidal fringe illuminated images are used to simulate the structured illumination in horizontal and vertical orientations, respectively. Each set of images contains three images representing the structured illumination in three phases at a gap of $2\pi/3$. Then, each simulated structured illumination image is multiplied by a sample image to achieve the fluorescence emission process under different orientations and phases. Convoluted with PSF of the microscope, the corresponding two sets of simulated recording images can be obtained. The next step is to transform these raw images into frequency space through a fast Fourier transform operation (FFT). For two-orientation illumination,

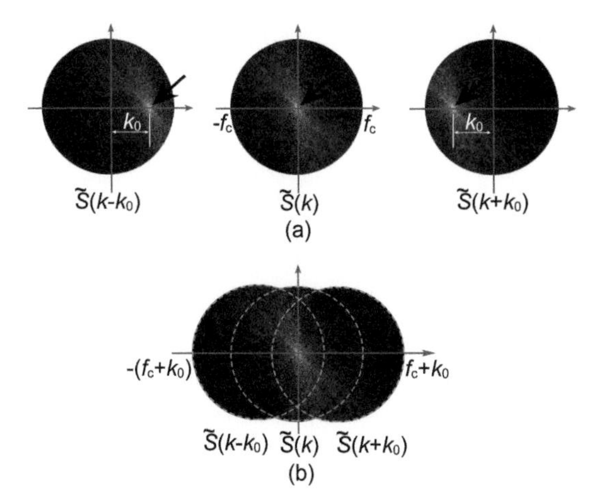

Figure 2.2 Principle of spatial frequency spectrum extension in single-orientation for linear SIM-SR. f_c is the cut-off frequency of OTF, and k_0 is the spatial frequency of the sinusoidal fringe illumination pattern. The arrows in (a) point out the central frequency of the spectrum.

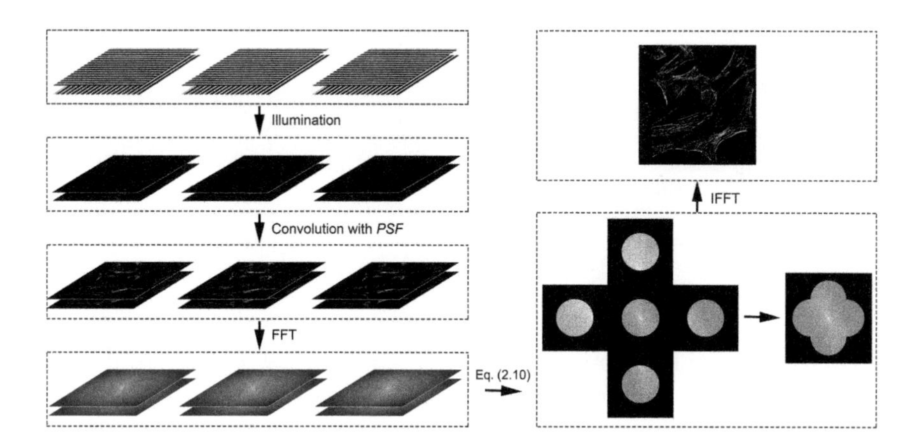

Figure 2.3 Procedure of SIM for super-resolution in horizontal and vertical directions.

according to Equation (2.10), five frequency spectra can be solved and allocated to their proper positions in the frequency space. Finally, a combined and enlarged frequency spectrum of the sample is obtained. Through IFFT (inverse FFT), a super-resolution image can be reconstructed in the real space.

2.2.2 Nonlinear SIM for super-resolution

In the above discussion, the precondition is that the emission intensity $E_m(x)$ is linearly proportional to the illumination intensity $I(x)$. If there is a nonlinear fluorescence effect, such as two-photon absorption, stimulated emission, or ground state depletion, Equation (2.3) will become:

$$E_m(x) = F[\mathrm{I}(x)] \cdot S(x) \tag{2.11}$$

Here F is a function of the illumination intensity, which depends on the nonlinear type of the sample in response to the illumination light. Normally, it can be expanded as a power series of the intensity with an infinite number of terms:

$$F[\mathrm{I}(x)] = a_0 + a_1 I(x) + a_2 I^2(x) + a_3 I^3(x) + \cdots \tag{2.12}$$

Substituting Equation (2.7) into Equation (2.12) and making a mathematical derivation, we obtain:

$$F[I(x)] = \sum_{n=0}^{\infty} b_n \cdot \cos[n(2\pi k_0 \cdot x + \varphi)] \tag{2.13}$$

where b_n ($n = 0,1,2,\cdots$) are constants. After transforming Equation (2.13) into the frequency domain and substituting the result into Equation (2.5), we find that the spectrum takes the form of a weighted sum of an infinite number of components:

$$\tilde{D}(k) = OTF(k) \sum_{n=-\infty}^{\infty} b_n \tilde{S}(k - nk_0) e^{-in\varphi} \tag{2.14}$$

Comparing this equation to the linear Equation (2.9), we can see that all frequency components from both the conventional *OTF* and the extended linear structured illumination *OTF* are involved when taking the terms of $n = -1$, 0, and 1. Importantly, Equation (2.14) also contains an infinite number of higher-order harmonics of the fundamental frequency k_0. In practice, only a few numbers, N, of the higher-order frequency components are able to be extracted with high enough SNR. Similar to the separation of the linear structured illumination raw data, it is possible to separate the higher-order frequency components if $(2N + 1)$ images are taken at different phases of the illumination pattern. Again, the single-orientation sinusoidal fringe pattern needs to be rotated in a sufficient number of orientations in order to fill the entire two-dimensional space isotropically.

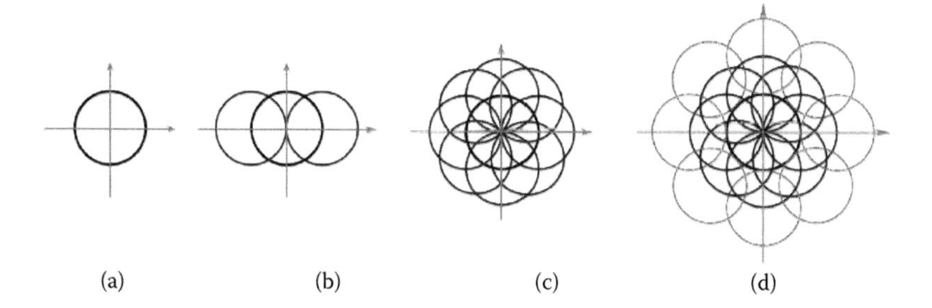

Figure 2.4 Schematic of frequency spectrum extension from linear SIM-SR to nonlinear SIM-SR. (a) Frequency spectrum for conventional microscopy; (b) frequency spectrum for linear SIM-SR in one-orientation extension; (c) frequency spectrum for linear SIM-SR in four-orientation extension; (d) frequency spectrum for nonlinear SIM-SR in four-orientation extension.

The simulation process of nonlinear SIM-SR is similar to that of the linear SIM-SR. The difference is the increased number of shifted phases and the number of separated high-frequency components. Figure 2.4 illustrates the more extended frequency spectrum from the conventional microscopy to the linear SIM-SR, and then to the nonlinear SIM-SR.

2.2.3 *SIM for optical sectioning*

On a conventional wide-field microscope, the targeted in-focus information is always blended with unwanted out-of-focus information, which degrades the imaging quality severely. Optical sectioning is a process of distinguishing the in-focus information from the merged data. Optical sectioning, often referred to as minimal depth or z resolution, allows the three-dimensional reconstruction of a sample from images captured at different focal planes.

To understand the principle of SIM for optical sectioning, we proceed from the discussion on the permitted spatial frequency intensity distribution with defocus on the conventional microscope. The normalized spatial frequency intensity distribution $H(z,\upsilon)$ can be expressed in the following formula (2.15) with the Stokseth's approximation (Neil et al. 1997):

$$H(z,\upsilon) = 2f(\upsilon)\left|\frac{J_1[4u\upsilon(1-\upsilon)]}{4u\upsilon(1-\upsilon)}\right| \tag{2.15}$$

with

$$u = 8(\pi/\lambda)\sin^2(\alpha/2)\cdot z \tag{2.16}$$

Here $J_1(-)$ represents the first-order Bessel function, $f(\upsilon) = 1 - 1.38\upsilon + 0.0304\upsilon^2 + 0.344\upsilon^3$, υ is the normalized spatial frequency, whose original value falls in the scope of $[0,f_c]$, $f_c = 2NA/\lambda$ is the cut-off frequency, $NA = n \cdot \sin\alpha$ (n is the refraction index, α is the half aperture angle of the objective lens) is the numerical aperture, and λ denotes the wavelength. z is the defocus variable, that is, the axial range apart from the focal plane ($z = 0$). In the case of $\upsilon = 0$, $H(z,\upsilon)$ gets its maximum value of $H_{max} = 1$. In the limit of $\upsilon = 1$, $H(z,\upsilon)$ reaches the minimum value of $H_{min} = 0$. Figure 2.5(a) and (b) shows the normalized spatial frequency intensity distribution with respect to the defocused range and the normalized spatial frequency according to Equation (2.15). It can be seen that the intensity of the higher-frequency component attenuates much more rapidly than the lower-frequency component with increasing defocus. So, if a proper high frequency sinusoidal fringe light illuminates the sample, only the in-focus part but not

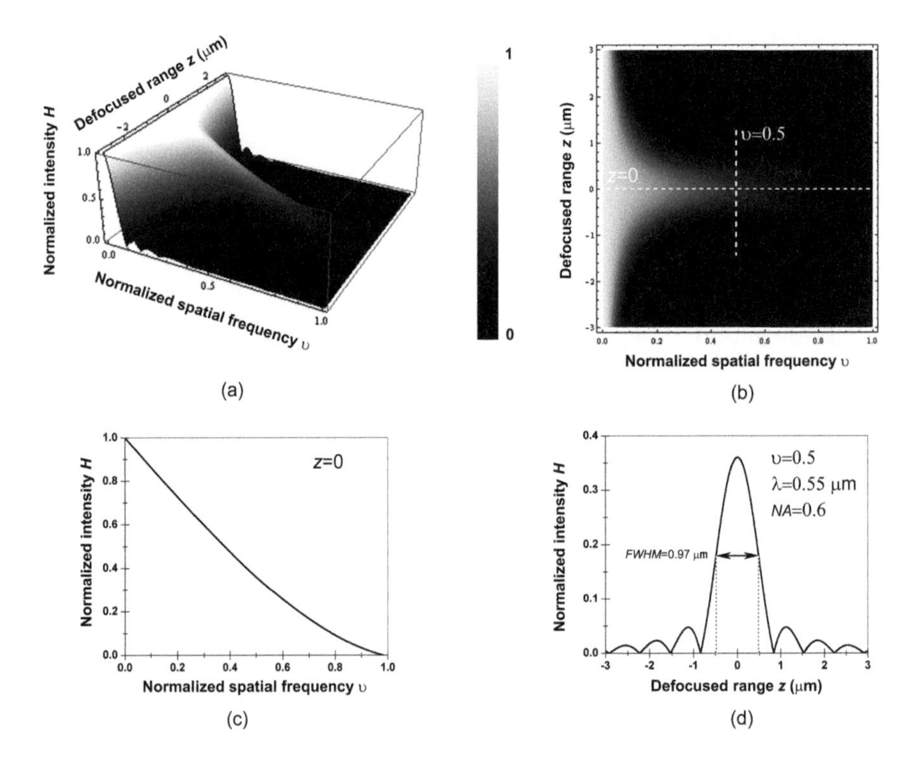

Figure 2.5 The normalized spatial frequency intensity distribution of a microscope with defocus. (a) and (b) show the normalized spatial frequency intensity distribution with respect to the defocused range and the normalized spatial frequency in 3D and 2D graphs, respectively. (c) Indicates the attenuation curve of $H(z,\upsilon)$ with respect to υ in the focal plane ($z = 0$). (d) Shows the $H(z,\upsilon)$ profile with respect to the defocused range at $\upsilon = 0.5$, and the full width at half maximum (FWHM) of the curve is calculated at the given parameters, which measures the optical sectioning strength of the SIM.

the out-of-focus part will be modulated by the structured illumination. Generally, a higher frequency of illumination will result in thinner depth of optical sectioning. But considering the modulation depth and SNR will degrade with increasing illumination frequency, in practice there is a trade-off between using fringe illumination and proper spatial frequency.

In order to specify the optical sectioning capability of Neil-SIM under a fringe illumination with a fixed spatial frequency υ, the full width at half maximum (FWHM) of the *H*(z, υ) curve is defined as the optical sectioning strength. According to Equation (2.15), the FWHM can be obtained by solving the following Equation (2.17) to get two roots z_1, z_2 and then FWHM = $|z_1\text{-}z_2|$.

$$2\left|\frac{J_1[4u\upsilon(1-\upsilon)]}{4u\upsilon(1-\upsilon)}\right|=\frac{1}{2} \tag{2.17}$$

$$\text{FWHM}=\frac{0.044407\lambda}{\upsilon(1-\upsilon)\sin^2[0.5\arcsin(NA/n)]} \tag{2.18}$$

As an example, the intensity profile at υ = 0.5 is plotted in Figure 2.5(d), and the FWHM is calculated to be 0.97 μm under the conditions of λ = 0.55 μm, *NA* = 0.6, which means the optical sectioning strength in this case is 0.97 μm. From Equation (2.18), the dependence of optical sectioning strength FWHM on the normalized spatial frequency and the numerical aperture is plotted in Figure 2.6. It can be seen that despite the value of

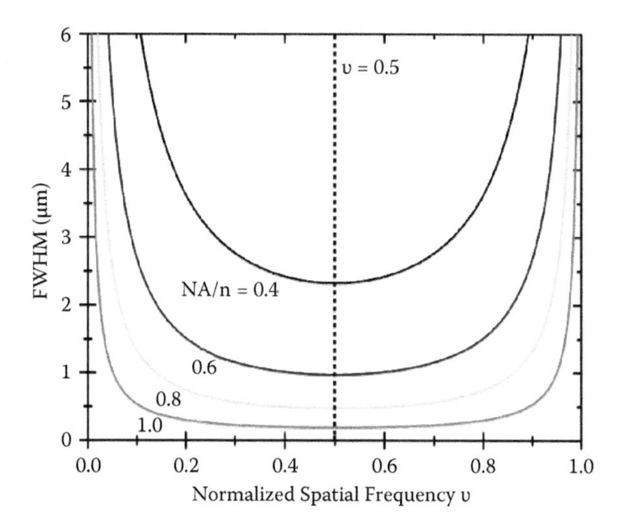

Figure 2.6 The dependence of optical sectioning strength (FWHM) on the normalized spatial frequency and the numerical aperture.

NA that is adopted, there exists an optimal normalized spatial frequency $\upsilon = 0.5$, at which the FWHM gets the minimum value. Of course, a higher NA will give a narrow FWHM. This means that, to obtain the best optical sectioning effect, the optimal spatial frequency of the illumination pattern at a given NA is half of the cut-off frequency.

Benefiting from the defocus feature of the high spatial frequency illumination near the focal plane, we can simply decompose the merged imaging data into two parts, that is, the in-focus component $D_{in}(x,y)$ and the out-of-focus component $D_{out}(x,y)$. Then, the problem remaining is how to eliminate the residuary fringe in the captured images to extract the wanted in-focus information. The distribution of photons on the detector in a conventional wide-field microscope can be expressed in the form:

$$D(x,y) = D_{out}(x,y) + D_{in}(x,y) \tag{2.19}$$

Under a one-dimensional sinusoidal structured illumination in the form of Equation (2.7), only $D_{in}(x,y)$ is modulated while $D_{out}(x,y)$ is not. For simplicity, assuming $I_0 = 1$ and $m = 1$, $D(x,y)$ can be rewritten as:

$$D(x,y) = D_{out}(x,y) + D_{in}(x,y) \cdot [1 + \cos(2\pi k_0 \cdot x + \varphi)] \tag{2.20}$$

To solve the targeted $D_{in}(x,y)$, three-step phase-shifting with interval of $2\pi/3$ is usually used to get the following equations:

$$D_{0^0}(x,y) = D_{out}(x,y) + D_{in}(x,y) \cdot [1 + \cos(2\pi k_0 \cdot x + \varphi)] \tag{2.21}$$

$$D_{120^0}(x,y) = D_{out}(x,y) + D_{in}(x,y) \cdot \left[1 + \cos\left(2\pi k_0 \cdot x + \varphi + \frac{2\pi}{3}\right)\right] \tag{2.22}$$

$$D_{240^0}(x,y) = D_{out}(x,y) + D_{in}(x,y) \cdot \left[1 + \cos\left(2\pi k_0 \cdot x + \varphi + \frac{4\pi}{3}\right)\right] \tag{2.23}$$

Then, $D_{in}(x,y)$ is solved as:

$$D_{in}(x,y) = \frac{\sqrt{2}}{3} \tag{2.24}$$

$$\sqrt{[(D_{120^0}(x,y) - D_{0^0}(x,y)]^2 + [(D_{240^0}(x,y) - D_{120^0}(x,y)]^2 + [(D_{240^0}(x,y) - D_{0^0}(x,y)]^2}$$

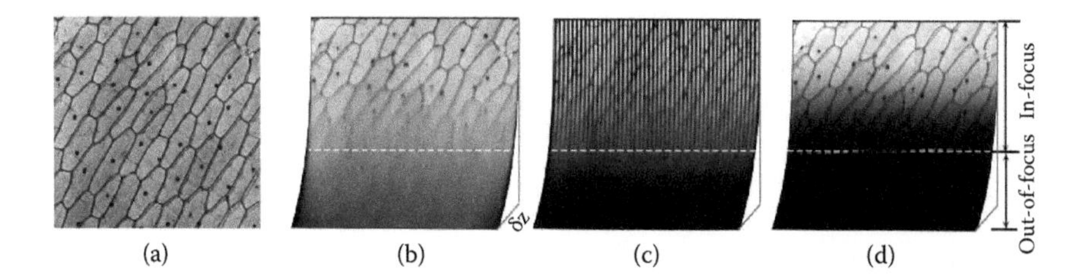

Figure 2.7 Simulation of SIM for optical sectioning effect. (a) Sample image in a single plane; (b) the image is tilted from the bottom using a defocus algorithm so that it becomes increasingly out-of-focus toward the bottom. (c) The tilted sample image under structured illumination; (d) SIM optical sectioning image where only the in-focus part of the sample is retained.

Meanwhile, the conventional wide-field image can also be obtained with

$$D_{out}(x,y)+D_{in}(x,y)=\frac{1}{3}[D_{0^0}(x,y)+D_{120^0}(x,y)+D_{240^0}(x,y)] \quad (2.25)$$

This means that the conventional wide-field image and the optical sectioning SIM image can be obtained simultaneously from the same data.

In the following, we simulate the procedure of Neil-SIM to show the optical sectioning effect. The simulated sample is shown in Figure 2.7(a). Figure 2.7(b) gives a defocus model used to simulate the spatial frequency attenuating with defocus in the conventional wide-field microscopy. The image is tilted so that it becomes increasingly out of focus toward the bottom. Figure 2.7(d) shows the optical sectioning result using the phase-shifting algorithm after three-step phase-shifting with SIM (one of the three images is shown in Figure 2.7c). Due to increased blurring toward the bottom, much of the defocused components of the image are eliminated and are dim; only the in-focus part remains and is bright.

2.3 Generation of structured illumination patterns

The generation of high frequency and high contrast of structured illumination patterns is essential to obtain the best quantity of images for both SIM-SR and Neil-SIM. In the following, we will present some approaches to generation of the structured illumination and their virtues and defects.

2.3.1 Interference

It is well known that interference of two or multiple laser beams can form one-, two-, or three-dimensional structured light patterns (Ryu et al. 2006; Littleton et al. 2007). In the most usual case, two coherent plane waves, normally formed by beam splitters, can generate a one-dimensional sinusoidal fringe pattern. The spatial frequency of the fringe is determined by the intersection angle of the two wave vectors, that is, $k_0 = 2n \cdot sin(\alpha)/\lambda$, where α is the half intersection angle, λ is the wavelength, and n is the refractive index of the surrounding medium. The phase of the fringe can be controlled by adjusting the relative optical path difference of the two beams. For this approach, the mechanical stability is a major concern, because the relative length of the two beam paths formed by the beam splitters must remain stable to sub-wavelength precision during the time needed to record an entire set of images.

When working with microscopes that possess high aperture angles in the sample plane, it is important to keep the vector nature of light in mind. Beams converging onto the focus plane at an angle of 90° with respect to each other do not generate an intensity interference pattern in *p*-polarization (i.e., the polarization vector is in the plane defined by the two interfering beams), whereas the interference of *s*-polarization (the vector perpendicular to the plane that contains the two interfering beams) results in optimal fringe contrast (Heintzmann 2006). Therefore, the illuminating light must be a coherent laser source for effective structural illumination.

2.3.2 *Diffraction plus interference*

As described above, beam splitters can divide one beam into two or multiple coherent beams. But the drawback is the mechanical stability due to the distributed arrangement of the optical elements. As we know, gratings can diffract incident beams into multiple-order beams that have relatively fixed phases and intensity ratios. Because the diffracted beams pass through the same optical elements, the influence of the environmental disturbance on the optical path difference among them can be reduced to a minimum. Thus, the mechanical stability can be improved a lot by using gratings instead of beam splitters. For example, a one-dimensional grating, which diffracts the incident beam into multiple-order beams in the grating vector direction, can be used for generation of one-dimensional structured illumination. To achieve the maximum degree of modulation, often the zero-order and other higher diffraction orders in the optical path are blocked to leave only the +1st order and the −1st order beams (which usually possess larger power than other orders) to interfere for producing the sinusoidal fringe pattern through objective lens. The phase shift of the fringe can be implemented easily by moving the grating in the grating vector direction.

To make flexible structured illumination patterns with fast speed of phase-shifting and fringe orientation rotation, the state-of-the-art programmable spatial light modulator (SLM) is broadly used as a dynamic diffractive optical element instead of the etched grating. There are a few types of SLMs in the market, such as liquid crystal SLM, liquid crystal on silicon (LCOS), digital micromirror device (DMD), etc. The best choice for diffraction grating is to use the pure phase-modulation SLM (e.g., LCOS), which is able to obtain the highest diffraction efficiency, thus increasing the utilizing efficiency of the illumination power.

2.3.3 *Projection*

Placing an illumination mask in the beam path of the illumination source is a direct approach to the generation of structured illumination. Through lenses (including objective lens), the structured illumination light can be projected to a small area. The mask can be a real grating or a programmable SLM, either of which works in transmission or reflection modes. Although the choice of light source is not restricted to a coherent or an incoherent one, the use of incoherent illumination light, which can completely suppress the speckle-noise that occurs in coherent illumination, is a good choice.

A simple arrangement on an available conventional microscope is the largest advantage of the fringe projection method. ApoTome from Carl Zeiss Inc. and OptiGrid module from Leica Inc. are such devices. As a plug-in module, a grating is employed to generate structured illumination upon various conventional upright microscopes. However, a trade-off in precision and speed is inevitable due to the mechanical movement of grating for phase-shifting. Usually, the phase-shifting must be in the precision of sub-wavelength. The mechanically slow movement speed harms the imaging speed in real-time and also may lead to fluorescence photobleaching and phototoxicity. This trade-off problem can be solved by using the programmable SLM with no moving parts, which will be shown in the next section in detail.

2.4 *DMD-based LED-illumination SIM*

As a concrete example, in this section we present an approach of structured illumination microscopy by using a digital micromirror device (DMD) for fringe projection and a low-coherence LED light for illumination. A 90-nm in-plane resolution was achieved with gold nanoparticles and bovine pulmonary artery endothelial (BPAE) cell samples. The optical sectioning capability of the microscope is demonstrated with mixed pollen grains and Golgi-stained mouse brain neurons, and the penetration depth of 120 μm and sectioning strength of 0.93 μm are obtained. The maximum acquisition speed for 3D imaging in the optical sectioning mode is 1.6×10^7 pixels/second, which is mainly limited by the sensitivity and speed of the CCD camera utilized. In contrast to other SIM techniques, the

DMD-based LED-illumination SIM is cost-effective, affords ease of wavelength switching, and is free of speckle-noise. The 2D super-resolution and 3D optical sectioning modalities can be easily switched and applied to either fluorescent or nonfluorescent specimens.

2.4.1 System configuration

The schematic diagram of the proposed DMD-based LED-illumination SIM system is shown in Figure 2.8. High brightness LEDs with switchable wavelengths of 365 nm, 405 nm, 450 nm, and 530 nm serve as excitation sources. LED-illumination has the advantages of low cost, ease of use,

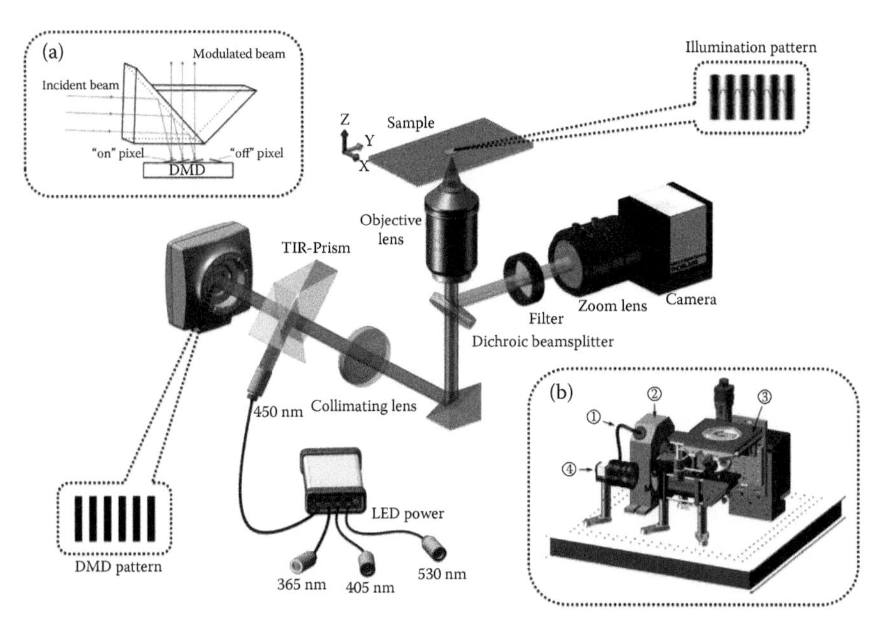

Figure 2.8 **(See color insert.)** Scheme of the DMD-based LED-illumination SIM microscope. A low-coherence LED light is introduced to the DMD via a TIR prism. The binary fringe pattern on the DMD is demagnified and projected onto the specimen through a collimating lens and a microscope objective lens. Higher orders of spatial frequencies of the binary fringe are naturally blocked by the optics, leading to a sinusoidal fringe illumination in the sample plane. Fluorescence or scattered light from the specimen is directly imaged onto the CCD camera. Inset (a) shows the principle of the TIR prism. The illumination light enters the prism and reflects through total internal reflection (TIR) within the prism and illuminates the DMD. Each micromirror of the DMD has an "on" position of +12° and an "off" position of −12° from the normal of the DMD. As a result, the illumination light is introduced at 24° from the normal so that the mirrors reflect the light at 0° when working at the "on" state. Inset (b) gives the configuration of the system, including the LED light guide ①, the DMD unit ②, the sample stage ③, and the CCD camera ④.

and being free of speckle-noise. Benefiting from the broad reflection spectrum of the DMD chip, multi-wavelength excitation can be implemented. Because the DMD chip is reflective and works at small incident angle, the illumination and projection lights have their paths very close to each other in front of the chip. It needs a long path to separate them well, which increases the overall path length of the system. In order to make the microscope system more compact, we employed a total internal reflection (TIR) prism to separate the illumination and the projection paths in minimal space. As shown in Figure 2.8, the nearly collimated LED light enters the TIR-prism, undergoes total internal reflection within the prism, and illuminates the DMD chip. Light modulated by the DMD passes through the TIR-prism, is collimated by an achromatic collimating lens, before being reflected by a dichroic beam splitter into the back aperture of a 100× objective (Apo TIRF, NA 1.49, Nikon Inc., Japan) or a 20× objective (EO M Plan HR, NA 0.6, 13-mm working distance, Edmund Optics Inc., USA). Sliding the collimating lens will slightly change the divergence of the illumination beam. This ensures that the fringe patterns on the DMD chip are precisely projected on the focal plane. The demagnification of the projection system is 176 for the 100× objective. The sample is mounted on a manual X-, Y-, and Z-axis motorized translation stage (M-405.PG, Physik Instrumente Inc., Germany) that can be moved axially in step precision of 50 nm. A USB CCD camera (DCU223M, 1024 × 768 pixels with pixel size of 4.65 × 4.65 μm^2, Thorlabs Inc., USA) is employed to capture the 2D image with a zoom lens (70–300 mm, F/4–5.6, Nikon Inc., Japan), which incorporated with the objective lens produces a variable magnification of the image. The CCD camera has a maximum full-frame rate of 27 fps. A long-pass filter is inserted in front of the zoom lens to block the excitation beam for fluorescence imaging. Automated data collection, DMD pattern generation, and motorized stage movement are implemented by custom software programmed in C++.

2.4.2 Fringe projection by DMD

A DLP Discovery D4100 kit from Texas Instruments Inc. (Dallas, Texas) is used as an amplitude-only SLM for the fringe projection. This DMD chip can switch binary images of two grayscale levels 0 and 255 up to 32 kHz with a resolution of 1024 × 768 pixels with a pixel size of 10.8 × 10.8 μm^2. In common use of DMD fringe projection, the generation of a sinusoidal fringe of 256 grayscale levels requires at least nine pixels for one period. This will apparently reduce the carrier frequency of the fringe and also limit the refreshing rate to 100 Hz. To solve this problem, a binary grating with a period of only four pixels is applied in our scheme, which enables fast transfer of patterns to the DMD up to a 32-kHz refreshing rate. A sinusoidal intensity distribution projected in the focal plane of the objective

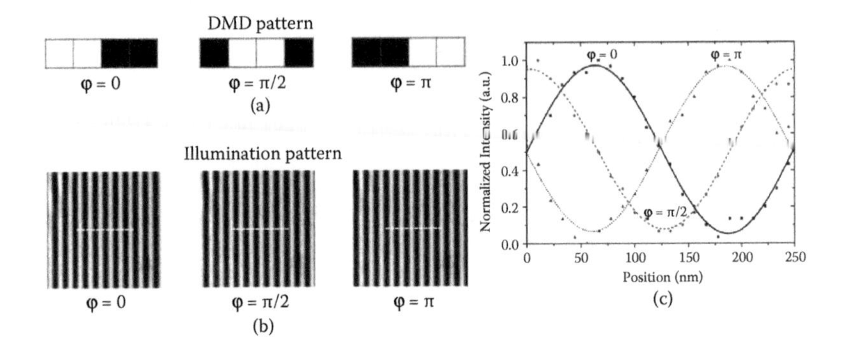

Figure 2.9 Principle of generation of high-frequency phase-shifted sinusoidal fringe illumination with the DMD. (a) Binary patterns loaded on the DMD in one period of four pixels at three different phases; (b) corresponding illumination patterns captured in the focal plane of the objective lens (100×); (c) the measured intensity distribution curves of the illumination patterns along the cut-lines of (b) for the three phases of $\varphi = 0$, $\pi/2$, and π, which show they are ideal sinusoidal fringes with a period of 245 nm and correct phase relations.

is naturally obtained, because the objective lens automatically blocks the higher-order spatial frequencies of the binary grating due to its limited bandwidth. Figure 2.9(a) shows the binary patterns loaded on the DMD chip in one period of four pixels at three phases. The corresponding illumination patterns captured in the focal plane of the objective are shown in Figure 2.9(b). The period of the sinusoidal fringe is measured to be a minimum of 245 nm in minimum (at 100× objective) for the illumination patterns. From the intensity profile analysis of Figure 2.9(c), it can be seen that the fringes are ideal sinusoidal and have the correct phase relations.

2.4.3 *Image process and data analysis*

The Fourier-based image reconstruction is processed according to the method of Gustafsson (2000). It should be noted that the initial phase is crucial for extraction of the high frequency components. It changes significantly with the sample location in the geometry of laser beams interference. Gustafsson et al. (2000) and Chang et al. (2009) both used a complex linear regression algorithm to determine the initial phase for each orientation pattern, which is precise but rather time-consuming. In the geometry of fringe projection, the initial phase can be preset by the DMD for different orientations and remains unchanged for different sample location, which simplifies the image processing.

In the fringe projection by DMD, the commonly used three-step phase-shifting at intervals of $2\pi/3$ cannot be performed, because only four pixels per period of binary pattern are used. Therefore, an alternate three-step phase-shifting at intervals of $\pi/2$ is adopted according to the same principle as described in Section 2.3. Here, for each slice, three sub-images (denoted by $D_{0°}$, $D_{90°}$, and $D_{180°}$) with a phase-shift of $\pi/2$ between two adjacent images are acquired. After the recording process, each image triple is then processed according to Equation (2.26) to obtain the sectioned images:

$$D_{in}(x,y) = \frac{1}{2}\sqrt{(2D_{90^0} - D_{0^0} - D_{180^0})^2 + (D_{180^0} - D_{0^0})^2} \quad (2.26)$$

The full widths at half maximum (FWHM) of gold nanoparticles are measured by fitting their intensity distributions to a Gaussian function with the ImageJ plugin (Point Spread Function Estimation Tool, Computational Biophysics Lab, ETH Zurich, Switzerland). For optical sectioning, the specimen is scanned axially by the motorized stage.

2.4.4 *Sample preparation*

Gold nanoparticles with a diameter of 80 nm (Nanocs Inc., New York) are immersed in a PMMA film. The slice of mixed pollen grains is purchased from Carolina Biological Supply Inc. (Burlington, Vermont). BPAE cell slices stained with Mito Tracker Red CMXRos are purchased from Molecular Probes Inc. (Eugene, USA).

The preparation of Golgi-Cox staining of mouse neuron cells is described in the following. Male Sprague-Dawley rats weighting between 200 and 230 g were provided by the Animal Center of the Fourth Military Medical University (FMMU), China. All experimental protocols were in accordance with the Guidelines for the Animal Care and Use Committee for Research and Education of the FMMU. The rats were housed at 24 ± 2°C with a 12-h light/12-h dark cycle, and food and water were available ad libitum. All efforts were made to minimize the number of animals used and their suffering. To prepare the material for modified rapid Golgi-Cox staining, rats were first anesthetized with sodium pentobarbital (100 mg/kg body weight, i.p.) and perfused with 150 ml of 0.01 M phosphate-buffered saline (PBS; pH 7.3), followed by 500 ml of 0.1 M phosphate buffer (PB; pH 7.3) containing 0.5% (w/v) paraformaldehyde. The modified rapid Golgi-Cox staining was conducted according to a protocol reported previously (Ranjan and Mallick 2010) with slight modifications. In brief, the whole brain was dissected out and put into the Golgi-Cox solution (prepared at least 5 days before) at 37°C for 24 h in darkness. Then the brain was transferred into 30% sucrose solution until it sank. Sections of thickness 200 μm were cut with a vibratome (Microslicer DTM-1000; Dosaka EM, Kyoto, Japan), mounted on the gelatinized slides and air dried. Then the sections were processed as follows:

1. Rinsed twice (5 min each) in distilled water to remove traces of impregnating solution.
2. Dehydrated in 50% alcohol for 5 min.
3. Kept in ammonia solution (3:1, ammonia:distilled water) for 5–10 min.
4. Rinsed twice (5 min each) in distilled water.
5. Kept in 5% sodium thiosulfate for 10 min in darkness.
6. Rinsed twice for 2 min each in distilled water.
7. Dehydrated twice (5–10 min each) in 70%, 80%, 95%, and 100% ethanol, followed by transpiration with xylene.
8. Finally the slides were mounted on a coverslip with Permount and lied laid out to dry for at least 7 days before observation.

2.4.5 *Super-resolution imaging*

To evaluate the spatial resolution of the DMD-based SIM microscope, gold nanoparticles (80 nm diameter) deposited on 170 μm thick coverslips are used as test samples. The Abbe resolution limit is ~151 nm calculated for the 100× objective of NA = 1.49 at the wavelength of 450 nm. The FWHMs are measured and averaged from 50 gold nanoparticles. The resultant FWHM for the SIM is 90 ± 10 nm, which is much better than that of conventional microscopy, 225 ± 5 nm. Figure 2.10 presents the images obtained from both conventional illumination mode and the SIM with the 450-nm LED.

Similar results are obtained with the BPAE cells slices (Figure 2.11). The intracellular mitochondrial network is stained with the fluorescent dye MitoTracker Red CMXRos. Under the excitation of 530-nm LED, the fluorescence image of the structure of mitochondria is clearly visible under

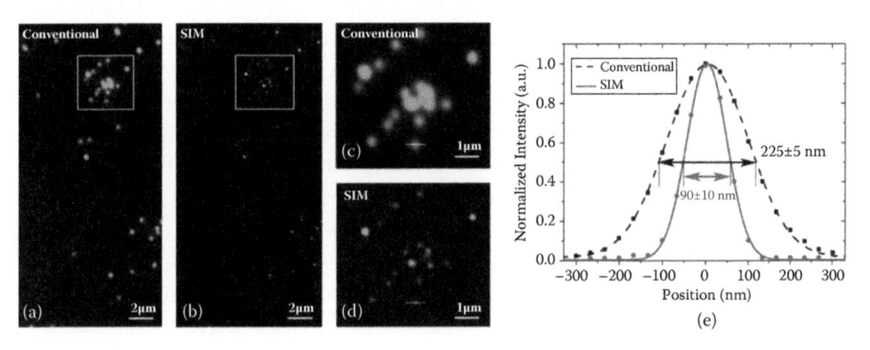

Figure 2.10 **(See color insert.)** Determination of the lateral SIM resolution using 80-nm-diameter gold nanoparticles. As a comparison, image (a) is captured in the conventional illumination mode, and (b) is the corresponding SIM image. Zooms are shown in (c) and (d). The line-scans of a selected nanoparticle in (c) and (d) are plotted in (e). The experiment was performed with the 100× objective of NA = 1.49 with LED wavelength of 450 nm.

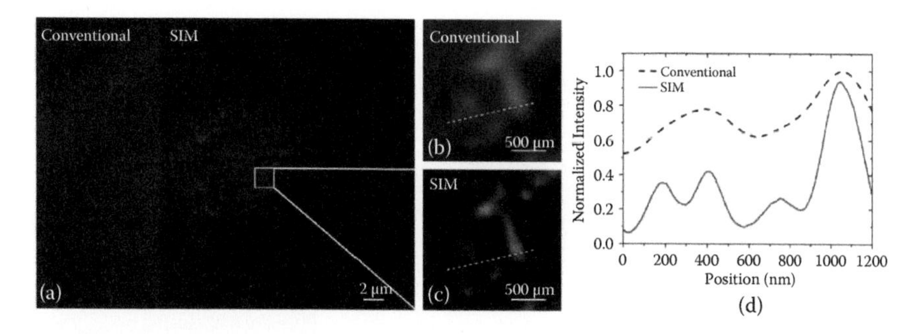

Figure 2.11 Images obtained by SIM and conventional wide-field microscopy for BPAE cells stained with MitoTracker Red CMXRos. The left portion of the entire image (a) is obtained with conventional illumination, and the right portion is the result of a SIM recording. Zooms are shown in (b) and (c). The line-scans of selected parts in (b) and (c) are plotted in (d).The experiment was performed with the 100× objective of NA = 1.49. The fluorescence signal from the BPAE cells was excited with a 530-nm LED, and detected by the CCD camera with a long-pass filter (LP 570 nm).

the 100× objective of NA = 1.49, with a long-pass filter (LP 570 nm) in front of the CCD camera. The SIM image again features much improved lateral resolution over the conventional wide-field microscopy.

To obtain a near isotropic resolution, multi-orientation illuminations are generally required. In the experiments, we applied two perpendicular illumination fringe orientations, that is, X and Y directions (0° and 90°) to achieve higher imaging speed and simplify the pattern design and data analysis. Nevertheless, it is possible to generate three-orientation illumination (0°, 60°, 120°) (Shao et al. 2011) or four-orientation illumination (0°, 45°, 90°, 135°) (Chang et al. 2009) with specifically designed illumination patterns.

2.4.6 Optical sectioning 3D imaging

As a benefit of the fringe projection and low-coherence LED-illumination, DMD-based SIM also has optical sectioning capability in free of speckle-noise. As an example, we use the setup for imaging a pollen grain specimen, which exhibits strong autofluorescence under the excitation with 450-nm LED light. Figure 2.12(a) shows a stack of images of the thick volume structure of the pollen grains with the 20×, NA = 0.6, long working distance objective. We slice 303 layers of the volume and capture 909 raw images in 1024 × 768 pixels. The axial slice interval is 400 nm, thus the slicing depth is 120.8 μm. The volume acquisition time of the SIM volume is 58 s, including 24 ms exposure time plus 36 ms readout time of the CCD camera for each raw image, 10 ms Z-stage settle time for each

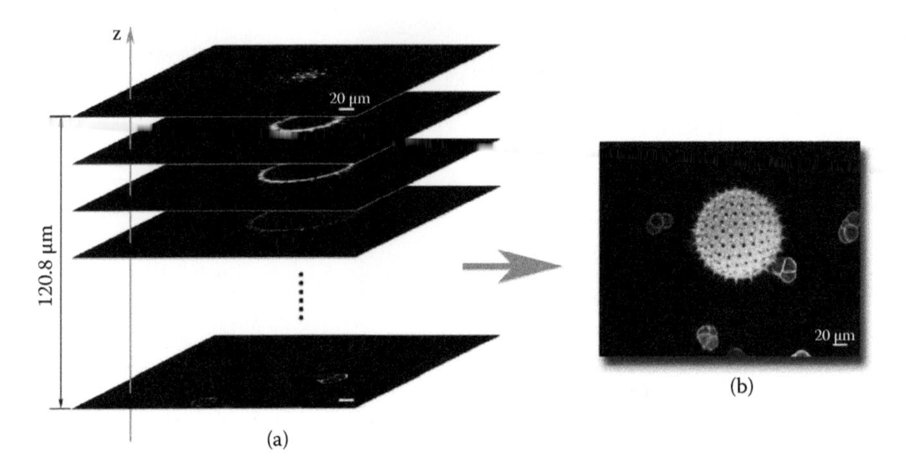

Figure 2.12 Optical sectioning images of the thick volume structure of the mixed pollen grains. The signal comes from autofluorescence of the sample with excitation using a 450-nm LED. (a) A stack of optically sectioned images comprised of 303 axial layers. (b) The maximum-intensity projection of the 303 planes along the Z-axis over a 120.8-μm thickness.

slicing layer, and 31 μs loading time for each DMD illumination fringe pattern. Figure 2.12(b) presents the maximum-intensity projection of the 303 planes along the Z-axis. The dimension of the biggest pollen grain is about 100 μm in diameter. This Figure 2.12 demonstrates the capability of out-of-focus light rejection and optical sectioning of SIM.

Another merit of the DMD-based LED-illumination SIM worthwhile mentioning is that, it can be used for nonfluorescent specimens. For instance, the apparatus is used for obtaining the 3D structure of Golgi-stained mouse brain neurons. Figure 2.13(a) shows a stack of Golgi-stained neuron cells with the 20×, NA = 0.6 objective under the illumination of a 450 -nm LED light. Here the dichroic beam splitter in Figure 2.8 is replaced by a 50:50 broadband beam splitter and the filter in front of the zoom lens is removed. One hundred eighty-six slicing layers consisting of 558 raw images in 1024 × 768 pixels are captured with an axial slice interval of 400 nm. Since the contrast of the Golgi-stained specimen coming from the back-reflection and scattering of illumination light by the microcrystallization of silver chromate is in the nervous tissue, which is much stronger than the fluorescence signal, the exposure time of the CCD camera can be reduced, thus increasing the acquisition speed. The acquisition time of 186 slicing layers is 27 s, including 9 ms exposure time plus 36 ms readout time of the CCD camera for each raw image, 10 ms Z-stage settle time for each slicing layer, and 31 μs loading time for each DMD illumination fringe pattern. Figure 2.13(b) presents the maximum-intensity projection of the 186 planes along the Z-axis. Figure 2.13(c) shows the fine structure of the

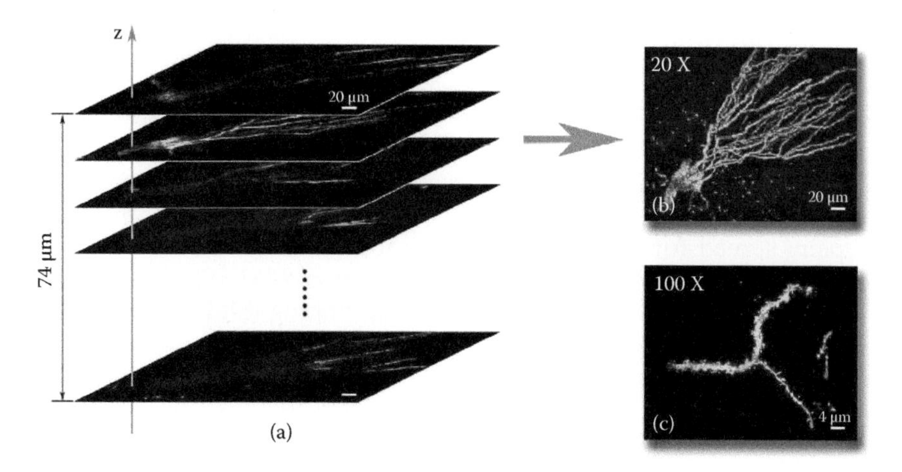

Figure 2.13 Optically sectioned images of Golgi-stained mouse brain neuron cells. The signal is from the back-reflection and scattering of the microcrystallization of silver chromate under the illumination of a 450-nm LED light. (a) A stack of optically sectioned images comprised of 186 axial layers obtained with 20×, NA = 0.6 objective. (b) The maximum-intensity projection of the 186 planes along the Z-axis over a 74-μm thickness. (c) The fine structure of the dendrite enlarged using a 100×, NA = 1.49 objective.

dendrite enlarged by using 100×, $NA = 1.49$ objective. So far, we could realize maximal 200 μm penetration depth without obvious degradation of the image quality. In order to describe the sectioning strength of the SIM microscopy, a normalized spatial frequency $\upsilon = \lambda/(2NA \cdot T)$ is introduced, where λ is the wavelength and T is the spatial periodicity of the projected fringe. The case of $\upsilon = 0$ corresponds to the conventional wide-field microscope, while $\upsilon = 0.5$ corresponds to the maximum sectioning strength. In the experimental setup, the period of the illumination fringe is $T = 1225$ nm at the 20X×, $NA = 0.6$ objective, which corresponds to $\upsilon = 0.306$, from which the FWHM of 930 nm of sectioning strength is obtained.

As for the speed of data acquisition in the optical sectioning mode of the SIM system, it is mainly limited by the speed and sensitivity of the CCD camera, because both the switching time of DMD and the axial translation stage settle time are negligible. In the present version, we use a 1024 × 768 pixel commercial CCD camera with a maximum full-frame rate of 27 fps at its minimal exposure time. The maximum acquisition speed of 1.6×10^7 pixels/second is reached at the imaging of the Golgi-stained neuron cells. Higher acquisition speed could be achieved by using faster and more sensitive cameras. For example, existing 4M (2048 × 2048) pixel scientific complementary metal-oxide semiconductor (SCMOS) cameras with a maximum full-frame rate of 100 fps will allow for about 3.14×10^8 pixels/second (7.5 s per volume for 186 slicing layers).

2.5 Conclusion

Super-resolution far-field optical microscopy based on special illumination schemes has been developed dramatically over the past decade. Of them, structured illumination microscopy (SIM) is easily compatible with the conventional optical microscope due to its wide-field geometry. In this chapter, we have introduced the principle of the SIM in terms of point spread function (PSF) and optical transform function (OTF) of the optical system. The SIM for super-resolution (SIM-SR) proposed by Gustafsson et al. and the SIM for optical sectioning (SIM-OS) proposed by Neil et al. are the most popular ones in the research community of microscopy. They have the same optical configuration but with different data post-processing algorithms. We mathematically described the basic theories for both of the SIMs, respectively, and presented some numerical simulations to show the effects of super-resolution and optical sectioning. Various approaches to generation of structured illumination patterns were reviewed. As an example, a SIM system based on DMD-modulation and LED-illumination was demonstrated. In this SIM setup, a 90-nm in-plane resolution was achieved with gold nanoparticles and BPAE cell samples. The optical sectioning capability of the microscope was demonstrated with mixed pollen grains and Golgi-stained mouse brain neurons, and the penetration depth of 120 μm and sectioning strength of 930 nm were obtained. The current maximum acquisition speed for 3D imaging in the optical sectioning mode was 1.6×10^7 pixels/second.

Since the structured illumination pattern is also limited by the passband of OTF, the super-resolution ability of linear SIM is only beyond the diffraction limit by a factor of two at most. To further increase the resolution, the saturated structured illumination microscopy (SSIM) is proposed, which exploits the nonlinear dependence of the emission rate of fluorophores on the intensity of the excitation light. By applying a sinusoidal illumination pattern with peak intensity close to that needed in order to saturate the fluorophores in their fluorescent state, a complex Moiré fringe emerges. The fringes contain the high order spatial frequencies information that may be extracted by computational techniques. Recently, a resolution of ~50 nm was reported by using specific fluorophores in SSIM.

Although the lateral resolution of 90 nm in the XY-plane was achieved in our system, the axial resolution is still ultimately limited by the PSF of the objective lens due to 2D fringe projection. Doubling the resolution in both lateral and vertical directions can be achieved by a 3D-SIM scheme that needs to illuminate the sample with 3D-structured light pattern. The 3D-structured illumination pattern is usually produced through interference of three coherent beams. The 3D-SIM can naturally eliminate the out-of-focus blur to get optical sectioning. It allows complete 3D visualization of structures inside cells at the level down to the ~100 nm range.

References

Abbe, E. (1873). "Beiträge zur Theorie des Mikroskops und der mikroskopischen Wahrnehmung." *Archiv für Mikroskopische Anatomie* **9**(1): 56.

Ansari, Z. et al. (2002). "Wide-field, real-time depth-resolved imaging using structured illumination with photorefractive holography." *Applied Physics Letters* **81**(12): 2148–2150.

Barlow, A. L. and C. J. Guerin (2007). "Quantization of widefield fluorescence images using structured illumination and image analysis software." *Microscopy Research and Technique* **70**(1): 76–84.

Bates, M. et al. (2007). "Multicolor super-resolution imaging with photo-switchable fluorescent probes." *Science* **317**(5845): 1749–1753.

Bélanger, S. et al. (2010). "Real-time diffuse optical tomography based on structured illumination." *Journal of Biomedical Optics* **15**(1): 016006–016007.

Best, G. et al. (2011). "Structured illumination microscopy of autofluorescent aggregations in human tissue." *Micron* **42**(4): 330–335.

Betzig, E. et al. (2006). "Imaging intracellular fluorescent proteins at nanometer resolution." *Science* **313**(5793): 1642–1645.

Beversluis, M. R. et al. (2008). "Effects of inhomogeneous fields in superresolving structured-illumination microscopy." *JOSA A* **25**(6): 1371–1377.

Blamey, N. et al. (2008). "The application of structured-light illumination microscopy to hydrocarbon-bearing fluid inclusions." *Geofluids* **8**(2): 102–112.

Bozinovic, N. et al. (2008). "Fluorescence endomicroscopy with structured illumination." *Optics Express* **16**(11): 8016–8025.

Bullen, A. (2008). "Microscopic imaging techniques for drug discovery." *Nature Reviews Drug Discovery* **7**(1): 54–67.

Caspi, D. et al. (1998). "Range imaging with adaptive color structured light." *IEEE Transactions on Pattern Analysis and Machine Intelligence* **20**(5): 470–480.

Chang, B.-J. et al. (2009). "Isotropic image in structured illumination microscopy patterned with a spatial light modulator." *Optics Express* **17**: 14710–14721.

Chang, B.-J. et al. (2011). "Subdiffraction scattered light imaging of gold nanoparticles using structured illumination." *Optics Letters* **36**(24): 4773–4775.

Chasles, F. et al. (2007). "Optimization and characterization of a structured illumination microscope." **15**(24): 16130–16140.

Chen, J. et al. (2013). "Super-resolution differential interference contrast microscopy by structured illumination." *Optics Express* **21**(1): 112–121.

Choi, Y. et al. (2011). "Full-field and single-shot quantitative phase microscopy using dynamic speckle illumination." *Optics Letters* **36**(13): 2465.

Cogger, V. C. et al. (2010). "Three-dimensional structured illumination microscopy of liver sinusoidal endothelial cell fenestrations." *Journal of Structural Biology* **171**(3): 382–388.

Cole, M. et al. (2001). "Time-domain whole-field fluorescence lifetime imaging with optical sectioning." *Journal of Microscopy* **203**(3): 246–257.

Cole, M. et al. (2000). "Whole-field optically sectioned fluorescence lifetime imaging." *Optics Letters* **25**(18): 1361–1363.

Dal Maschio, M. et al. (2010). "Simultaneous two-photon imaging and photostimulation with structured light illumination." *Optics Express* **18**(18): 18720–18731.

Dan, D. et al. (2013). "DMD-based LED-illumination Super-resolution and optical sectioning microscopy." *Scientific Reports* **3**.

Delica, S. and C. M. Blanca (2007). "Wide-field depth-sectioning fluorescence microscopy using projector-generated patterned illumination." *Applied Optics* **46**(29): 7237–7243.

Docter, M. W. et al. (2007). "Structured illumination microscopy using extraordinary transmission through sub-wavelength hole arrays." *Journal of Nanophotonics* **1**(1): 011665–011610.

Ducros, N. et al. (2013). "Fluorescence molecular tomography of an animal model using structured light rotating view acquisition." *Journal of Biomedical Optics* **18**(2): 020503.

Edouard, B. and R. Mattias (2008). "Application of structured illumination for multiple scattering suppression in planar laser imaging of dense sprays." *Optics Express* **16**(22): 17870–17881.

Fedosseev, R. et al. (2005). "Structured light illumination for extended resolution in fluorescence microscopy." *Optics and Lasers in Engineering* **43**(3): 403–414.

Fernández-Suárez, M. and A. Y. Ting (2008). "Fluorescent probes for super-resolution imaging in living cells." *Nature Reviews Molecular Cell Biology* **9**(12): 929–943.

Fiolka, R. et al. (2008). "Structured illumination in total internal reflection fluorescence microscopy using a spatial light modulator." *Optics Letters* **33**(14): 1629–1631.

Fiolka, R. et al. (2012). "Time-lapse two-color 3D imaging of live cells with doubled resolution using structured illumination." *Proceedings of the National Academy of Sciences* **109**(14): 5311–5315.

Fitzgibbon, J. et al. (2010). "Super-resolution imaging of plasmodesmata using three-dimensional structured illumination microscopy." *Plant Physiology* **153**(4): 1453–1463.

Frohn, J. et al. (2001). "Three-dimensional resolution enhancement in fluorescence microscopy by harmonic excitation." *Optics Letters* **26**(11): 828–830.

Frohn, J. T. et al. (2000). "True optical resolution beyond the Rayleigh limit achieved by standing wave illumination." *Proceedings of the National Academy of Sciences* **97**(13): 7232–7236.

Fukano, T. and A. Miyawaki (2003). "Whole-field fluorescence microscope with digital micromirror device: imaging of biological samples." *Applied Optics* **42**(19): 4119–4124.

Gao, L. et al. (2011). "Depth-resolved image mapping spectrometer (IMS) with structured illumination." *Optics Express* **19**(18): 17439–17452.

Gao, L. et al. (2012). "Noninvasive imaging beyond the diffraction limit of 3D dynamics in thickly fluorescent specimens." *Cell* **151**(6): 1370–1385.

Gardeazábal Rodríguez, P. F. et al. (2008). "Axial coding in full-field microscopy using three-dimensional structured illumination implemented with no moving parts." *Optics Letters* **33**(14): 1617–1619.

Gonçalves, M. S. T. (2008). "Fluorescent labeling of biomolecules with organic probes." *Chemical Reviews* **109**(1): 190–212.

Gruppetta, S. and S. Chetty (2011). "Theoretical study of multispectral structured illumination for depth resolved imaging of non-stationary objects: focus on retinal imaging." *Biomedical Optics Express* **2**(2): 255.

Gur, A. et al. (2011). "The limitations of nonlinear fluorescence effect in super resolution saturated structured illumination microscopy system." *Journal of Fluorescence* **21**(3): 1075–1082.

Gustafsson, M. G. (2000). "Surpassing the lateral resolution limit by a factor of two using structured illumination microscopy." *Journal of Microscopy* **198**(2): 82–87.

Gustafsson, M. G. (2005). "Nonlinear structured-illumination microscopy: wide-field fluorescence imaging with theoretically unlimited resolution." *Proceedings of the National Academy of Sciences USA* **102**(37): 13081–13086.

Gustafsson, M. G. et al. (2000). "Doubling the lateral resolution of wide-field fluorescence microscopy using structured illumination." Presented at BiOS 2000 The International Symposium on Biomedical Optics, International Society for Optics and Photonics.

Gustafsson, M. G. et al. (2008). "Three-dimensional resolution doubling in wide-field fluorescence microscopy by structured illumination." *Biophysical Journal* **94**(12): 4957–4970.

Hagen, N. et al. (2012). "Quantitative sectioning and noise analysis for structured illumination microscopy." *Optics Express* **20**(1): 403.

Hao, X. et al. (2013). "Nanomorphology of polythiophene–fullerene bulk-heterojunction films investigated by structured illumination optical imaging and time-resolved confocal microscopy." *Methods and Applications in Fluorescence* **1**(1): 015004.

Heintzmann, R. (2003). "Saturated patterned excitation microscopy with two-dimensional excitation patterns." *Micron* **34**(6): 283–291.

Heintzmann, R. (2006). "Structured illumination methods." In *Handbook of Biological Confocal Microscopy*, edited by J. Pawley, 15. New York: Springer.

Heintzmann, R. and P. A. Benedetti (2006). "High-resolution image reconstruction in fluorescence microscopy with patterned excitation." *Applied Optics* **45**(20): 5037–5045.

Heintzmann, R. and C. G. Cremer (1999). "Laterally modulated excitation microscopy: improvement of resolution by using a diffraction grating." Presented at BiOS Europe '98, International Society for Optics and Photonics.

Hell, S. W. and J. Wichmann (1994). "Breaking the diffraction resolution limit by stimulated emission: stimulated-emission-depletion fluorescence microscopy." *Optics Letters* **19**(11): 780–782.

Hess, S. T. et al. (2006). "Ultra-high resolution imaging by fluorescence photoactivation localization microscopy." *Biophysical Journal* **91**(11): 4258–4272.

Hirvonen, L. et al. (2008). "Structured illumination microscopy using photoswitchable fluorescent proteins." Presented at Biomedical Optics (BiOS) 2008, International Society for Optics and Photonics.

Hirvonen, L. M. et al. (2009). "Structured illumination microscopy of a living cell." *European Biophysics Journal* **38**(6): 807–812.

Huang, B. et al. (2010). "Breaking the diffraction barrier: super-resolution imaging of cells." *Cell* **143**(7): 1047–1058.

Huang, B. et al. (2008). "Three-dimensional super-resolution imaging by stochastic optical reconstruction microscopy." *Science* **319**(5864): 810–813.

Huisken, J. et al. (2004). "Optical sectioning deep inside live embryos by selective plane illumination microscopy." *Science* **305**(5686): 1007–1009.

Jiang, S.-H. and J. G. Walker (2004). "Experimental confirmation of non-scanning fluorescence confocal microscopy using speckle illumination." *Optics Communications* **238**(1): 1–12.

Jiang, S.-H. and J. G. Walker (2005). "Non-scanning fluorescence confocal microscopy using speckle illumination and optical data processing." *Optics Communications* **256**(1): 35–45.

Jiang, S.-H. and J. G. Walker (2009). "Speckle-illuminated fluorescence confocal microscopy, using a digital micro mirror device." *Measurement Science and Technology* **20**(6): 065501.

Karadaglić, D. (2008). "Image formation in conventional brightfield reflection microscopes with optical sectioning property via structured illumination." *Micron* **39**(3): 302–310.

Karadaglić, D. et al. (2002). "Confocal endoscopy via structured illumination." *Scanning* **24**(6): 301–304.

Karadaglić, D. and T. Wilson (2008). "Image formation in structured illumination wide-field fluorescence microscopy." *Micron* **39**(7): 808–818.

Keller, P. J. et al. (2008). "Reconstruction of zebrafish early embryonic development by scanned light sheet microscopy." *Science* **322**(5904): 1065–1069.

Klar, T. A. and S. W. Hell (1999). "Subdiffraction resolution in far-field fluorescence microscopy." *Optics Letters* **24**(14): 954–956.

Kner, P. et al. (2009). "Super-resolution video microscopy of live cells by structured illumination." *Nature Methods* **6**(5): 339–342.

Köhler, A. K. J. V. (1893). "Ein neues Beleuchtungsverfahren für mikrophotographische Zwecke." *Zeitschrift für Wissenschaftliche mikroskopie und für Mikroskopische Technik* **10**(4): 8.

Kristensson, E. et al. (2008a). "High-speed structured planar laser illumination for contrast improvement of two-phase flow images." *Optics Letters* **33**(23): 2752–2754.

Kristensson, E., et al. (2008b). "Spatially resolved, single-ended two-dimensional visualization of gas flow phenomena using structured illumination." *Applied Optics* **47**(21): 3927–3931.

Křížek, P. and G. M. Hagen (2012). Chapter: "Current optical sectioning systems in florescence microscopy." Current Microscopy Contributions to Advances in Science and Technology. Formatex Research Center, Spain. 826–832.

Křížek, P. et al. (2012). "Flexible structured illumination microscope with a programmable illumination array." *Optics Express* **20**(22): 24585–24599.

Krzewina, L. G. and M. K. Kim (2006). "Single-exposure optical sectioning by color structured illumination microscopy." *Optics Letters* **31**(4): 477–479.

Langhorst, M. F. et al. (2009). "Structure brings clarity: structured illumination microscopy in cell biology." *Biotechnology Journal* **4**(6): 858–865.

Lefman, J. et al. (2011). "Live, video-rate super-resolution microscopy using structured illumination and rapid GPU-based parallel processing." *Microscopy and Microanalysis* **17**(2): 191.

Lim, D. et al. (2008). "Wide-field fluorescence sectioning with hybrid speckle and uniform-illumination microscopy." *Optics Letters* **33**(16): 1819–1821.

Lim, D. et al. (2011). "Optically sectioned in vivo imaging with speckle illumination HiLo microscopy." *Journal of Biomedical Optics* **16**(1): 016014–016018.

Littleton, B. et al. (2007). "Coherent super-resolution microscopy via laterally structured illumination." *Micron* **38**(2): 150–157.

Lu, C.-H. et al. (2013). "Flow-based structured illumination." *Applied Physics Letters* **102**: 161115.

Lukic, V. et al. (2009). "Optical tomography with structured illumination." *Optics Letters* **34**(7): 983–985.

Mandula, O. et al. (2012). "Line scan-structured illumination microscopy super-resolution imaging in thick fluorescent samples." *Optics Express* **20**(22): 24167–24174.
Markaki, Y. et al. (2012). "The potential of 3D-FISH and super-resolution structured illumination microscopy for studies of 3D nuclear architecture." *Bioessays* **34**(5): 412–426.
Mazhar, A. et al. (2010). "Structured illumination enhances resolution and contrast in thick tissue fluorescence imaging." *Journal of Biomedical Optics* **15**(1): 010503–010506.
Monneret, S. et al. (2006). "Highly flexible whole-field sectioning microscope with liquid-crystal light modulator." *Journal of Optics A: Pure and Applied Optics* **8**(7): S461.
Mudry, E. et al. (2012). "Structured illumination microscopy using unknown speckle patterns." *Nature Photonics* **6**(5): 312–315.
Neil, M. et al. (1997). "Method of obtaining optical sectioning by using structured light in a conventional microscope." *Optics Letters* **22**(24): 1905–1907.
Neil, M. et al. (1998). "A light efficient optically sectioning microscope." *Journal of Microscopy* **189**(2): 114–117.
Neumann, A. et al. (2008). "Structured illumination for the extension of imaging interferometric microscopy." *Optics Express* **16**: 6785–6793.
Orieux, F. et al. (2012). "Bayesian estimation for optimized structured illumination microscopy." *IEEE Transactions on Image Processing* **21**(2): 601–614.
Patterson, G. H. et al. (1997). "Use of the green fluorescent protein and its mutants in quantitative fluorescence microscopy." *Biophysical Journal* **73**(5): 2782–2790.
Planchon, T. A. et al. (2011). "Rapid three-dimensional isotropic imaging of living cells using Bessel beam plane illumination." *Nature Methods* **8**(5): 417–423.
Ranjan, A. and B. N. Mallick (2010). "A modified method for consistent and reliable Golgi–Cox staining in significantly reduced time." *Frontiers in Neurology* **1**: 157.
Rego, E. H. et al. (2012). "Nonlinear structured-illumination microscopy with a photoswitchable protein reveals cellular structures at 50-nm resolution." *Proceedings of the National Academy of Sciences* **109**(3): E135–E143.
Rocchini, C. et al. (2001). "A low cost 3D scanner based on structured light." Computer Graphics Forum, Wiley Online Library.
Rust, M. J. et al. (2006). "Sub-diffraction-limit imaging by stochastic optical reconstruction microscopy (STORM)." *Nature Methods* **3**(10): 793–796.
Ryu, J. et al. (2006). "Multibeam interferometric illumination as the primary source of resolution in optical microscopy." *Applied Physics Letters* **88**(17): 171112–171113.
Ryu, J. et al. (2003). "Application of structured illumination in nano-scale vision." Presented at Computer Vision and Pattern Recognition Workshop, 2003. CVPRW'03, Institute of Electrical and Electronic Engineers.
Schermelleh, L. et al. (2008). "Subdiffraction multicolor imaging of the nuclear periphery with 3D structured illumination microscopy." *Science* **320**(5881): 1332–1336.
Schröter, T. J. et al. (2012). "Scanning thin-sheet laser imaging microscopy (STSLIM) with structured illumination and HiLo background rejection." *Biomedical Optics Express* **3**(1): 170.
Shao, L. et al. (2008). "I^5 S: wide-field light microscopy with 100-nm-scale resolution in three dimensions." *Biophysical Journal* **94**(12): 4971–4983.

Shao, L. et al. (2011). "Super-resolution 3D microscopy of live whole cells using structured illumination." *Nature Methods* **8**(12): 1044–1046.

Shao, L. et al. (2012). "Interferometer-based structured-illumination microscopy utilizing complementary phase relationship through constructive and destructive image detection by two cameras." *Journal of Microscopy* **246**(3): 229–236.

Shroff, S. A. et al. (2007). "Estimation of phase shifts in structured illumination for high resolution imaging." Frontiers in Optics, Optical Society of America.

Shroff, S. A. et al. (2008). "OTF compensation in structured illumination superresolution images." In Proc. SPIE.

Shroff, S. A. et al. (2010). "Lateral superresolution using a posteriori phase shift estimation for a moving object: experimental results." *JOSA A* **27**(8): 1770–1782.

Siegel, J. et al. (2001). "Whole-field five-dimensional fluorescence microscopy combining lifetime and spectral resolution with optical sectioning." *Optics Letters* **26**(17): 1338–1340.

So, P. T. (2004). Chapter: "Multi-photon excitation fluorescence microscopy." *Frontiers in Biomedical Engineering*, Springer: Germany. 529–544.

So, P. T. et al. (2000). "Two-photon excitation fluorescence microscopy." *Annual Review of Biomedical Engineering* **2**(1): 399–429.

Somekh, M. G. et al. (2008). "Resolution in structured illumination microscopy: a probabilistic approach." *JOSA A* **25**(6): 1319–1329.

Somekh, M. G. et al. (2009). "Stochastic transfer function for structured illumination microscopy." *JOSA A* **26**(7): 1630–1637.

Somekh, M. G. et al. (2011). "Effect of processing strategies on the stochastic transfer function in structured illumination microscopy." *JOSA A* **28**(9): 1925–1934.

Sonnen, K. F. et al. (2012). "3D-structured illumination microscopy provides novel insight into architecture of human centrosomes." *Biology Open* **1**(10): 965–976.

Strauss, M. P. et al. (2012). "3D-SIM super resolution microscopy reveals a bead-like arrangement for FtsZ and the division machinery: implications for triggering cytokinesis." *PLoS Biology* **10**(9): e1001389.

Su, X. and Q. Zhang (2010). "Dynamic 3-D shape measurement method: a review." *Optics and Lasers in Engineering* **48**(2): 191–204.

Ventalon, C. et al. (2007). "Dynamic speckle illumination microscopy with wavelet prefiltering." *Optics Letters* **32**(11): 1417–1419.

Ventalon, C. and J. Mertz (2005). "Quasi-confocal fluorescence sectioning with dynamic speckle illumination." *Optics Letters* **30**(24): 3350–3352.

Ventalon, C. and J. Mertz (2006). "Dynamic speckle illumination microscopy with translated versus randomized speckle patterns." *Optics Express* **14**(16): 7198–7209.

Walker, J. G. (2001). "Non-scanning confocal fluorescence microscopy using speckle illumination." *Optics Communications* **189**(4): 221–226.

Wang, L. et al. (2011). "Wide-field high-resolution structured illumination solid immersion fluorescence microscopy." *Optics Letters* **36**(15): 2794–2796.

Waterman-Storer, C. M. et al. (1998). "Fluorescent speckle microscopy, a method to visualize the dynamics of protein assemblies in living cells." *Current Biology* **8**(22): 1227–1230.

Webb, S. et al. (2002). "A wide-field time-domain fluorescence lifetime imaging microscope with optical sectioning." *Review of Scientific Instruments* **73**(4): 1898–1907.

Wicker, K. and R. Heintzmann (2010). "Single-shot optical sectioning using polarization-coded structured illumination." *Journal of Optics* **12**(8): 084010.

Wicker, K. et al. (2013). "Phase optimisation for structured illumination microscopy." *Optics Express* **21**(2): 2032–2049.
Wilson, T. (1990). *Confocal Microscopy*. Academic Press, London **426**: 1-64.
Wong, C. et al. (2006). "Study on potential of structured illumination microscopy utilizing digital micromirror device for endoscopy purpose." Presented at IEEE International Symposium on Biophotonics, Nanophotonics and Metamaterials.
York, A. G. et al. (2012). "Resolution doubling in live, multicellular organisms via multifocal structured illumination microscopy." *Nature Methods* **9**(7): 749–754.
Zhang, H. et al. (2011). "Nonlinear structured illumination microscopy by surface plasmon enhanced stimulated emission depletion." *Optics Express* **19**(24): 24783–24794.

Problems

1. What is a Moiré pattern? How does it occur?
2. Describe the basic principles of structured illumination microscopy for both super-resolution (SIM-SR) and optical sectioning (SIM-OS).
3. What is the role of a Moiré pattern in super-resolution SIM? Describe how to separate the in-focus and out-of-focus portions in wide-field imaging by structured illumination in optical sectioning SIM?
4. What are the advantage and disadvantage of SIMs compared with other microscopies (for example, STED, PALM/STORM in super-resolution, and confocal, light sheet, two-photon microscopy in optical sectioning)?
5. Why is three-step phase-shifting required in both SIMs?
6. How is frequency scope expanded in SR-SIM?
7. Why is it essential to employ high spatial frequency and high contrast structured illumination fringes in SIMs? How is structured illumination generated efficiently?
8. How does the DMD-based LED-illumination SIM work? What are the advantages of this method?

chapter three

Super-resolution imaging with stochastic optical reconstruction microscopy (STORM) and photoactivated localization microscopy (PALM)

Yujie Sun

Contents

3.1 Introduction

Modern cell biology covers a large scale, extending from a single macromolecule to a multicell system. In order to fully understand the molecular mechanisms of a cellular process, one needs to know not only what molecules are involved but often their spatial-temporal behavior, which requires molecular imaging. Compared to many other tools, fluorescence microscopy has several virtues that make it the most prevalent imaging strategy in cell biology. These include noninvasiveness, high specificity, and temporal-spatial detection.

Major technological breakthroughs during the past two decades, such as the ultra-sensitive imaging detector (EMCCD), stable compact laser sources (solid state lasers), high performance objectives, and fluorescent proteins, along with many other incremental advancements, have largely improved the performance of fluorescence microscopy and made it even routine to image dynamic single molecules in live cells. Despite the tremendous improvements, the application of fluorescence microscopy is still hindered due to its limited spatial resolution, imposed by the diffraction of light.

Optical diffraction originates from the wave nature of light. For any pointed light source such as a fluorophore, its image formed via a focusing lens is not a sharp "focal point," but rather a pattern with a finite size called "Airy disk" due to optical diffraction. When two point sources are so close that their Airy disks overlap significantly, optical imaging is not able to resolve them. This was originally described by Abbe in 1873 and thus is named the Abbe limit. Chapter 1 describes the optical diffraction limit in detail.

In fluorescence microscopy, optical diffraction is practically defined by the emission wavelength of the fluorophore and the numerical aperture (NA) of the objective. Several definitions have been developed to describe the resolution imposed by diffraction. Here we use the Abbe limit to define the resolution:

$$d = \frac{\lambda}{2NA} \tag{3.1}$$

where λ is the emission wavelength of the fluorophore and NA is the numerical aperture of the objective. Given that most fluorophores used in fluorescence microscopy emit in the visible band, that is, 400 to 700 nm, and objectives with good correction for chromatic and spherical aberrations usually have an NA less than 1.5, the theoretical Abbe resolution would be 133 nm. Nonetheless, considering the UV damage to the cell as well as better penetration efficiency of light with longer wavelength, we usually estimate the spatial resolution using an emission wavelength of 600 nm, which corresponds to ~200 nm for Abbe resolution. Compared to the lateral resolution, the axial resolution of fluorescence microscopy is even worse, on the order of 500 nm, due to the nonsymmetrical wavefront emerging from the objective.

The incompatibility between the common length scales in the cell and the limited imaging resolution has been a constraint of fluorescence microscopy as a descriptive or half-quantitative tool in cell biology (Figure 3.1). First, many macromolecular complexes, intracellular organelles, and even subcellular structures are on the scale of tens of nanometers, smaller than the optical resolution. Second, cells are super-crowded environments and

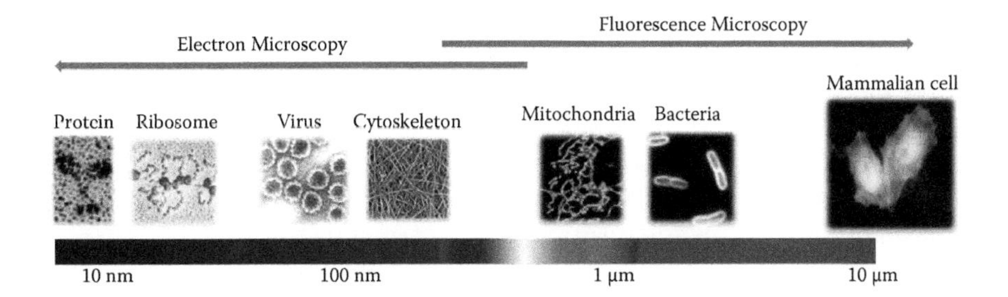

Figure 3.1 Common scale in biology and application ranges of conventional fluorescence microscopy and electron microscopy.

even for a target molecule with a concentration as low as 1 μM, which, by a quick calculation, corresponds to the five molecules in an optical diffraction defined cube with a 200-nm side length. Therefore, conventional fluorescence microscopy is generally inadequate for resolving the subcellular organization of proteins or dissecting the architecture of many intracellular structures. Some immediate examples that are beyond the resolving power of conventional fluorescence microscopy are the cytoskeleton and cytoskeleton-related proteins, synaptic structures, and nuclear pore architectures.

In fact, ultra-structure study in cell biology has been the realm of electron microscopy (EM). This is not surprising as an electron beam has a much shorter wavelength and higher spatial resolution even though it also obeys diffraction limitation. Under optimal conditions, EM can achieve subnanometer resolving power, hundreds of times better than conventional fluorescence microscopy. However, EM usually requires cells that are stained and frozen for enhanced contrast and structure maintenance, so it is unsuitable for dynamic imaging in living cells. In addition, it is generally hard to achieve multiple specific labeling in EM, making multimolecule imaging rare in EM studies. Therefore, it seems that we are facing a dilemma to trade off between resolution and specificity, as well as between fixed and living cells.

In the field of molecular optical imaging, people have actually never stopped trying to "break" the diffraction barrier and increase the spatial resolution. The past two decades have proven to be an exciting era witnessing the emergence and development of several novel fluorescence microscopy approaches that "break" the diffraction barrier. These so called super-resolution imaging techniques dramatically improve the optical resolution, approaching the resolving power of the EM while still retaining the advantages of fluorescence microscopy.

There were originally three types of major super-resolution methods that are based on distinct working mechanisms: stimulated emission depletion microscopy (STED; Hell and Wichmann 1994; Hell 2007), structured illumination microscopy (SIM; Gustafsson 2000), and stochastic optical reconstruction microscopy (Rust et al. 2006) or (fluorescence) photoactivated localization microscopy (STORM/(F)PALM; Betzig et al. 2006; Hess et al. 2006). Recent developments of new super-resolution imaging modalities have obscured the difference. For example, reversible saturable optically linear fluorescent transitions (RESOLFT; Hofmann et al. 2005) and saturated structured illumination microscopy (SSIM) (Heintzmann et al. 2002; Gustafsson 2005) demonstrate both higher resolution and better biological compatibility.

Currently, most of these super-resolution methods are commercially available. While they have demonstrated attractive promise in cell biology research, these techniques all have individual strengths and weaknesses. For instance, SIM is probably the most user-friendly to cell biologists because it requires almost no changes in sample preparation nor special skill sets for data analysis. However, SIM only improves the resolution to 100 nm, by a factor of two compared with conventional imaging. In contrast, other super-resolution techniques such as STED and STORM/(F)PALM offer much higher spatial resolution but require rather sophisticated instrumentation, fluorophores with special photophysical properties, and relatively advanced data analysis routines. Therefore, it would be great if a cell biologist could combine SIM with either STED or STORM/(F)PALM, so both efficiency and spatial resolution can be taken care of in the research.

STORM/(F)PALM are usually categorized as single-molecule localization-based super-resolution imaging techniques. Taking advantage of photocontrollable properties of fluorophores, STORM/(F)PALM are able to stochastically activate and image individual fluorophore molecules within a diffraction-limited region. The single-molecule localization precision is generally much higher than the diffraction limit. Therefore, superimposition of these precisely localized molecules can reconstruct a super-resolution image that "breaks" the diffraction limit. Since single fluorescent molecules were first visualized at room temperature nearly 20 years ago, single-molecule imaging techniques have seen dramatic development both in methodology and applications, and gradually become a routine tool set in cell biology. As an extension of single-molecule imaging, STORM/(F)PALM inherit several advantages that make them probably the most implemented super-resolution methods for cell biologists. First, the instrumentation of STORM/(F)PALM is relatively simple and affordable, essentially the same as the widely used totally internal reflection microscope (TIRFM); second, STORM/(F)PALM demonstrate the highest spatial resolution due to their best photon usage efficiency; third, single-molecule localization allows STORM/(F)PALM to count molecules in a subdiffraction space; fourth, photoswitchable fluorophores give STORM/(F)PALM the ability to track single-molecule dynamics in live cells. Finally, STORM/(F)PALM demonstrate the best three-dimensional (3D) resolving capability.

This chapter focuses on STORM/(F)PALM super-resolution imaging techniques. I will start with an introduction to single-molecule localization, then discuss some basics of STORM/(F)PALM, including fluorophores and labeling for super-resolution imaging. I will also provide basic notes for instrumentation, multicolor and three-dimensional imaging, and live-cell super-resolution imaging.

3.2 Single-molecule localization microscopy (SMLM)

In the introduction to Chapter 1 and this chapter, we showed that in optical microscopy, diffraction transforms any point source into an Airy disk pattern, known as the point-spread function (PSF). The emission wavelength and the numerical aperture (NA) of the microscope objective together define the size of PSF (Equation 3.1), which is approximately 200 nm in the lateral direction and 500 nm in the axial direction.

Despite the large size of the PSF spot, we can actually identify its center with precision that is much higher than the diffraction limit. This is based on the fact that photons emitted from a fluorophore have a particular distribution in the PSF, which can be approximately described by a two-dimensional (2D) Gaussian function (Figure 3.2).

Therefore, just like finding the peak of a mountain, the center of a PSF can be determined by fitting the photon distribution using a 2D Gaussian function (Equation 3.2).

$$P(x, y, z_0, A, x_0, y_0, s_x, s_y) = z_0 + A \cdot exp\left[-\frac{1}{2}\left[\left(\frac{x-x_0}{s_x}\right)^2 + \left(\frac{y-y_0}{s_y}\right)^2\right]\right] \quad (3.2)$$

where z_0 is a constant reflecting the contribution of the background fluorescence, detector noise, and EMCCD baseline reading; *A* is the peak height,

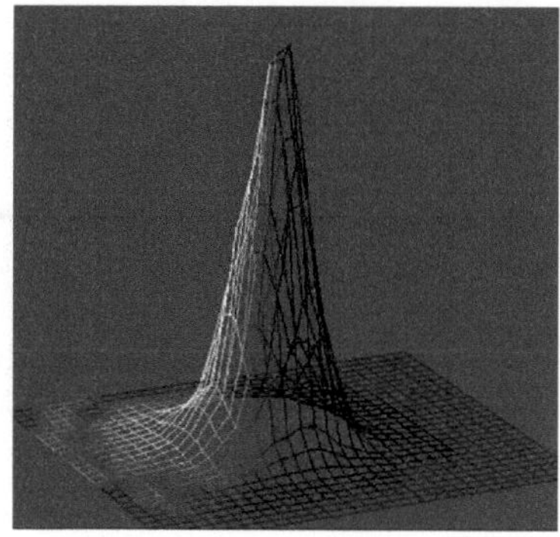

Figure 3.2 Point spread function (PSF) of a single fluorescent molecule.

x_0 and y_0 are the coordinates of the center, and s_x and s_y are the standard deviations of the 2D Gaussian fit in the lateral directions. The localization precision of the center, that is, the standard error of the mean, is given by:

$$\sigma_{\mu i} = \sqrt{\frac{S_i^2}{N} + \frac{a^2/12}{N} + \frac{8\pi S_i^4 b^2}{a^2 N^2}} \tag{3.3}$$

where μ stands for the mean value, *i* represents the *x* and *y* direction, *N* is the detected photon number, *a* is the effective pixel size of the detector, and *b* is the background noise (Thompson et al. 2002; Yildiz et al. 2003). The first term accounts for the shot noise, reflecting the stochastic nature of photon counting. The second term is called pixelation noise due to the finite pixel size in the image. The last term accounts for the rest of the background noise such as out-of-focus fluorescence, EMCCD readout noise, dark current, etc. It is apparent that the uncertainty in determining the molecule's position scales with the inverse square root of photon number emitted from the molecule that are detected on the camera. Therefore in fluorescence microscopy, given other conditions unchanged, options that can increase detected photons such as brighter fluorophores, a more sensitive detector, and even longer exposure time, would help improve the localization precision. We can make a rough estimation to see how each parameter affects the localization precision. In a typical fluorescence experiment, s_i is usually 125 nm, *a* is 150 nm for an EMCCD camera equipped with a 100× objective, *b* is usually less than 5 photons for optimized imaging condition, and *N* is about 5000 photons for the most used STORM fluorophore Alexa647 (see discussion later). This set of conditions yields a localization precision value of 1.9 nm. The localization precision is very sensitive to the background as it is inverse linearly related to *b*. Therefore, in SMLM experiments, it is important to minimize the background fluorescence. Common imaging modalities for background reduction include confocal, multiphoton, TIRFM, and light sheet microscopy. Figure 3.3 illustrates how localization precision is dependent on the photon number with a range that covers most fluorophores used in STORM/(F)PALM. We can see that this quadratic dependence sets ~1000 photons as a critical number to evaluate the localization precision. Interestingly, most of the fluorescent proteins are generally dim with only a few hundred photons on average, while organic dye molecules usually emit a few thousand photons, so result in much better localization precision. The effect would become more obvious with a larger fluorescence background.

A few notes are worth mentioning before we move to the next section. First, all the parameters in Equation (3.3) are related. For instance, smaller pixel size *a* usually introduces larger background *b*, while too large pixel size leads to a loss of spatial information. For high localization precision,

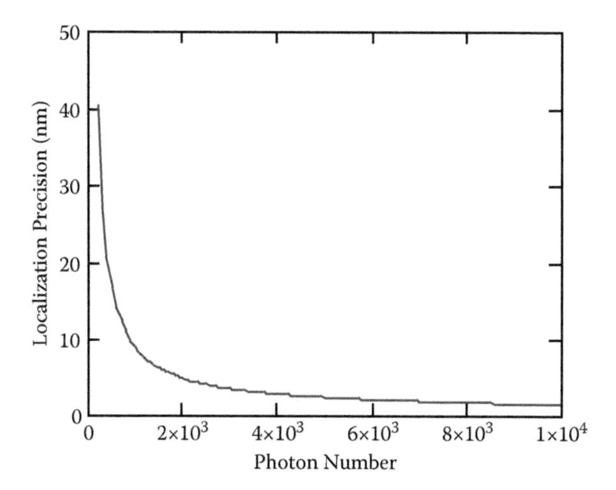

Figure 3.3 Localization precision dependence on photon number.

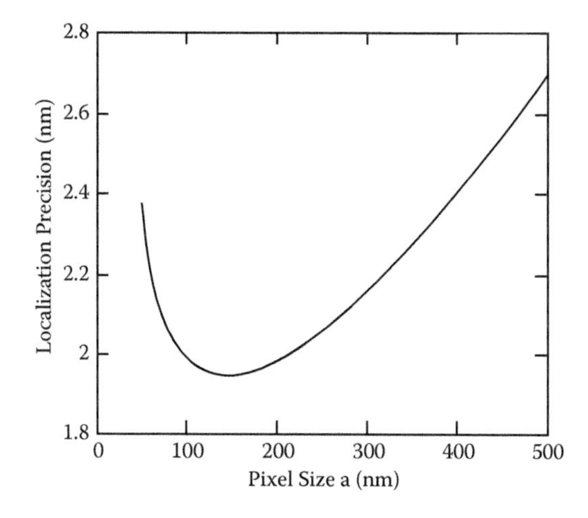

Figure 3.4 Localization precision dependence on the pixel size.

one needs to test various conditions. For a quick rule of thumb, it is better to have the effective pixel size close to or slightly smaller than the standard deviation. Figure 3.4 shows how localization precision depends on the pixel size with other parameters fixed. Second, 2D Gaussian fit is only suitable for PSF that is free of the polarization effect. Most fluorophores are dipoles and their absorption and emission profiles are dependent on the relative orientation between the excitation polarization and the dipole orientation. Fortunately, fluorophores used in STORM/(F)PALM experiments are usually dangling on the targeted molecule. In this case, the thermal motion of the fluorophore is so fast that it can sample the whole space during a polarized excitation duration and thus average out the polarization effect.

3.3 *Localization precision and spatial resolution*

With enough photons and sensitive detectors, we can now use SMLM to localize single fluorophores at a very high precision. However, it is actually just one step toward the super-resolution imaging to resolve subcellular structures with high spatial resolution.

In fact, spatial resolution is determined by both localization precision and sampling/labeling density. Imagine that we have a cellular structure with a dimension near the diffraction limit, say 1 μm in the long axis, as given in Figure 3.5a. Under a conventional fluorescence microscope, the PSF of the fluorophores would overlap severely, resulting in a featureless blurry image (Figure 3.5b). Apparently, it is not possible to localize individual fluorophores in the image, not even to mention the underlying structure. STORM/(F)PALM super-resolution techniques offer a clever way to see individual fluorophores, the working mechanism of which will be addressed later. For now, let us assume that we have identified the fluorophores in the structure with a resolution ten times higher than the conventional fluorescence imaging. The localization precision is $\sigma_{\mu i}$ as defined in Equation 3.3. There are several ways to estimate the spatial resolution based on the localization precision. Here we use the full width at half maximum (FWHM) of a Gaussian profile to define the spatial resolution $R_{localization}$ as given in Equation (3.4):

$$R_{localization} = 2.355 \cdot \sigma_{\mu i} \tag{3.4}$$

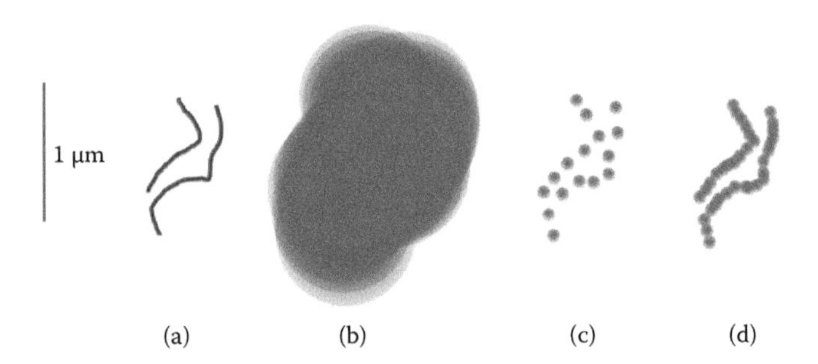

Figure 3.5 The Nyquist-Shannon theorem defines the spatial resolution that is labeling density dependent.

Thus, with the typical condition described in the previous section, where $\sigma_{\mu i} = 1.9$ nm, we know the spatial resolution would be about 4.5 nm. If we assume Figure 3.5c is the super-resolution image of the structure, we immediately notice the image is problematic, that is, ambiguous structure assembly due to insufficient fluorescent spots, which can be due either to poor fluorescent labeling density or incomplete imaging of fluorophores in the structure. Actually, this kind of problem is very common in signal processing and quantitatively described by the Nyquist-Shannon theorem. The theorem points out that in order to resolve a spatial feature of frequency f, the spatial sampling frequency needs to be no smaller than $2f$. In other words, the mean distance between fluorescent labels must be more than twice as dense as the desired resolution. The spatial resolution $R_{sampling}$ limited by the Nyquist-Shannon theorem is given below:

$$R_{sampling} = \frac{2}{d^{1/dim}} \tag{3.5}$$

where d is the localization density and dim is the dimensionality. The effective spatial resolution R_{eff} is then estimated by Equation (3.6):

$$R_{eff} = \sqrt{R_{localization}{}^2 + R_{sampling}{}^2} \tag{3.6}$$

For a 2D super-resolution image with $R_{localization} = 4.5$ nm, it is better to have a labeling density higher than 500 fluorophores in a 100×100 nm^2 square. This high labeling density is sufficient to reveal the detailed structural organization (Figure 3.5d).

It is also important to note that in practice, there are other factors affecting the spatial resolution of the target structure, such as the finite size of the fluorescent protein, and the primary and secondary antibody used in immuno-labeling.

3.4 *Basics of STORM/(F)PALM: Sequential imaging of fluorophores*

The previous discussion seems to put us in a paradoxical situation for super-resolution imaging. On the one hand, the labeling density needs to be high enough for sufficient sampling frequency; on the other hand, fluorophores need to be separated far enough so that their PSFs do not have much overlap and each individual fluorophore can be precisely localized. In 2006, three groups solved this paradox independently (Rust et al. 2006; Betzig et al. 2006; Hess et al. 2006). The trick is to sequentially image photocontrolled fluorophores.

Returning to Figure 3.5, while the fundamental reason for the inability to resolve the structure is that most fluorophores are densely distributed in a diffraction-limited region, the practical problem is actually because all fluorophores emit light simultaneously, with their PSFs largely overlapping so that no single molecules can be identified. In principle, any cellular structure or complex is composed of individual constituent molecules. We have shown that a single molecule can be localized with a precision well beyond the diffraction limit. Thus, if the coordinate of each molecule can be determined with high precision, a high-resolution image of the structure can be reconstructed by stacking all coordinates up in one image.

The key is that, instead of exciting all fluorophores at once, we only image a sparse subset of all molecules in each frame so that individual PSFs have a very low chance to overlap spatially, allowing single molecules to be localized with nanometer precision. This process can be repeated many times in order to accumulate enough single-molecule localizations to fulfill the Nyquist spatial sampling requirement (Figure 3.5). Finally, a complete image can be obtained by adding all the images of the subsets together. Since the localization of each individual molecule is much better than the diffraction limit, the final image composed of all individual molecules is a super-resolution image that "breaks" the diffraction limit.

The realization of sparse subset activation in the whole fluorophore population is based on the fact that a fluorophore may have multiple fluorescent states under certain circumstances. Imagine that we have a fluorophore that has a dark state besides its fluorescent state and permanent damaged state due to photobleach (Figure 3.6), which is actually true for most fluorophores—although it may not be often seen in ensemble fluorescence imaging, blinking of fluorophores is almost inevitable in single-molecule fluorescence experiments. Unlike the molecules in the permanent damaged state, fluorophore molecules in dark states only lose their capability of fluorescing temporarily. A major source of dark states is the electron triplet state with a lifetime in the millisecond range. Sometimes, fluorophore molecules in the triplet state may be converted to radical cations or anions if reacting with oxidant or reductant. By chance

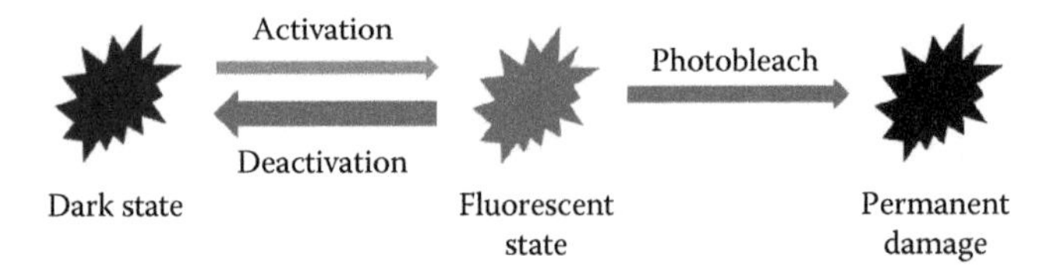

Figure 3.6 A fluorophore may be able to switch multiple times between a fluorescent state and some dark state before it is permanently photobleached.

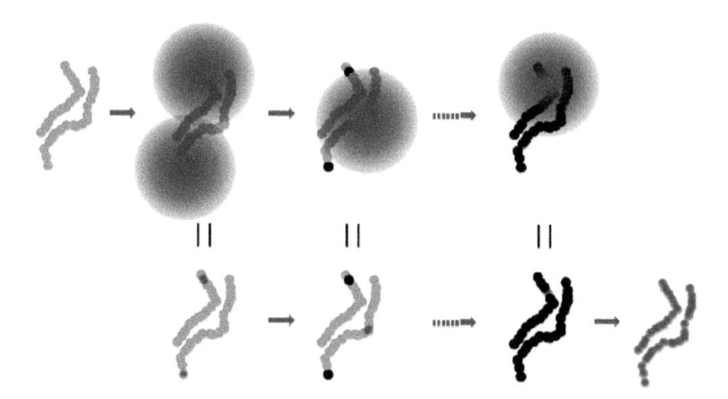

Figure 3.7 A super-resolution image is obtained by iteratively activating, imaging, and photobleaching sparse subsets of the fluorophores in the structure.

or by certain treatments, molecules in dark states are able to recover back to the fluorescent state, and the activation and deactivation may cycle many times (Figure 3.6).

Now, if we can first put all fluorophores in their dark states (Figure 3.7), they somehow control their activation to the fluorescent state at a slow rate such that during each imaging frame, there are only one or two being activated to the fluorescent state, the PSF centers of the activated fluorophores can be precisely localized in each frame (lower row in Figure 3.7). All activated fluorophores are photobleached quickly. By iterating the process, we can image all fluorophores in the structure sequentially, and the accumulation of all localizations reconstructs a super-resolution image of the structure.

Apparently, in order to realize the super-resolution scheme based on single-molecule localization, we first need to have a reliable way to modulate the states of a fluorophore. Although many properties such as pH and temperature can be used for fluorophore modulation, light is obviously better for its invasiveness and temporal-spatial specificity.

Actually, STORM/(F)PALM share the same idea to use photomodulatable fluorophores to achieve controlled activation of the fluorescent molecules. Generally, the target structure is first fluorescently labeled at high density with either photomodulatable organic dye molecules or fluorescent proteins. Then the temporal control of activation and emission is achieved by modulation of two lasers, which are used as the activation and excitation illumination, respectively. The activation laser, usually with a shorter wavelength, is used to activate the fluorophore from dark states into the fluorescent state. The excitation laser, usually with a longer wavelength, is used to excite the fluorophore to fluoresce and/or to a permanently photobleached state or back to dark states. The "on/off" switching rates are both dependent on the power of the corresponding illumination, which offers a way to fine-tune the number of activated fluorophores, that is, a sparse subset, in each imaging frame (Figure 3.7). Generally, the activation power is set so low that only a sparse subset of the fluorophores in dark states can be activated. This activation for each individual molecule is rather stochastic. Therefore, even though the fraction of activated molecules is proportional to the activation power, locations of activated molecules are random in the sample. In contrast, the power of the excitation laser is usually set very high so that the activated fluorophores would on the one hand emit enough photons in a short exposure, and on the other hand be quickly photobleached or switched to dark states.

Originally, the difference between STORM and (F)PALM lay in the choice of fluorophores used for super-resolution imaging. STORM was based primarily on organic dye molecules, which may be induced to switch between the fluorescent state and long-lived dark states under appropriate redox conditions. STORM was invented by the Zhuang group in 2006 (Rust et al. 2006). They found the photoswitching property of the organic dye Cy5 can be well controlled if it is in close proximity to another organic dye with a shorter excitation wavelength, which in their case was Cy3. Under appropriate buffer conditions, Cy5 can be quickly turned into dark states with strong 647-nm illumination and switched back to the fluorescent state with 532-nm illumination. It was found that Cy5 was able to switch between the two states many times before getting permanently photobleached. In the demonstration, the Zhuang group showed a resolution of 20 nm for RecA-coated circular plasmid DNA. The paired-dye scheme in STORM is very powerful as one can combine the "reporter" dye molecule Cy5 with several types of "activator" dye molecules with different wavelengths, such as Alexa405, Alexa488, and Cy3, to achieve multicolor super-resolution imaging, with no need to worry about chromatic aberration. In the follow-up technical development, the Zhuang group implemented multicolor, live–cell, and 3D STORM super-resolution imaging. Their effort has helped to promote the SMLM-based super-resolution imaging techniques widely used in cell biology. Recently, the Zhuang group implemented a dual-objective scheme in the STORM setup, which increases the spatial resolution to 7 nm and allows super-resolution imaging of actin cytoskeletal structures. It was also found that organic dye molecules can be activated directly by UV laser and switched between dark states and fluorescent states. Based on this property, Heilemann et al. introduced direct STORM (dSTORM) without the need of an activator molecule (Heilmann et al. 2008).

PALM and FPALM are similar to STORM except that they were developed using fluorescent proteins (FPs). As FPs are genetically encoded fluorophores, PALM is intrinsically more compatible with live-cell super-resolution imaging. PALM and FPALM were first carried out using photoconvertible EosFP and photoactivable GFP, and achieved spatial resolution of about 30 nm.

Because STORM and (F)PALM share the same mechanism, the boundary between them has become less and less clear. These days, the acronyms have actually become the symbols of SMLM-based super-resolution imaging techniques.

3.5 Photomodulatable fluorophores

Over the past few years, the palette of photomodulatable fluorophores has expanded dramatically. Both organic dye molecules and fluorescent proteins have been engineered to carry different modulation modes by light, including photoactivatable, photoconvertible, and photoswitchable.

Photoactivatable fluorophores are natively in dark states and can be irreversibly activated to the fluorescent state using an appropriate illumination (typically UV light). Photoconvertible fluorophores have multiple fluorescent states, and upon certain illumination (also typically UV light) their fluorescence emission can be converted from one wavelength to another, irreversibly. Photoswitchable fluorophores can be reversibly switched between their dark and fluorescent states upon exposure to an appropriate combination of illumination. Currently, most organic dye molecules used in STORM/(F)PALM belong to the photoswitchable category, while fluorescent proteins have more variable modes of modulation.

As the properties of fluorophores are vital for STORM/(F)PALM, it is worth noting some important considerations for selection and design of fluorophores. First, in order to achieve high resolution in STORM/(F)PALM, the fluorophores need to be localized at a very high precision. According to Equation (3.3), the fluorophore should be sufficiently bright, that is, emitting a large number of photons per photomodulation event. In addition, according to Equation (3.5), the labeling density of the fluorophore should be sufficiently high to match the Nyquist resolution requirement. For organic fluorophores, high labeling density usually means high specificity and lower steric hindrance in immunolabeling, whereas for fluorescent proteins, it usually means a high success rate of protein folding and maturation. Second, as STORM/(F)PALM are all based on two-state switching, the contrast ratio between the two states needs to be as high as possible. Some fluorophores in dark states still have residual fluorescence, so its contrast ratio in the bright and dark states is finite. Using PA-GFP as an example, with a contrast ratio of 40, the number of PA-GFP in a diffraction-limited area (250×250 nm^2) cannot be more than 40 because otherwise the sum of the residual fluorescence of the dark PA-GFP would be as bright as an activated PA-GFP in this area. This labeling density corresponds to a Nyquist resolution of ~80 nm (Equation 3.5). Third, the super-resolution fluorophores need to have a low duty cycle that is, a small fraction of time spent in the fluorescent state versus the dark states. It is also proportional to the probability of spontaneous activation by thermal energy. A low duty cycle offers good control of the number of photoactivated molecules and reduces the number of activated molecules that remain for an extended period, which is critical for fast super-resolution imaging.

With the tremendous amount of work on testing and optimization during the past few years, researchers in this field have identified or developed a large number of photomodulatable fluorophores for STORM/(F)PALM imaging, including both organic dye molecules and fluorescent proteins (Huang et al. 2010).

The major advantage of organic fluorophores is their outstanding brightness and photostability, rendering high single-molecule localization precision. Other merits result from their varied structures and modifications, which provide labeling solutions that are hard to achieve with fluorescent proteins. For instance, fluorescent labeling of DNA and RNA can be easily done with intercalating dyes or labeled nucleic acid (Lubeck and Cai 2012). Easy chemical modification can make a particular dye molecule more membrane permeable; and far-red organics dyes with excellent performance are available for deep tissue imaging. For applications in super-resolution imaging, the disadvantages and limitations of organic fluorophores are as obvious. First of all, the photoswitching mechanisms of organic fluorophores are not well understood. Although organic fluorophore-based super-resolution imaging generally requires removal of oxygen and addition of reductive reagents such as β-mercaptoethylamine (MEA) or β-mercaptoethanol (βME), the exact buffer conditions vary a lot and usually take effort to optimize. In addition, as immunolabeling is generally used for protein labeling in the cell, the labeling specificity has to be a concern because with a ten-fold better resolution in STORM/(F)PALM, false targeting could cause misleading artifacts and absurd conclusions. It usually takes a lot of effort to optimize antibody types, concentrations, incubation times, and temperatures for immunofluorescence. In addition, cell fixation may damage the ultrastructures. The large size of primary or secondary antibody is also a concern as it limits the effective resolution. Another obvious problem with immunolabeling is its incompatibility with live cell imaging. Regarding these major limitations, researchers have developed some approaches to mitigate the problems. For instance, Dempsey et al. (2011) have done a comprehensive evaluation of 26 commercially available organic fluorophores using different buffer conditions. This survey identified several fluorophores with distinct spectral properties that are suitable for multicolor super-resolution imaging. In the current list of options, the best-performing organic fluorophore is Alexa647 and its structural analog Cy5, both of which emit in the far-red spectrum. The second best option is probably the near-infrared Dylight 750 or Cy7. Good options in the red and green spectrum are Cy3B and Alexa488, respectively. Regarding the labeling of different cellular structures, Shima et al. (2012) evaluated many membrane-specific, lipophilic organic fluorophores that are suitable for super-resolution imaging of particular membrane systems in live cells. To avoid the bulky size of antibodies in immunolabeling, one can use small peptide tags such as SNAP-tag (Keppler et al. 2003), which can react with a fluorophore substrate (for SNAP-tag, BG-Alexa647, New England Biolabs). This combines the specificity advantage of genetic encoding and superior fluorophore properties of organic dyes (Jones et al. 2011). Another option is to use a nanobody, which is a variable domain of antibody heavy chain and much smaller than an antibody (Ries et al. 2012). Many efforts have also been made to label live cells for super-resolution imaging. Specific labeling strategies such as FlAsH, ReAsH, CHoAsH, and Click chemistry are all found to be suitable for live-cell imaging. In fact, reductive glutathione thiol is naturally present at a few mM in live cells, so organic fluorophores such as rhodamine and oxazine that need a reductive environment can be used in live-cell super-resolution imaging without an exogenous reductive reagent.

In contrast to organic fluorophores, the obvious disadvantage of fluorescent proteins (FP) is the relatively low brightness. For instance, one of the best super-resolution FP, mEos3.1, can emit ~1000 photons, about one-quarter the number of photons emitted by Cy5. According to Equation (3.3), with other conditions equivalent, the highest resolution using mEos3.1 is only half of that using Cy5. FPs also have several other drawbacks. Many FPs tend to dimerize or oligomerize, which can introduce artifacts in target protein localization and function. Some FPs have very long maturation time, incompatible with live-cell imaging. It also takes some time to optimize the expression level of transiently transfected cell lines to avoid artifacts in super-resolution imaging. Despite these limitations, FPs offer many advantages for super-resolution imaging. First of all, FPs can be genetically encoded with the target protein at a one-to-one ratio, not only intrinsically compatible with live-cell imaging, but also valuable for molecule counting. We previously mentioned that FPs for super-resolution imaging can be grouped in three categories: photoactivatable, photoswitchable, and photoconvertible. As photoactivatable fluorescent proteins (PA-FPs) rely on UV-induced activation from the dark states to the bright state, the fluorescent contrast is critical for their performance in super-resolution imaging. For example, the first PA-FP, PA-GFP, shows a contrast ratio of 100 with 405-nm activation and 488-nm excitation. This is not ideal for super-resolution imaging. PAmCherry and PATagRFP are PA-FPs with red emission. Both of them exhibit good contrast, three- to fivefold higher than that of PA-GFP. Note that when using PA-FPs for super-resolution imaging, it is hard to check the cell condition with conventional imaging methods because the cell is invisible before FPs are photoactivated. Photoswitchable fluorescent proteins (PS-FPs) are used often in super-resolution imaging. Among many PS-FPs, Dronpa is well studied. It is excited at 503 nm and emits at 518 nm. When illuminated by a strong 488-nm laser, Dronpa turns into a dark state, which can be switched back to the bright state by a 405-nm illumination. This process can be repeated multiple times. Although Dronpa seems to work similarly to Cy5, its brightness is about 10 times lower. Chang et al. recently developed a series of novel PS-FPs, mGeos, derived from the EosFP family (Chang et al. 2012). These PS-FPs exhibit a very good photon budget. A few more PS-FPs with different colors have been also developed, including mTFP, KFP1, rsCherry, rsTagRFP, and mApple. PS-FPs generally offer multiple localization of individual molecules, which increases the effective localization precision. Nevertheless, the brightness of PS-FPs still needs improvement. Photoconvertible fluorescent proteins (PC-FPs) convert their fluorescence from one wavelength to another by specific illumination, most often 405 nm, which induces either chromophore backbone cleavage or oxidation and emission wavelength shift. The first PC-FP used in super-resolution imaging is EosFP, which emits at 516 nm by 488 nm excitation and when illuminated by 405 nm, converts to 581-nm emission. Several other PC-FPs that switch to different colors have been developed, including PS-CFP2 (cyan to green), mIrisFP, NijiFP, mClavGR2, mMaple, mKikGR (green to red), and PSmOrange (orange to far red). Among them, mIrisFP and NijiFP are both photoconvertable and photoswitchable, which is particularly useful in live-cell pulse-chase PALM experiments. In the family of PC-FPs, EosFP is so far the best for its excellent brightness and contrast ratio. Its disadvantages are slow maturation and potential artifacts in fusion protein localization and function due to the oligomerization of EosFP. Recently, Zhang et al. (2012) developed mEos3.2, which is a true monomer and matures quickly with high labeling density. It is currently considered the most appropriate FP choice for super-resolution imaging. Compared with PA-FPs and PS-FPs, PC-FPs are more suitable for super-resolution imaging for two main reasons. One is that the native fluorescence before photoconversion can be used to evaluate cells for transformation efficiency; the other is that the irreversibility of photoconversion provides the possibility for protein counting. Compared to organic fluorophores, PC-FPs such as mEos3.2 do not require a special buffer. It can be imaged in a standard physiological buffer, which is essential for live-cell super-resolution imaging. Therefore, for single-color (F) PALM imaging, the best choice would be mEos3.2. For two-color imaging, the second FP option can be PS-CFP2.

In summary, we provide a brief introduction to organic fluorophores and fluorescent proteins for super-resolution imaging. To choose the proper fluorophores for specific experiments, one needs to consider a variety of factors. We recommend these factors be evaluated in the following order: live or fixed cells, required resolution for the target structure (brightness and labeling density), whether quantification of the number of molecules is required, and single- or multicolor imaging.

3.6 Basic microscope setup of STORM/(F)PALM

As an extension of SMLM, STORM/(F)PALM are in principle compatible with most microscope setup for single-molecule imaging, which generally requires good mechanical stability, high photon collection capability, and superior detection sensitivity.

Most STORM/(F)PALM setups are built around a research-level inverted fluorescence microscope. The excitation light sources are usually solid-state lasers with 50 to 500 mW power, which have become fairly affordable. A typical set of lasers used in STORM/(F)PALM experiments are a 405-nm laser (~50 mW) for photoactivation of most fluorophores, a 561-nm laser (~100 mW) for imaging EosFP or Cy3B, and a 647-nm (~200 mW) laser for imaging Cy5/Alexa647. Additional lasers can be chosen based on the recommended fluorophore list of STOMR/(F)PALM. The intensity and timing of lasers can be controlled by several mechanisms. Very often, an acousto-optic tunable filter (AOTF) is used to achieve laser modulation. AOTF can provide rapid wavelength selection and intensity controls at microsecond temporal resolution, sufficient for milliseconds exposure time in STORM/(F)PALM. Currently, the working wavelength range of commercially available AOTF is either 400–650 nm or 450–700 nm, neither of which covers the broad spectral range of super-resolution fluorophores (405–750 nm). In this case, one can choose to control a 405-nm LED laser by an external TTL signal or modulate a 750-nm laser using an ultrafast mechanical shutter and a tunable neutral density filter. Note that a combination of a half-wave plate and a polarizing beam splitter is commonly used to adjust the output intensity of lasers. The lasers can be combined or coupled either in free space or through an optical fiber and sent into the back illumination port of the microscope.

STORM/(F)PALM have unsatisfactory time resolution because they take thousands of frames with sparsely activated and imaged fluorophores to reconstruct a super-resolution image. In order to reduce the data collection time, we need high laser power density to excite the activated fluorophores with enough photons and also quickly switch off or bleach them in each frame. It sounds pretty straightforward to use high-power laser sources. However, high-power lasers are much more expensive and lasers can lose as high as 70% power between the laser source and the output of the objective. In this case, one can shrink the size of the collimated laser beam before the focusing lens for the back-focal plane of the objective, which will proportionally reduce the illumination area and increase the power density. Nonetheless, there should be an upper limit of the laser power that can be used for super-resolution imaging because excess heat from the laser would destroy live cells and even for fixed cells it would deteriorate the image quality. It would take some titration to figure out the right dose of laser. For imaging of EosFP (561 nm) and Alexa647 (647 nm), a good starting point may be a few kW/cm^2.

To maximize the signal-to-noise ratio, the illumination of STORM/(F)PALM is usually set in total internal reflection fluorescence (TIRF) mode, which can restrict the illumination depth only ~200 nm above the substrate and reduce background fluorescence as much as 95%. Nevertheless, it is obvious that TIRF illumination is not suitable for imaging deep inside a cell or tissue samples. STORM/(F)PALM setup is generally equipped with a high-NA objective (NA ≥ 1.40), which is not only a requirement for realizing the TIRF mode through an objective, but also essential for collecting enough photons for high-resolution localization. High NA objectives also generate smaller PSFs, allowing higher activation density per frame and thus better time resolution. Before imaged on the detector, the fluorescence emission needs to be separated from the excitation and activation laser light. This is usually achieved by a combination of dichroic mirrors and fluorescence filters. SMLM-based STORM/(F)PALM experiments require high-end filter sets, which are sometimes custom made. A very useful filter set is the TIRF-level flatness 405/488/561/647 poly-chroic mirrors and Quad-band emission filters. This filter set provides multi-band excitation, activation, and emission, simultaneously. To optimize the performance of this filter set, extra emission filters and Dual-view or Quad-view imaging accessories are recommended.

On the detection side, a cooled electron-multiplying charge-coupled device (EMCCD) camera is usually used for STORM/(F)PALM. EMCCD cameras have several features that make them qualified as single-photon imaging devices. First, the CDD chip can be chilled down to −70°C and lower, significantly suppressing the dark current of the chip. Second, the camera has an on-chip electron-multiplying design, which allows specific amplification of the real photon-electron but not the readout noise. Third, the back-illumination on the thinned CCD provides high quantum efficiency. Most EMCCD cameras are 512 × 512 pixels with 10 MHz pixel readout rate, which restricts the maximum frame rate of about 30 frames per second (fps). This frame rate is sometimes not fast enough for STORM/(F)PALM experiments that run at 20-ms exposure time, not to mention live-cell super-resolution imaging mode. This problem is usually circumvented by using a smaller region of interest. Recently, a new generation of CMOS camera, scientific CMOS (sCMOS), is emerging as a great option for super-resolution imaging as it carries a much higher frame rate and illumination area while retaining similar sensitivity as an EMCCD camera. We previously pointed out that for SMLM, the effective pixel size should be close to or slightly smaller than the PSF of the fluorophore. For a typical EMCCD pixel size of 16 μm, one would need a magnification of 100× to 200×, which can be achieved by combining a 100×objective with additional magnification lenses in the detection light path.

Finally, sample drift compensation is very important for STORM/(F)PALM experiments because of the high spatial resolution and low time resolution. Generally, the microscope should be set on a vibration-isolated optical table. If possible, one can build a rack around the optical table and place anything contributing vibrational noise on the rack. A common method of drift correction is to use fiduciary markers such as fluorescent beads, quantum dots, or gold particles that are stuck on the surface. It is also useful to use image-correlation analysis of a constant feature to correct drift.

3.7 Multicolor STORM/(F)PALM imaging

Multicolor fluorescent imaging is crucial for cell biology study as we often need to investigate protein-protein interaction, macromolecule colocalization, etc. It becomes much more challenging for STORM/(F)PALM imaging because such high resolution reveals more details of the structure. Sometimes a colocalized image would appear un-colocalized under super-resolution imaging, which could be real or just simply due to chromatic aberration in multicolor imaging. Therefore, high-quality apochromatic objectives are required. While the microscopic setup seems good enough for super-resolution imaging, the real challenge is good fluorophores for STORM/(F)PALM. Currently, multicolor STORM/(F)PALM suffer the problems of either cross-channel artifacts or lack of fluorophores with good performance. STORM can perform multicolor imaging via "reporter"-based or "activator"-based approaches.

The reporter-based approach simply uses multiple spectrally distinct fluorophores to achieve super-resolution imaging in each channel. Alexa647, Alexa750/Cy7/DyLight750, Cy3B, and Atto488/Alexa488 are good options for organic fluorophores. The advantage of this approach is the low cross-channel signal, which is essential for resolving a multicomponent complex structure or colocalized features where the molecule copy number of each molecule type is highly different. However, it should be noted that Alexa750, Cy3B or Alexa488 performs less ideally than Alexa647, so the resulting super-resolution images would have lower resolution. More importantly, one would need to do a very careful chromatic aberration correction because under STORM/(F)PALM, even minor aberrations can become strikingly apparent. FPs are also widely used in multicolor super-resolution imaging. There are a couple of concerns with FPs. One is their less ideal brightness than organic fluorophores. The other is that FPs usually have relatively large spectral profiles, resulting in more severe cross-talk. Actually, it is also possible to combine Alexa647 with mEos3.2 for multicolor STORM/(F)PALM imaging. However, one must pay attention to the optimal buffer condition to achieve good performance for both organic fluorophores and FPs in the same sample.

The activator-based approach is based on the observation that a shorter-wavelength fluorophore can facilitate switching of a nearby longer-wavelength fluorophore in addition to its spontaneous activation. In most cases, Alexa647/Cy5 is used as the reporter fluorophore, which is paired up with different activator fluorophores, such as Alexa405, Alexa488, Cy3 etc., and labeled on different types of target molecules. Different channels are assigned by recording the temporal sequence of activator lasers. The primary advantage of this approach is that it is free of chromatic aberration. The great performance of Alexa647/Cy5 would also result in the best super-resolution images. The obvious drawback of this approach is the cross-talk as this approach actually images a single color, so it is very hard to distinguish nonspecific activation and spontaneous activation events.

3.8 3D super-resolution imaging

While 3D super-resolution imaging is an inherent need in cell biology study, it is not easy to achieve good 3D super-resolution imaging in a diffraction-limited regime as objectives have relatively poor axial sectioning capability. The most popular approach to do 3D super-resolution imaging is probably the astigmatic imaging using a cylindrical lens to form the image (Huang et al. 2008). Through a cylindrical lens, a PSF becomes elliptical when it is out-of-focus and the ellipticity and orientation are dependent on the axial position of the object. There are many other ways to achieve 3D super-resolution imaging. For instance, dual-plane imaging captures defocused images at two different focal planes to encode the axial position (Juette et al. 2008). The PSF can also be transformed into a double-helical shape using a spatial light modulator (Pavani et al. 2009), which offers a larger (~2.5 μm) axial range.

The aforementioned 3D imaging methods all need to illuminate a large part of the sample, which either generates much background from off-focus volume or photobleaches many fluorophores before they are supposed to be imaged. To overcome this problem, two-photon imaging and light sheet microscopy have been implemented and demonstrated a depth of several tens of microns.

3.9 *Fast and live cell imaging*

Cells are dynamic systems. It is apparently exciting if super-resolution imaging can become live. However, as STORM/(F)PALM are essentially imaging approaches that trade time for spatial resolution, it is inherently hard to achieve high temporal resolution with STORM/(F)PALM. First of all, the desired Nyquist resolution defines how many molecules need to be accumulated to reconstruct a super-resolution image. Typically, one needs to gather at least a few hundred images containing sparsely activated fluorophores for a satisfactory Nyquist resolution. To compete with the cell motion, the data collection has to be as fast as possible. With the emergence of super-fast sCMOS and high-power laser sources, the hardware is getting ready for the task. The rate-limiting factor is probably going to be lack of fast-switching fluorophores. Organic fluorophores such as Alexa647 and Cy5 are found to be able to switch so quickly that live-cell super-resolution imaging can be done at a 0.5 second time resolution and a spatial resolution of 25 and 50 nm in the lateral and axial dimensions, respectively (Jones et al. 2011). However, the high excitation laser intensities (~15 kW/cm^2) may cause phototoxicity for live cells. Furthermore, it is not trivial to label specific cellular structures in living cells with organic fluorophores. Although FPs are inherently compatible with live-cell imaging, they are now limited to their slow switching rates. Simultaneously with searching for good fluorophores, scientists have also been working on algorithms that can speed up the process. For instance, multiple emitter fitting algorithms, such as DAOSTORM and compressed sensing, can fit overlapping molecules at a density significantly higher than a conventional single-molecule fitting algorithm, which can improve the data acquisition speed by an order of magnitude.

3.10 *Conclusion and outlook*

Since they were invented in 2006, STORM/(F)PALM have made dramatic progress in not only in technology technological improvements but also scientific applications, covering many biological structures and processes in cell biology, microbiology, and neurobiology, etc. Obviously, pursuing of better super-resolution imaging will never stop because biology viewed under higher temporal and spatial resolution keeps surprising us. Despite these breakthroughs, STORM/(F)PALM still have much room for improvement. First, multi-color, higher-resolution, and live imaging all seek for fluorophores with high photon number, fast switching rate, and reliable labeling specificity. Second, live-cell super-resolution imaging requires new algorithms that allow fast data processing. Third, higher spatial resolution calls for new optical methods, such as the dual-objective approach and correlative light electron microscopy (CLEM), which offers unprecedented resolution that eventually bridges the gap between fluorescence microscopy and EM.

References

Betzig, E. et al. (2006). "Imaging intracellular fluorescent proteins at nanometer resolution." *Science* **313**(5793): 1642–1645.

Chang, H. et al. (2012). "A unique series of reversibly switchable fluorescent proteins with beneficial properties for various applications." *Proc. Natl. Acad. Sci. USA* **109**(12): 4455–4460.

Dempsey, G. T. et al. (2011). "Evaluation of fluorophores for optimal performance in localization-based super-resolution imaging." *Nat. Methods* **8**: 1027–1036.

Gustafsson, M. G. L. (2000). "Surpassing the lateral resolution limit by a factor of two using structured illumination microscopy." *J. Microsc.* **198**: 82–87.

Gustafsson, M. G. L. (2005). "Nonlinear structured-illumination microscopy: wide-field fluorescence imaging with theoretically unlimited resolution." *Proc. Natl. Acad. Sci. USA* **102**: 13081–13086.

Heilemann, M., S. van de Linde, M. Schüttpelz, R. Kasper, B. Seefeldt, A. Mukherjee, P. Tinnefeld, and M. Sauer. 2008. Subdiffraction-Resolution Fluorescence Imaging with Conventional Fluorsecent Probes13. *Angewandte Chemie International Edition* **47** (33):6172–6176.

Heintzmann, R. et al. (2002). "Saturated patterned excitation microscopy: a concept for optical resolution improvement." *J. Opt. Soc. Am. A* **19**: 1599–1609.

Hell, S. W. (2007). "Far-field optical nanoscopy." *Science* **316**: 1153–1158.

Hell, S. W. and J. Wichmann (1994). "Breaking the diffraction resolution limit by stimulated emission: stimulated-emission-depletion fluorescence microscopy." *Opt. Lett.* **19**: 780–782.

Hess, S. T. et al. (2006). "Ultra-high resolution imaging by fluorescence photoactivation localization microscopy." *Biophys. J.* **91**(11): 4258–4272.

Hofmann, M. et al. (2005). "Breaking the diffraction barrier in fluorescence microscopy at low light intensities by using reversibly photoswitchable proteins." *Proc. Natl. Acad. Sci. USA* **102**: 17565–17569.

Huang, B. et al. (2010). "Breaking the diffraction barrier: super-resolution imaging of cells." *Cell* **143**: 1048–1058.

Huang, B. et al. (2008). "Three-dimensional super-resolution imaging by stochastic optical reconstruction microscopy." *Science* **319**: 810–813.

Jones, S. A. et al. (2011). "Fast, three-dimensional super-resolution imaging of live cells." *Nat. Methods* **8**: 499–505.

Juette, M. F. et al. (2008). "Three-dimensional sub-100 nm resolution fluorescence microscopy of thick samples." *Nat. Methods* **5**(6): 527–529.

Keppler, A. et al. (2003). "A general method for the covalent labeling of fusion proteins with small molecules in vivo." *Nat. Biotechnol.* **21**(1): 86–89.

Lubeck, E. and L. Cai (2012). "Single-cell systems biology by super-resolution imaging and combinatorial labeling." *Nat. Methods* **9**: 743–748.

Pavani, S. R. P. et al. (2009). "Imaging beyond the diffraction limit by using a double-helix point spread function." *Proc. Natl. Acad. Sci. USA* **106**(9): 2995–2999.

Ries, J. et al. (2012). "A simple, versatile method for GFP-based super-resolution microscopy via nanobodies." *Nat. Methods* **9**: 582–584.

Rust, M. J. et al. (2006). *Nat. Methods* **3**: 793–795.
Shima, S.-H. et al. (2012). "Super-resolution fluorescence imaging of organelles in live cells with photoswitchable membrane probes." *Proc. Natl. Acad. Sci. USA* **109**(35): 13978–13983.
Thompson, R. E. et al. (2002). "Precise nanometer localization analysis for individual fluorescent probes." *Biophys J.* **82**: 2775–2783.
Yildiz, A. et al. (2003). "Myosin V walks hand-over-hand: single fluorophore imaging with 1.5-nm localization." *Science* **300**: 2061–2065.
Zhang, M. et al. (2012). "Rational design of true monomeric and bright photoactivatable fluorescent proteins." *Nat. Methods* **9**(7): 727–729.

Problems

1. Name as many factors as you can that contribute to the background in single-molecule fluorescence imaging experiments.
2. Design one or more routines to measure the background noise of your microscope setup.
3. Name factors that determine the localization precision in single-molecule fluorescence imaging experiments.
4. Design one or more routines to measure the single-molecule localization precision of your microscope setup.
5. Based on this chapter and additional reading, describe a few mechanisms for photoactivation of organic fluorophore.
6. Based on this chapter and additional reading, describe a few mechanisms for photoactivation of fluorescent proteins.
7. Discuss the direction for developing live-cell super-resolution imaging.

chapter four

Small-molecule labeling probes

Jie Wang, Jie Li, Yi Yang, Maiyun Yang, and Peng R. Chen

Contents

4.1 Introduction

Dissecting biological processes frequently requires the ability to label and visualize various biomolecules in the context of living cells. Fluorescent imaging is becoming a powerful strategy in illuminating biological systems, which in many cases remain as a black box to us. Fluorescent microscopy, in conjunction with fluorescent probes, provides a valuable toolkit in lighting up these diverse biological systems, allowing us to monitor and image biomolecules at the cellular, tissue, or even whole-animal levels with increasing precision (Roger and Tsien 1995). For example, fluorescent imaging of proteins, the most abundant biomolecules within a cell, which participate in essentially all life processes, has revolutionized our ability to study the expression, interaction, localization, and dynamics of proteins in living species. Numerous fluorescent probes have been developed that either covalently or noncovalently bind to biological macromolecules (e.g., proteins or nucleic acids), or are located within an intracellular region or compartment such as the cytoskeleton, mitochondria, Golgi apparatus, endoplasmic reticulum, and the nucleus. Besides these versatile fluorophores, many biocompatible conjugation methods allow labeling of biomacromolecules of interest with high specificity (Wysocki and Lavis 2011).

In this chapter, we will first introduce the properties of various fluorescent probes, including probes suitable for two-photon, near-infrared, or super-resolution imaging. The labeling methods as well as the underlying bioorthogonal reactions will next be discussed, followed by some examples in which small-molecule labeling probes play a pivotal role in studying and solving biological problems.

4.2 General view of small-molecule labeling probes

In general, a small-molecule labeling probe contains a chromophore for fluorescent imaging and a targeting group for labeling purposes. In addition, a modifying group tethered to the chromophore and a linker between the chromophore and the targeting groups are usually necessary. The modifying group can fine-tune the optical physical property of the chromophore, and the linker is to eliminate the potential influences of the targeting group in the chromophore. In Figure 4.1, two examples of small-molecule-based fluorescent probes are shown with the aforementioned four components labeled in different colors (Mandato and Bement 2003; Nolting et al. 2011). These fluorescent probes are usually small, robust, versatile, and easy to modify, which play an irreplaceable role in labeling and imaging biomolecules in diverse settings. In this chapter we discuss these four components of the small molecule fluorescent probe in details.

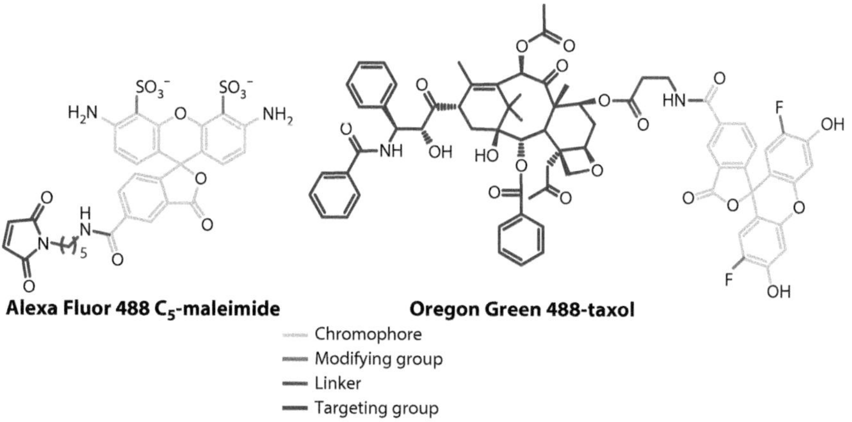

Alexa Fluor 488 C_5-maleimide

Oregon Green 488-taxol

Chromophore
Modifying group
Linker
Targeting group

Figure 4.1 Two examples of small molecule fluorescent probes with the four components labeled in different colors.

4.2.1 Chromophore and the modification group

A chromophore is central to a small-molecule-based fluorescent probe. The chromophore is responsible for the photophysical properties of the molecule, including the excitation and emission wavelength, fluorescent quantum yield, and photostability. The structure and emission wavelength of many commonly used fluorophores are shown in Figure 4.2 (Chan et al. 2012). The synthesis and modification of coumarin are straightforward. However, its excitation wavelength is below 400 nm, which can cause the excitation of many endogeneous molecules inside cells that will result in a high fluorescence background. The high quantum yield of fluorescein and rhodamine makes them very popular fluorescent dyes. Most of the modifications on these two fluorophores are centered in the benzene ring distal from the chromophore, which generally gives a rather small influence to their fluorescence properties. Bodipy exhibits excellent membrane

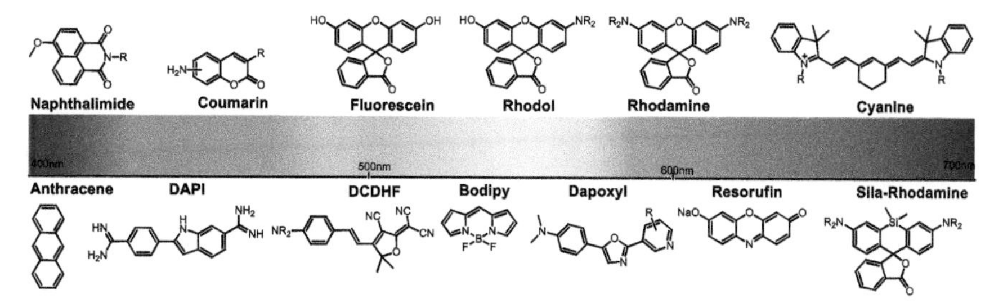

Figure 4.2 The name and structure of commonly used fluorophores and their approximate emission wavelengths.

permeability and can be readily modified to meet various needs. The cyanine type of fluorophores feature good photostability and high fluorescent quantum yield, making them highly valuable dyes for super-resolution imaging. In addition, Cy3 (excitation = 550 nm, emission = 570 nm) and Cy5 (excitation = 650 nm, emission = 670 nm) are a pair of cyanine dyes widely used for FRET (fluorescence resonance energy transfer) analysis (Mishra et al. 2000).

Modifying groups play a crucial role in modulating the property of a given fluorophore. By directly modifying the fluorophore or tethering a chemical moiety at a periphery region, the modifying groups are frequently employed to alter the property of fluorophores to satisfy different needs. For instance, introduction of an auxochrome can shift the fluorophore's absorption peak toward the red region and increase its absorption strength; the addition of a carboxyl group can increase the solubility of the dye in water' and addition of a sulphonate group can enhance the water solubility, fluorescent quantum yield, and ability to resist photobleaching and/or pH changes. We next show a few examples of commercially available fluorophores in which the modifying group significantly improves or alters their physical properties (Figure 4.3; Lain Johnson 2010).

Many commonly used commercial fluorescent probes are derivatives of fluorescein or rhodamine with certain modifying groups. For example, difluorofluorescein (Figure 4.4), a famous pH-sensitive probe, which is commonly referred to as Oregon Green, is a functionalized fluorescein derivative carrying two fluorine groups. When compared with the original fluorescein compound, Oregon Green has higher quantum yield and greater resistance to photobleaching. Moreover, Oregon Green has a lower pKa (pKa = 4.7 versus 6.4 for fluorescein), making its fluorescence pH insensitive in the physiological pH range. In addition, the pH sensitivity of Oregon Green in weakly acidic conditions (pH 4 to 6) also makes it a useful pH indicator for acidic organelles of a live cell. Another popular commercial fluorophore, Alexa Fluor 488, is the original rhodamine110 molecule functionalized with two sulfo-groups, which has significant improvements in photostability, pH insensitivity, and water solubility.

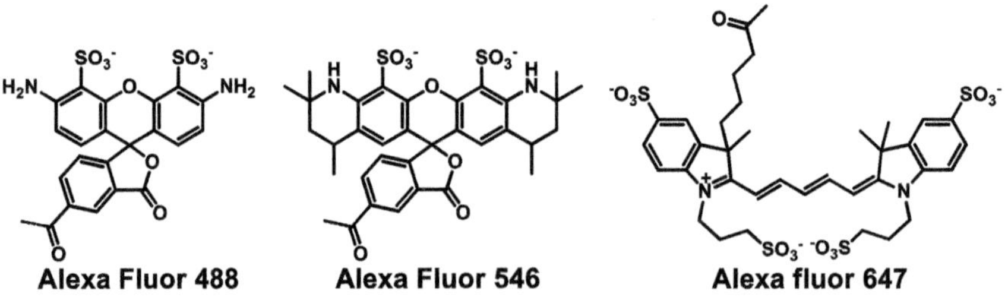

Figure 4.3 Some commercially available dyes carrying sulfonic modifying groups.

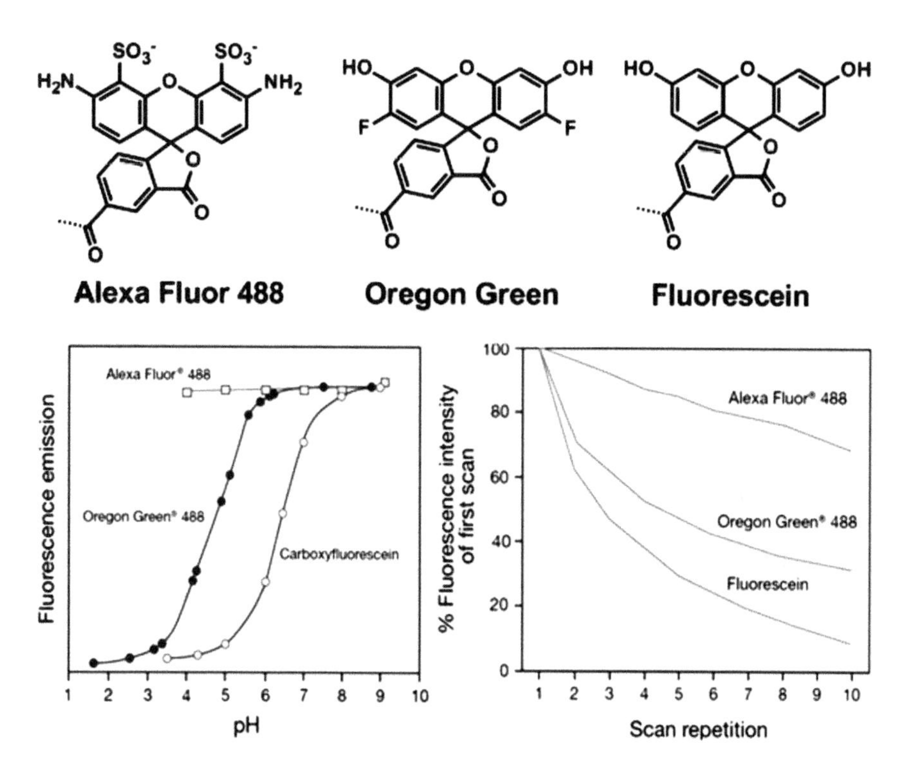

Figure 4.4 Modifications and improvements of the Green fluorescent dyes.

Indeed, Alexa Fluor 488 is perhaps the first choice of a fluorescein substitute in most applications. From these two examples, we can see that the modifying groups are crucial for a small-molecule fluorescent probe.

4.2.2 Targeting group and the linker

A key advantage for synthetic fluorophores is the ability to use various chemical means to modulate their property as well as their location within a biological system (Wysocki and Lavis 2011). The chromophore and modifying groups determine the photophysical properties of a probe, whereas the targeting group enables the labeling of a specific biological target by the probe. A targeting group here refers to a chemical moiety that allows the covalent or noncovalent attachment of the fluorophore to a biological target. The covalent attachment is usually carried out via the biocompatible conjugation reaction between a pair of chemical handles on the targeting group and the biomolecular target, respectively, while the noncovalent attachment is typically achieved through a small molecule capable of specific binding to the target biomolecule or cellular structure. For example, the Alexa Fluor 488 C5-maleimide (Figure 4.1) can covalently label a protein by reacting with the sulfydryl group on proteins, whereas the Oregon Green 488-taxol (Figure 4.1) can label the cytoskeleton by the specific interaction between taxol and the cytoskeleton. The labeling method is discussed in more detail in this chapter (Schuler et al. 2002; Mandato and Bement 2003).

The property of small molecule fluorescent probes may in many cases be affected by the modifying or targeting group. To circumvent this problem, a linkage group, typically a short PEG (polyethyleneglycol) chain or an alkyl chain is often inserted to separate the chromophore and additional functional groups so that these independent parts can work well without interference from each other.

In summary, various fluorophores and modifying groups, in conjunction with targeting groups and linkers, provide a great many ways to combine those four parts together, which can make a powerful toolkit to satisfy different application needs.

4.3 *Small molecule fluorescent probes for special purposes*

In vivo optical fluorescent imaging techniques, in conjugation with diverse fluorescent labeling probes, have revolutionized our ability to visualize the biological processes at the molecular, cellular, and tissue levels. To make many essential cellular processes visible, however, people need to see deeper, clearer, and with more finely tuned probes by using fluorescent imaging. For example, how does a virus, or even a toxin protein alone, cause the infection of mammalian host cells, and what affects their intracellular distribution? Difficulties such as photon scattering, autofluorescence (prominent in tissue imaging), and diffraction limit make the normal fluorescent probes increasingly unsuitable for biological imaging. Recently, many special and high-quality small molecule fluorophores have been developed, including probes suitable for near-infrared (NIR) imaging, two-photon (2P) imaging, and super-resolution imaging.

4.3.1 *Near-infrared probes*

The NIR window that extends from 650 to 900 nm exhibits high penetration capability into living tissue without potential interference from endogenous fluorescence (Jöbsis-van derVliet 1999; Pastrana 2013. The combined absorption from water and hemoglobin are minimal in this region, while the tissue remains largely transparent to light (Nolting et al. 2011). Meanwhile, NIR fluorescent labels have excellent biocompatibility, because they avoid the use of UV-Vis light, which may cause phototoxicity or unwanted autofluorescence background. In recent years, remarkable

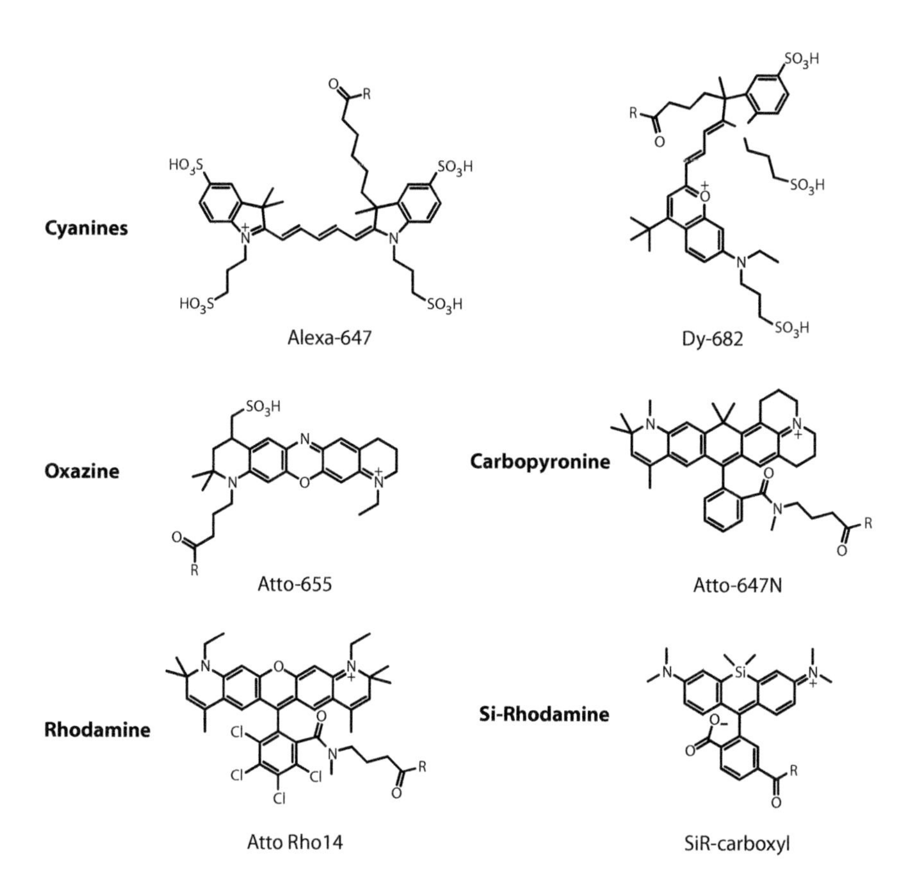

Figure 4.5 Chemical structures of some widely used near-infrared probes.

advances have been made in the development of GFP-like fluorescent proteins that absorb and emit in the low 600-nm range, but further red-shifting has proven difficult thus far (Fernandez-Suarez and Ting 2008). Fortunately, a series of small-molecule-based NIR dyes with redder excitation and emission wavelengths have been developed recently. The widely used, commercially available NIR fluorescent imaging probes can be divided into four classes: cyanine, carbopyronine, rhodamine, and oxazine (Nolting et al. 2011). The chemical structures of some widely used NIR fluorescent probes are shown in Figure 4.5. However, even though these NIR fluorophores possess remarkably high photostability and brightness, they tend to be membrane impermeable and show nonspecific binding to cellular components (Lukinavičius et al. 2013).

A class of fluorophores based on silicon-containing rhodamine derivatives was created lately, which showed excellent spectroscopic properties, and in most cases can be cell membrane permeable (Egawa et al. 2011).

Lukinavičius and coworkers (2013) reported a highly membrane permeable and biocompatible near-infrared silicon–rhodamine fluorophore (SiR-carboxyl; Figure 4.5) that can be specifically coupled with an intracellular protein of interest by using different labeling techniques. The highly efficient and selective protein labeling was achieved by introducing various targeting groups to SiR (e.g., SiR-SNAP, SiR-CLIP, and SiR-Halo). Another unique feature that distinguishes this dye from other NIR fluorophores is its fluorogenic property and the high brightness (signal-to-noise ratio), which permits live-cell imaging without cumbersome washing steps. A possible explanation for the observed fluorogenic property is that the reactions of these targeting group-bearing SiR variants with their respective protein tags keeps the fluorophores in their fluorescent zwitterionic form. Meanwhile, the two sources of background signal, aggregation of unreacted dye and unspecific binding to hydrophobic structures, may stay in the nonfluorescent spirolactone form. In addition, this new fluorophore proved to be ideally suited for use in live-cell super-resolution microscopy.

4.3.2 *Two-photon probes*

Another class of dyes that can be used in the NIR region for bioimaging is two-photon fluorescent probes. Two-photon laser scanning microscopy, in which fluorescence is induced by two-photon excitation (2PE), has become a powerful technique to obtain three-dimensional images of cells and diffuse tissues (Yao and Belfield 2012). Some commonly used fluorescent probes can be used for 2PFM bioimaging; however, their low two-photon absorption (2PA) or poor photostability limits their application in 2PFM. Rhodamine is one of the most widely used templates for creating specific two-photon dyes (Milojevich et al. 2013). For example, rhodamine and its derivatives have recently been used as two-photon fluorescence probes for monitoring mercury ions in live cells (Lee et al. 2010), as multiphoton probes for exploring mitochondrial zinc ions (Masanta et al. 2011), and as benchmark molecules in two-photon excitation systems (So et al. 2000). Development of more powerful probes for 2PFM and expansion of their applications in diverse biological systems are currently actively pursued in many laboratories worldwide (Yao and Belfield 2012).

4.3.3 *Probes and chemical concepts in super-resolution imaging*

The resolution limit in light microscopy is caused by the wave nature of light and the diffraction of light at small apertures (Huang et al. 2010). However, employing physical or chemical means to distinguish fluorescence emission of fluorophores in an additional dimension, such as time or spectroscopic characteristics, suggested a way to bypass the resolution limit. This can be realized either in a deterministic way, such as by generation of a light pattern, or in a stochastic way, such as by reduction of the number of simultaneously fluorescing fluorophores. Deterministic far-field super-resolution imaging methods (e.g., stimulated emission depletion, STED), and stochastic super-resolution imaging methods (e.g., photoactivated localization microscopy, PALM: stochastic optical reconstruction microscopy, STORM; and direct STORM, dSTORM), have both advantages and disadvantages (van de Linde et al. 2012). Stochastic super-resolution imaging methods all share the same underlying mechanism and need stochastic photoactivation, photoconversion, reversible photoswitching of fluorescent proteins (FPs) or synthetic organic fluorophores.

Three classes of synthetic organic fluorophores have been used for super-resolution imaging: (1) regular nonshifting, nonactivatable fluorescent dyes; (2) reversible photoactivatable molecules (also called photoswitchers); and (3) irreversible photoactivatable compounds (also called photocaged fluorophores). Among these regular fluorescent probes, ATTO dyes (Figure 4.5) have been widely used in deterministic super-resolution imaging methods such as STED, owing to their intense brightness, high photostability, and long fluorescence lifetimes (Fernandez-Suarez and Ting 2008).

However, stochastic super-resolution imaging methods such as STORM rely on modulatable fluorophores, and make use of reversible (photoswitching) and irreversible (photocaging) photoactivatable probes (Raymo et al. 2012). Within the class of photoswitchers, photochromic molecules such as rhodamine B can photoswitch via light-induced isomerization without the addition of other molecules (Fölling et al. 2007 and see Figure 4.6). Another class of photoswitchers involves reversible photoswitching of standard synthetic fluorophores that appeared in 2005

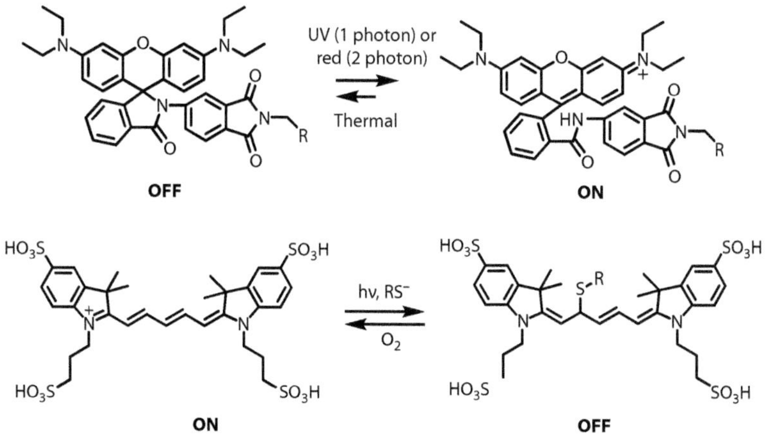

Figure 4.6 Chemical mechanisms of two reversible photoswitchers.

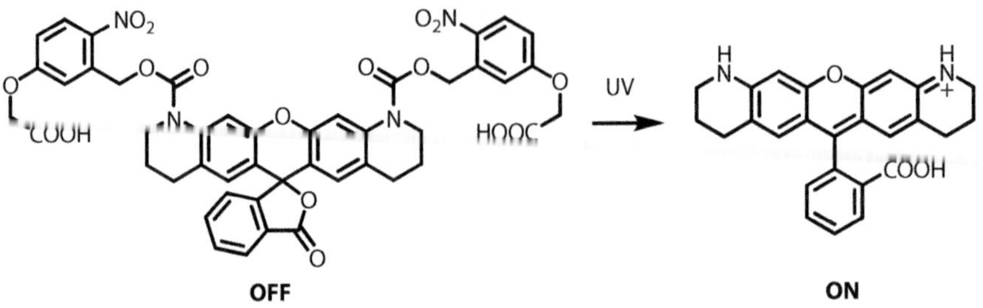

Figure 4.7 Photo-uncaging of caged Q-rhodamine.

(van de Linde et al. 2012). Single carbocyanine dyes such as Cy5 and Alexa Fluor 647 were used as efficient reversible single-molecule optical switches. Single-molecule photoswitching experiments in aqueous buffer in the presence of 100 mM β-mercaptoethylamine (MEA) and an oxygen scavenger demonstrated that single carbocyanine molecules can be cycled between a fluorescent and nonfluorescent state more than 100 times with a reproducibility of >90% at room temperature. The photoactivation of cyanine dyes can be facilitated by a second chromophore molecule. The underlying mechanism was then studied and showed that its dark-state photoconversion depends on both pH and thiol concentration. The product generated in the dark state is a cyanine-thiol adduct which can react with oxygen to repopulate the fluorescent form (Figure 4.6; Dempsey et al. 2009).

The intriguing photoswitching mechanism of carbocyanine dyes can be transferred to the majority of other standard synthetic fluorophores, most of which belong to the class of rhodamine dyes with a pronounced electron affinity and thus share a similar redox activity, to enable dSTORM experiments (Heilemann et al. 2009).

Among photocaged organic fluorophores, caged Q-rhodamines (Figure 4.7) and caged fluorescein have been used for super-resolution imaging of caged rhodamine-dextran that had been dried on glass cover slips (Fernandez-Suarez and Ting 2008). However, caged compounds have not yet been used for super-resolution imaging of biological samples. It must be noted that reversible photoswitchers are advantageous for single-molecule super-resolution imaging because the same fluorophore can be used multiple times.

4.4 *Labeling techniques*

The modification of biomolecules with small-molecule-based probes is an important method for elucidating and engineering biological functions both in vitro and in vivo. During recent years the progress in the development of labeling techniques has provided a broad palette of applications.

Here we take proteins as the example and describe the different labeling techniques and give the readers a general understanding of these chemical tools and how they work.

4.4.1 *Methods for protein labeling in vitro*

4.4.1.1 *Classic labeling methods for cysteine and lysine residues*

As one of the naturally occurring nucleophilic amino acid residues present in proteins, the thiol group of cysteine is often used for bioconjugation. It can readily undergo alkylation with iodoacetamide or Michael addition with maleimide, and through dithiol exchange it can also react with dithiol-modified probes. The ε-amino group of lysine is another frequently used target for protein modification. Lysine can react with succinimidyl esters, isocyanates, and sulfonyl chlorides, generating the corresponding amides, thioureas, and sulfonamides (Figure 4.8). It has to be noted that these reagents can also react with the N-termini of proteins.

4.4.1.2 *New methods for labeling cysteine and lysine residues*

Some new techniques have recently been developed for modification of lysine and cysteine residues (Figure 4.9). Iridium-catalyzed transfer hydrogenation for lysine-specific reductive alkylation provides high yield of the labeling product at neutral pH (McFarland and Francis 2005). Another lysine-specific labeling method was reported based on the 6π-aza-electrocyclization reaction that is rapid and robust under neutral pH conditions (Tanaka et al. 2008). A two-step method was also developed for cysteine modification: the first step involves the transformation of cysteine to dehydroalanine followed by the second step, the Micheal addition with thiol group-modification reagents to give specific labeling cysteines (Bernardes et al. 2008).

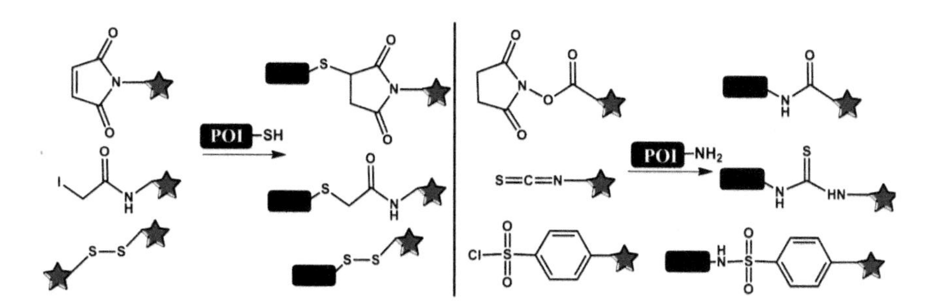

Figure 4.8 Classic strategies for labeling cysteine and lysine residues. POI, protein of interest.

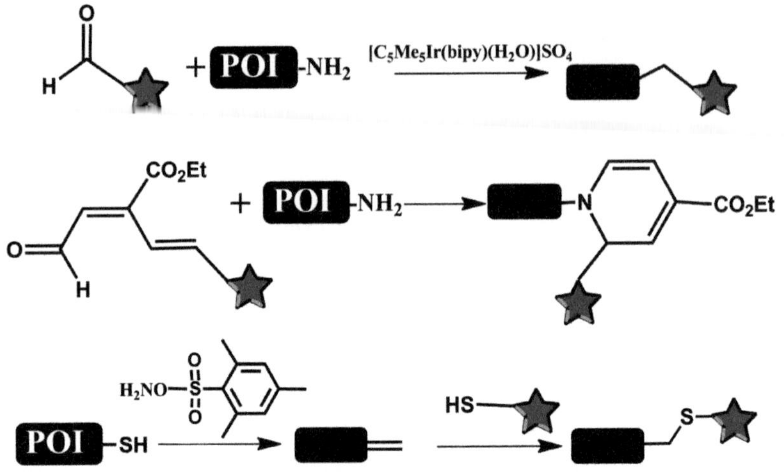

Figure 4.9 New methods for labeling lysine and cysteine residues. POI, protein of interest.

4.4.1.3 Methods for labeling tyrosine and tryptophane

Although not as extensively explored as cysteine and lysine, methods for selective labeling of other natural amino acid residues have also draw great interest. For example, tyrosine-based labeling was achieved via a three-component Mannich reaction with aldehyde and analine (Figure 4.10; Joshi et al. 2004). Another reaction explored for tyrosine modification is a palladium catalyst-mediated reaction involving the π-allyl species (Tilley

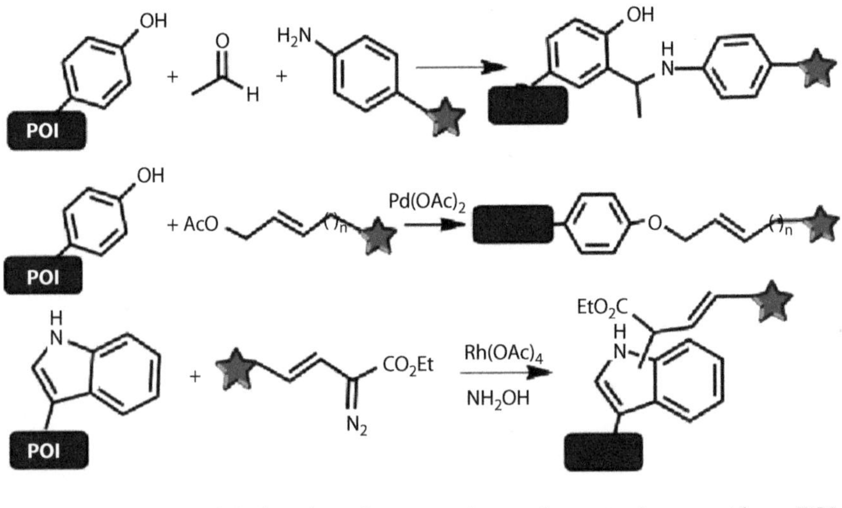

Figure 4.10 Protein labeling based on tyrosine and tryptophane residues. POI, protein of interest.

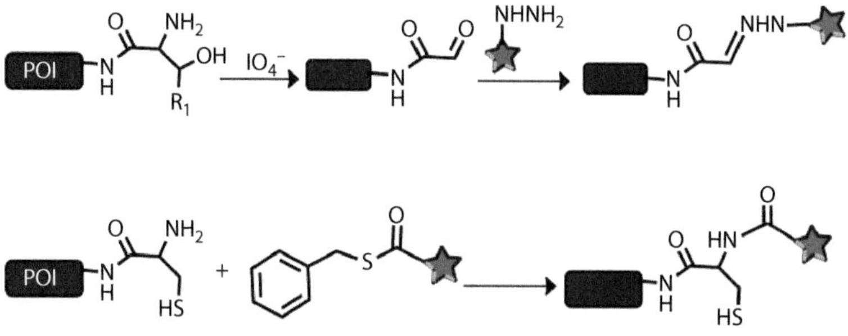

Figure 4.11 N-terminal-specific labeling of serine, threonine, and cysteine. POI, protein of interest.

and Francis 2006). Tryptophan can also be site specifically labeled using rhodium acetate in the presence of a diazo compound (Antos et al. 2009).

4.4.1.4 N-terminal-specific labeling methods

Proteins containing specific N-terminal residues such as serine, threonine, or cysteine can be selectively modified with different reagents (Ma et al. 2008). The N-terminal 1,2-amino alcohol structure of serine or threonine can be oxidized by IO_4^-, thus allowing its reaction with hydrazide or aminooxy modified probes. In addition, the benzyl thioester derivatives can be used for specific labeling of N-terminal cysteine (Figure 4.11).

Other methods have been developed to label N-terminal α-amino groups in case the reactive amino acid residues (serine, threonine, and cysteine) do not exist. To avoid interference of the ε-amino group of lysine, the pH of the reaction medium is often adjusted to 8.0 because the ε-amino group is protonated under these conditions. Another approach is guanidination of the lysine ε-amino group to prevent its labeling with the probes.

4.4.2 Methods for protein labeling in vivo

4.4.2.1 Bioorthogonal chemical reactions for protein labeling

Bioorthogonal reactions typically rely on a pair of non-natural functional groups that react exclusively with each other without perturbation of the biological system. Several bioorthogonal reactions have been developed in recent years with a high specificity and reactivity (Sletten and Bertozzi 2009; Hao et al. 2011).

The ketone or aldehyde group can react with aminooxy or hydrazide to form a stable oxime or hydrazine molecule. These two groups have been incorporated into proteins that served as a reactive handle for protein labeling. The Staudinger ligation involves the coupling of an azide group with triarylphosphine derivatives bearing an ester group on the aromatic ring, which results in an amide linkage as well as one equivalent of nitrogen gas. The Cu(I) catalyzed azide-alkye cycloaddition (CuAAC), commonly referred to as the "click reaction," is probably the most widely used bioorthogonal reaction in living systems. This cycloaddition reaction can be conducted in conditions resembling that in living species (e.g., room temperature, water, and neutral pH, highly reduced environment) to form a triazole compound with high specificity and efficiency. To avoid the cytotoxicity of Cu(I) ions, a strain-promoted, copper-free click reaction has been developed using the cyclooctyne-based reagents. In addition, various copper(I) ligands have been created to attenuate the Cu(I)-induced toxicity and accelerate CuAAC, which further enhanced the biocompatibility of click reactions with living organisms.

A recently emerging bioorthogonal labeling strategy is based on the inverse-electron demand Diels-Alder reaction between a tetrazine compound and a strained *trans*-cyclooctene or a norbornene dienophile. This reaction can be conducted at room temperature with an extremely fast reaction rate. In addition, the simple terminal alkene can react with tetrazole derivatives through photoinduced 1,3-dipolar cycloaddition. This reaction, termed the "photoinduced click reaction," requires UV light irradiation to form a nitrile-imine dipole which then undergoes a cycloaddition to afford the pyrazoline cycloadduct (Lim and Lin 2011).

In parallel with these strain-promoted and light-induced bioorthogonal reactions, investigators have also explored the repertoire of transition-metal-mediated bioorthogonal reactions. An emerging example is the palladium-mediated cross-coupling reactions that have been successfully demonstrated on exogenously delivered small molecules inside cells, paving the way for applying similar chemistry to label proteins. Indeed, Pd-mediated labeling reactions (Suzuki coupling and Sonogashira coupling) have recently been performed on purified proteins as well as on proteins displayed on an *Escherichia coli* cell surface without apparent toxicity. Very recently, our group reported the site-specific protein labeling inside pathogenic Gram-negative bacterial cells via a ligand-free palladium-mediated cross-coupling reaction. The identified simple compound $Pd(NO_3)_2$ exhibited high efficiency and biocompatibility for protein labeling in vitro and inside living enteric bacteria.

As a critical step for employing the aforementioned bioorthogonal reactions to label biomolecules, functional groups involved in these reactions have been incorporated into biomolecules via metabolic or genetic engineering strategies. For example, both residue-specific and site-specific methods have been developed for incorporation of unnatural amino acids (UAAs) carrying diverse bioorthogonal handles into proteins, the most abundant biomolecules within a cell. Among them, the recently emerged pyrrolysine-base genetic-code expansion system allowed UAAs to be introduced into proteins in prokaryotic and eukaryotic cells, and even in animals. Unnatural sugars carrying different bioorthogonal handles have

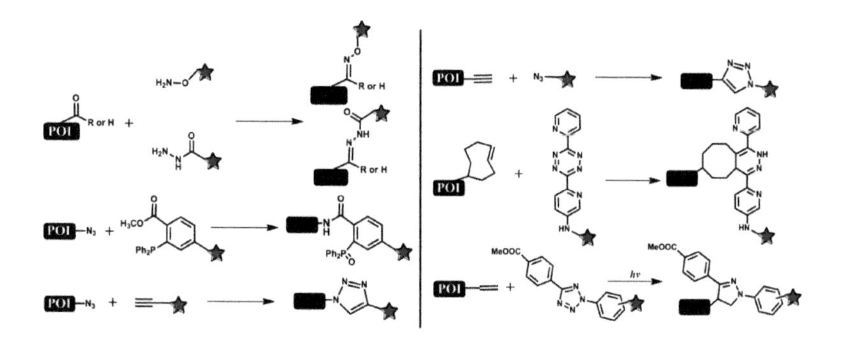

Figure 4.12 Bioorthogonal reactions for protein-specific labeling. POI, protein of interest.

also been metabolically incorporated into cell surface glycans as a means of engineering the cell surface with diverse chemical reactivity.

Taken together, these bioorthogonal reactions, in conjunction with various genetic or metabolic-incorporated bioorthogonal handles, provide a wide array of methods for labeling biomolecules in a native cellular setting (Figure 4.12).

4.4.2.2 *Protein labeling using enzyme-mediated reactions of specific peptide/nucleic acid sequences*

In nature, additional chemistries beyond the functional groups contained in the canonical 20 amino acids are often required to carry out protein's physiological functions. Cells often use enzyme cofactors and posttranslational modification (PTM) machineries to fulfill this requirement at the posttranslational stage. Mimicking PTM, researchers have developed a suite of enzyme-mediated protein modification methods. The target protein is fused or inserted with a peptide tag to which an enzyme can ligate its substrate with high specificity (Figure 4.13). For protein labeling, a substrate analog containing a bioorthogonal group is usually used, and the enzyme is chosen or engineered so that it can tolerate the substrate modification. One of the early works used a mutant form of human O^6-alkylguanine-DNA alkyltransferase (**hAGT**) as the enzyme, which can react with O^6-benzylguanine derivatives on a protein. **BirA**, the biotin ligase from *E. coli*, directs the biotinylate molecule to a protein containing a specific peptide sequence that can be labeled with a streptavidin-modified probe. Another bacterial enzyme, Sortase A (**StrA**), recognizes the peptide sequence (LPXTG) near the C-terminus of proteins and catalyzes its labeling with a polyglycine-containing probe. The lipoic acid ligase (**LplA**) can catalyze the attachment of an aryl azidolipoic acid derivative (or fluorophore-conjugated lipoic acid) to proteins containing the appropriate peptide. The "HaloTag technology," which

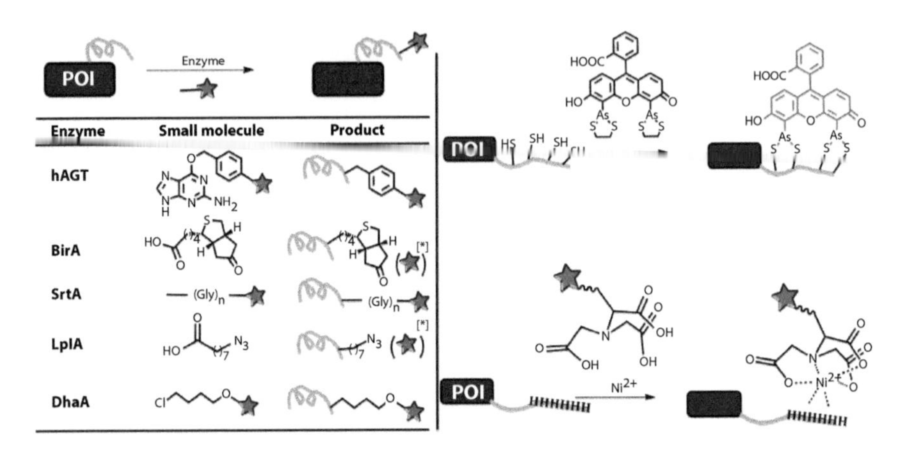

Figure 4.13 Enzyme tag and peptide tag for recombinant protein labeling. [*], the fluorescent probe was inserted through a subsequent chemical labeling step. POI, protein of interest.

was developed by Promega, uses the haloalkane dehalogenase (**DhaA**) to trap the alkyl chloride probes and has been widely used for specific protein labeling in living systems. Bertozzi and coworkers recently utilized formylglycine-generating enzyme (**FGE**) to convert the Cys residue in a 13- or 6-residue consensus sequence to aldehyde containing formylglycine in bacteria and mammalian cells. This approach eliminates the need for enzyme substrate engineering so that it might be advantageous for certain applications. Looking for additional enzymes capable of directly converting a natural amino acid side-chain to an unnatural moiety will be an exciting avenue to explore in the future.

Some peptide sequences can be modified with enzyme-free techniques. For example, the tetracysteine motif (CCXXCC) can be selectively labeled with a biarsenical-functionalized fluorescent probe (such as **FlAsH**—fluorescein arsenical hairpin binder) and **ReAsH** (resorufin arsenical hairpin binder). The hexahistidine tag, originally developed for protein purification, has also been applied for protein labeling by conjugation to a nickel nitrilotriacetic acid (Ni-NTA) probe. In another strategy, the nickel ion is replaced by zinc ion to overcome the fluorescence quenching effect and cell toxicity of Ni(II) ions. The zinc-containing multinuclear complex can selectively bind to a tetra-aspartate peptide and has been applied to cell surface protein labeling. Most recently, the zinc finger-mediated approach has been applied for cell surface protein labeling. The zinc finger DNA-binding domain is fused to the cell surface protein, thus enabling the site-specific labeling of a fluorophore-modified DNA probe (Mali et al. 2013).

4.4.3 *Other methods for protein labeling in living systems*

Several techniques based on affinity labeling have recently been demonstrated to specifically label the natural protein endogenously expressed in a living system without genetic manipulation. Probes used in these methods typically contain three components: (1) a ligand is used to specifically attach the entire probe to the target; (2) a reactive group that can covalently bind to the protein residue near the active site; (3) a reporter tag that is usually a fluorophore or a biotin molecule.

Ligand-directed chemical labeling (Figure 4.14) has been proved to be a successful strategy for endogenous protein modification (Takaoka et al. 2013). Researchers used phenylsulfonate linker as a reactive group, which, through an S_N-2-type reaction with a natural nucleophilic amino-acid residue, can transfer the labeling probe to the target protein. This method

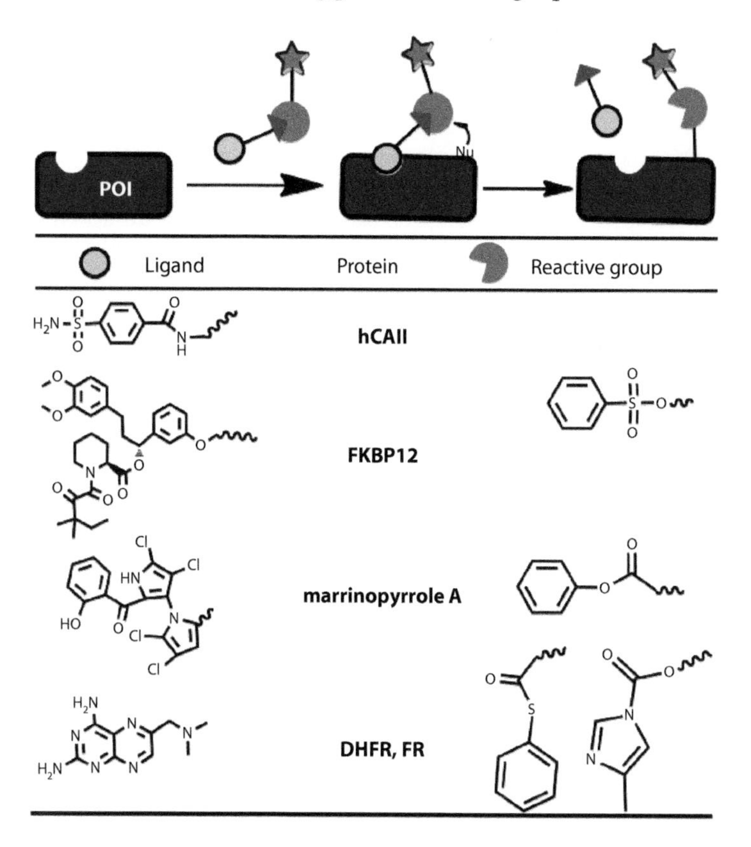

Figure 4.14 Ligand-directed traceless labeling of endogenous proteins. POI, protein of interest.

has been used for the labeling of endogenous proteins such as human carbonic anhydrase II (hCAII) and FK506-binding protein 12 (FKBP12). Other reactive groups, such as acylphenol, acyl imidazole, and thioester have also been demonstrated for effective labeling of specific proteins.

4.5 Applications of small molecule labeling probes

Fluorescent probes can be labeled on proteins and other biomolecules through various labeling methods mentioned above. These small-molecules-based labeling probes have been widely used in studying the function of biomolecules, detecting signal molecules, and identifying ultrastructure within a biological sample. Here we use a few examples to show their applications in actin ultrastructure identification, glycan imaging, and cell surface receptor trafficking.

4.5.1 Super-resolution imaging for ultrastructure identification

Fluorescence-labeling-based super-resolution imaging provides valuable information for identifying the ultrastructures of cell components with high spatial resolution and molecular specificity. Actin ultrastructures in different cell cytoskeletons have been recently demonstrated by the Zhuang group. The multicolor and three-dimensional (3D) super-resolution imaging method (STORM) they developed gave an unambiguous view of cell ultrastructures (Bates et al. 2007).

Actin is essential for diverse cellular processes, including the establishment of neuronal polarity, the transport of cargos, the growth of neurites, and the stabilization of synaptic structures in neuron cells (Dent and Gertler 2003; Cingolani and Goda 2008; Kapitein and Hoogenraad 2011). However, the organizations of actins in axons remain largely unknown. Zhuang and coworkers found a periodicity actin-spectrin-based cytoskeleton structure in axons by using a super-resolution imaging strategy based on ligands and immune labeling. They labeled actin filaments with phalloidin-Alexa Fluor 647 suitable for STORM imaging. The actin wrapped around the circumference of the axons to form a ring-like structure evenly spaced along the axonal shafts with a periodicity of ~180 to 190 nm. This topological arrangement is very different from that in dendritic shafts, where actin forms filaments running along the dendritic shafts. Next, βII-spectrin, a component of rod-like αII-βII spectrin tetramer, which was hypothesized as the linker between periodicity actin rings, was immune-labeled on the C-terminal. A similar periodic, ring-like structure was found in axons but not in dendrites. The spacing between adjacent rings is 182 nm, which is nearly identical to the values measured from the actin rings. Adducin, a protein that caps the growing end of actin filaments and promotes the binding of spectrin to actin in erythrocyte cytoskeleton, was then labeled, which yielded similar findings. To further prove the relevance, two-color STORM imaging was performed on actin/βII-spectrin, actin/adducin, and βII-spectrin/adducing combinations. The results indicated that actins and spectrins are indeed localized together and form a periodicity structure. In the end, an ultrastructure model of actin in neuron cells was demonstrated. The cortical cytoskeleton of axons is comprised of short actin filaments that are capped by adducin at one end and arranged into ring-like structures, which wrap around the circumference of the axon. Also, they hypothesized the periodic distribution of sodium channels in the axon initial segments is most likely coordinated by the underlying periodic cytoskeleton structure.

4.5.2 *In vivo glycan imaging*

Glycans are covalent assemblies of sugars (oligosaccharides and polysaccharides) that exist in either free form or in covalent complexes with proteins or lipids. Most classes of glycan exist as membrane-bound glycoconjugates or as secreted molecules, which become part of the extracellular matrix (ECM) that mediates cell adhesion and motility, as well as intracellular signaling events. Some other glycans like N-linked glycans mainly located on the endoplasmic reticulum that mediate protein folding. Changes in glycan structures often serve as markers indicating altered gene expression during development and disease progression (Haltiwanger and Lowe 2004; Ohtsubo and Marth 2006). However, in vivo visualization of glycans was largely hampered due to their incompatibility with genetic manipulations.

Bertozzi and coworkers exploited glycan metabolic pathways, such as the N-acetylgalactosamine (GalNAc) salvage pathway, to introduce bioorthogonal functional groups into glycans (Sletten and Bertozzi 2009), which provided an elegant procedure for glycan imaging on living cells. In particular, they incorporated an azide-bearing $Ac_4GalNAz$ into growing zebrafish embryo glycans, which allowed the subsequent strain-promoted click reaction with DIFO-fluorophore for imaging glycans in growing zebrafish embryos. This "strain-promoted cyclo-addition reaction" (Laughlin et al. 2008) has quick reaction kinetics and low reagent toxicity when compared with the Staudinger ligation or CuAAC reaction. Labeling zebrafish embryos with DIFO-647 at different incubation time of $Ac_4GalNAz$, a burst in fluorescence intensity in the jaw region, pectoral fins, and olfactory organs was found at 60 hpf (hours postfertilization) which was continued for more than 72 hpf.

In addition, to get more detailed information, two-color detection experiments were performed. Zebrafish embryos were labeled with DIFO-647 at 60 hpf to visualize the cell-surface glycans exposed at that time point. Because the dye is membrane-impenetrable, nascent azide-labeled glycan trafficking through the secretory pathway remained unreacted. Next, the embryos were labeled with DIFO-488 at several different hpfs. Intense labeling of the pharyngeal epidermis in the jaw region was observed which was derived from the second reaction (DIFO-488, 60~61 hpf), but not from the first fluorophore labeling (DIFO-647, 62~63 hpf). They also found that, in caudal regions of the pharyngeal epidermis that were labeled during both reactions, a corrugated distribution of glycan was observed: old glycan was restricted to peaks at the extreme ventral surface, whereas new glycan was produced in troughs projecting dorsally. Such a distribution was not found in any other regions in the entire zebrafish. In the end, the authors expanded their analysis to encompass the period from 60 to 72 hpf using three DIFO-fluorophore conjugates (DIFO-647, 60~61 hpf; DIFO-488, 62~63 hpf; DIFO-555, 72~73 hpf). After labeling, the kinocilia of mechanosensory hair cells surrounding the head of the embryo were robustly labeled with DIFO-555, but not with DIFO-647 or DIFO-488. In contrast, adjacent epithelial cells were labeled with the three DIFO-fluorophores, which indicates that these regions matured early during development.

Taken together, the spatial and temporal differences in the expression of cell surface glycans during zebrafish embryo development was elegantly depicted by this work, which is hardly feasible using traditional methods.

4.5.3 Detection of cell surface receptor localization and trafficking

Cell surface receptors are mainly divided into three classes: ion channel-linked receptors, enzyme-linked receptors, and G-protein-coupled receptors (GPCRs). They are essential in ion transport or intracellular-extracellular signal transduction. Receptors like the GPCRs form the largest family of membrane signaling molecules, which are also the major target for drug development. Receptors such as epidermal growth factor receptor (EGFR), which is one of the enzyme-linked receptors, mediate cell proliferation and differentiation during animal development. Malfunction of surface receptors may cause severe damage to cells, leading to many diseases such as cancer, neurodegenerative diseases, and cardiovascular diseases. Thus, cell surface receptors are important targets for drug development. Detection of localization and trafficking of surface receptors on a native cell may provide unequivocal information regarding their functions as well as pathological roles in many diseases.

Baskin et al. (2007) revealed the location of both neurokinin-1 receptor (NK1R) and prototypical (GPCR) in living cells by using the acyl carrier protein (ACP) labeling method. They simultaneously labeled ACP–NK1R fusion protein with Cy3 (donor) and Cy5 (acceptor) at different, well-defined ratios. When the expression level is close to physiological condition (≈25,000 receptors per cell), no FRET (fluorescence resonance energy transfer) signal was detected, which indicated NK1Rs exist as monomers in the physiological condition.

However, when the expression level of the ACP–NK1R fusion protein is higher (≈63,000 receptors per cell), they observed high FRET efficiency which was not linearly dependent on DA (donor/acceptor) ratios, excluding the presence of dimers. This was further proved through calculation. The FRET signal was further explained by high frequency of stochastic encounters between donors and acceptors. However, on the expression level (≈63,000 receptors per cell), stochastic encounters between donors and acceptors couldn't induce such high FRET efficiency if the NK1Rs were homogeneously distributed on the cell membrane. The high FRET efficiency would fit only when local concentration of NK1Rs was about 80 times higher than that from a homogeneous distribution. After further calculation and experiments, they demonstrated that NK1Rs tends to be concentrated in microdomains, which were found to constitute about 1% of the entire cell membrane.

Meyer et al. (2006) developed an O^6-alkylguanine-DNA-alkyltransferase (snap-tag) based "no wash" labeling method, allowing the monitoring of EGFR trafficking during cell migration. DRBG-488, an O^6-benzylguanine derivative which contained a quencher and a Fluor-488 moiety, was designed and synthesized. Before labeling, Fluor-488 was quenched by the quencher via the FRET mechanism. However, when ligated to SNAP-tag, the quencher group will be released, resulting in a 1000-fold fluorescence enhancement. This DRBG-488 facilitated labeling strategy enabled continuous imaging of cell surface receptors without a cumbersome washing procedure under living conditions. They found that, as cells began to migrate upon EGF stimulation, newly formed internalized vesicles were located in a significantly greater number at the rear than at the front region with respect to the direction of migration of the cell (Figure 4.15), which indicates

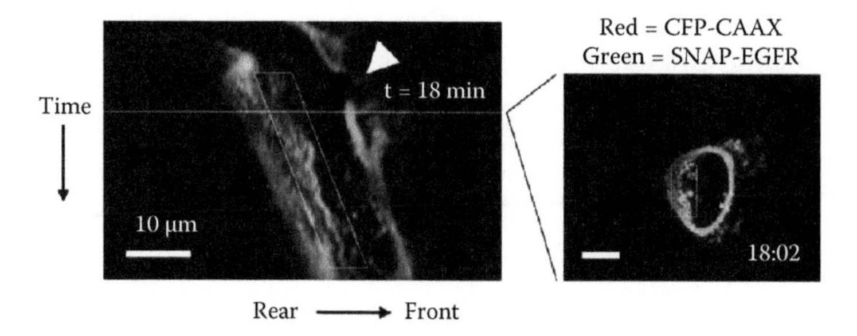

Figure 4.15 **(See color insert.)** Difference in internalization of SNAP-EGFR from rear to front during cell migration in epidermal growth factor-stimulated cells. (From T. Komatsu et al. (2011). *J. Am. Chem. Soc.* **133**(17): 6745–6751. Copyright American Chemical Society. Reprinted with permission.)

endocytosis of the EGFR in stimulated cells may be slowed down at the leading edge as compared to the rear during cell migration.

As shown above, small-molecule-based labeling probes provide powerful tools in ultrastructure identification and visualization of protein and glycan dynamics. However, some of the small molecular fluorescent probes are toxic to the cells. Some of the labeling systems may disturb protein function because of their big size. These drawbacks limited their application in studying protein functions in a native cellular setting. Development of more biocompatible labeling probes and chemical reactions is greatly needed in the years to come.

4.6 Conclusion and outlook

The revolutionary ability to visualize and monitor biomacromolecules in living systems, as a result of the development of various fluorescent probes, has dramatically expanded our understanding of nature. However, although the in vitro protein labeling strategy has been widely explored, expanding such methodologies for protein labeling in living cells is still in its infancy. In addition to more biocompatible labeling strategies, methods for more precise labeling are also urgently needed. One of the major advantages of site-specific protein labeling in comparison with the genetic encoded fluorescent proteins is the small size of the labeling probes that can be incorporated at virtually any desired site on the protein. This feature is extremely valuable when the bulky GFP induces significant perturbations to the target protein's structure and function. Introduction of bioorthogonal functionalities into proteins with high efficiency and specificity should open up the possibility of manipulating and monitoring biomolecular function and dynamics in living cells. Noteworthy, the pyrrolysine-based system has already demonstrated its versatility for encoding UAAs in prokaryotic cells, eukaryotic cells, and multicellular organisms, which is becoming a one-stop shop for site-specific incorporation of chemical reactivity into proteins in these diverse living species. We envision that an exciting future direction is the study of generation, trafficking, and dynamic conformational change of crucial cell surface proteins and intracellular proteins in their native cellular environment. Given these promising technology developments, we believe it should not be a long way to go.

References

Antos, J. M., et al. (2009). "Chemoselective tryptophan labeling with rhodium carbenoids at mild pH." *J. Am. Chem. Soc.* **131**(17): 6301–6308.

Baskin, J. M. et al. (2007). "Copper-free click chemistry for dynamic in vivo imaging." *Proc. Natl. Acad. Sci. U.S.A.* **104**(43): 16793–16797.

Bates, M. et al. (2007). "Multicolor super-resolution imaging with photo-switchable fluorescent probes." *Science* **317**(5845): 1749–1753.

Bernardes, G. J. L. et al. (2008). "Facile conversion of cysteine and alkyl cysteines to dehydroalanine on protein surfaces: versatile and switchable access to functionalized proteins." *J. Am. Chem. Soc.* **130**(15): 5052–5053.

Chan, J. et al. (2012). "Reaction-based small-molecule fluorescent probes for chemoselective bioimaging." *Nat. Chem.* **4**(12): 973–984.

Cingolani, L. A. and Y. Goda (2008). "Actin in action: the interplay between the actin cytoskeleton and synaptic efficacy." *Nat. Rev. Neurosci.* **9**(5): 344–356.

Dempsey, G. T. et al. (2009). "Photoswitching mechanism of cyanine dyes." *J. Am. Chem. Soc.* **131**(51): 18192–18193.

Dent, E. W. and F. B. Gertler (2003). "Cytoskeletal dynamics and transport in growth cone motility and axon guidance." *Neuron* **40**(2): 209–227.

Egawa, T. et al. (2011). "Development of a far-red to near-infrared fluorescence probe for calcium ion and its application to multicolor neuronal imaging." *J. Am. Chem. Soc.* **133**(36): 14157–14159.

Fernandez-Suarez, M. and A. Y. Ting (2008). "Fluorescent probes for super-resolution imaging in living cells." *Nat. Rev. Mol. Cell Biol.* **9**(12): 929–943.

Fölling, J. et al. (2007). "Photochromic rhodamines provide nanoscopy with optical sectioning." *Angew. Chem. Int. Ed.* **46**(33): 6266–6270.

Haltiwanger, R. S. and J. B. Lowe (2004). "Role of glycosylation in development." *Annu. Rev. Biochem.* **73**(1): 491–537.

Hao, Z. et al. (2011). "Introducing bioorthogonal functionalities into proteins in living cells." *Acc. Chem. Res.* **44**(9): 742–751.

Heilemann, M. et al. (2009). "Super-resolution imaging with small organic fluorophores." *Angew. Chem. Int. Ed.* **48**(37): 6903–6908.

Huang, B. et al. (2010). "Breaking the diffraction barrier: super-resolution imaging of cells." *Cell* **143**(7): 1047–1058.

Jöbsis-van der Vliet, F. F. (1999). "Discovery of the near-infrared window into the body and the early development of near-infrared spectroscopy." *J. Biomed. Opt.* **4**(4): 392–396.

Joshi, N. S. et al. (2004). "A three-component mannich-type reaction for selective tyrosine bioconjugation." *J. Am. Chem. Soc.* **126**(49): 15942–15943.

Kapitein, L. C. and C. C. Hoogenraad (2011). "Which way to go? Cytoskeletal organization and polarized transport in neurons." *Mol. Cell. Neurosci.* **46**(1): 9–20.

Komatsu, T. et al. (2011). "Real-time measurements of protein dynamics using fluorescence activation-coupled protein labeling method." *J. Am. Chem. Soc.* **133**(17): 6745–6751.

Lain Johnson, M. T. Z. S. (2010). T*he Molecular Probes® Handbook: A Guide to Fluorescent Probes and Labeling Technologies*. Life Technologies Corporation**[au: please add location of publisher]**.

Laughlin, S. T. et al. (2008). "In vivo imaging of membrane-associated glycans in developing zebrafish." *Science* **320**(5876): 664–667.

Lee, J. H. et al. (2010). "A two-photon fluorescent probe for thiols in live cells and tissues." *J. Am. Chem. Soc.* **132**(4): 1216–1217.

Lim, R. K. V. and Q. Lin (2011). "Photoinducible bioorthogonal chemistry: a spatiotemporally controllable tool to visualize and perturb proteins in live cells." *Acc. Chem. Res.* **44**(9): 828–839.

Lukinavičius, G. et al. (2013). "A near-infrared fluorophore for live-cell super-resolution microscopy of cellular proteins." *Nat. Chem.* **5**(2): 132–139.

Ma, H. et al. (2008). "N-Terminal specific fluorescence labeling and its use in local structure analysis of proteins (Invited Review)." *Curr. Chem. Biol.* **2**(3): 249–255.

Mali, P. et al. (2013). "Barcoding cells using cell-surface programmable DNA-binding domains." *Nat. Meth.* **10**(5): 403–406.

Mandato, C. A. and W. M. Bement (2003). "Actomyosin Transports microtubules and microtubules control actomyosin recruitment during xenopus oocyte wound healing." *Curr. Biol.* **13**(13): 1096–1105.

Masanta, G. et al. (2011). "A mitochondrial-targeted two-photon probe for zinc ion." *J. Am. Chem. Soc.* **133**(15): 5698–5700.

McFarland, J. M. and M. B. Francis (2005). "Reductive alkylation of proteins using iridium catalyzed transfer hydrogenation." *J. Am. Chem. Soc.* **127**(39): 13490–13491.

Meyer, B. H. et al. (2006). "FRET imaging reveals that functional neurokinin-1 receptors are monomeric and reside in membrane microdomains of live cells." *Proc. Natl. Acad. Sci. U.S.A.* **103**(7): 2138–2143.

Milojevich, C. B. et al. (2013). "Surface-enhanced hyper-raman scattering elucidates the two-photon absorption spectrum of rhodamine 6G." *J. Phys. Chem. C* **117**(6): 3046–3054.

Mishra, A. et al. (2000). "Cyanines during the 1990s: a review." *Chem. Rev.* **100**(6): 1973–2012.

Nolting, D. et al. (2011). "Near-infrared dyes: probe development and applications in optical molecular imaging." *Curr. Org. Synth.* **8**(4): 521–534.

Ohtsubo, K. and J. D. Marth (2006). "Glycosylation in cellular mechanisms of health and disease." *Cell* **126**(5): 855–867.

Pastrana, E. (2013). "Near-infrared probes." *Nat. Meth.* **10**(1): 36–36.

Raymo, F. et al. (2012). "Photoactivatable fluorophores." *ISRN Physical Chemistry* **2012**: 15.

Roger Y. and Tsien, A. W. (1995). *Handbook of Biological Confocal Microscopy.*

Schuler, B. et al. (2002). "Probing the free-energy surface for protein folding with single-molecule fluorescence spectroscopy." *Nature* **419**(6908): 743–747.

Sletten, E. M. and C. R. Bertozzi (2009). "Bioorthogonal chemistry: fishing for selectivity in a sea of functionality." *Angew. Chem. Int. Ed.* **48**(38): 6974–6998.

So, P. T. C. et al. (2000). "Two-photon excitation fluorescence microscopy." *Annu. Rev. Biomed. Eng.* **2**(1): 399–429.

Takaoka, Y. et al. (2013). "Protein organic chemistry and applications for labeling and engineering in live-cell systems." *Angew. Chem. Int. Ed.* **52**(15): 4088–4106.

Tanaka, K. et al. (2008). "A submicrogram-scale protocol for biomolecule-based PET imaging by rapid 6π-azaelectrocyclization: visualization of sialic acid dependent circulatory residence of glycoproteins." *Angew. Chem. Int. Ed.* **47**(1): 102–105.

Tilley, S. D. and M. B. Francis (2006). "Tyrosine-selective protein alkylation using π-allylpalladium complexes." *J. Am. Chem. Soc.* **128**(4): 1080–1081.

van de Linde, S. et al. (2012). "Live-cell super-resolution imaging with synthetic fluorophores." *Annu. Rev. Phys. Chem.* **63**(1): 519–540.

Wysocki, L. M. and L. D. Lavis (2011). "Advances in the chemistry of small molecule fluorescent probes." *Curr. Opin. Chem. Biol.* **15**(6): 752–759.

Xu, K. et al. (2013). "Actin, spectrin, and associated proteins form a periodic cytoskeletal structure in axons." *Science* **339**(6118): 452–456.

Yao, S. and K. D. Belfield (2012). "Two-photon fluorescent probes for bioimaging." *Eur. J. Org. Chem.* **2012**(17): 3199–3217.

Problems

1. What advantages do small molecular fluorescent probes have, compared to fluorescent protein technology?
2. How would you design a novel small molecular fluorescent probe for a specific experiment purpose? (For example, design a red fluorescent probe for super-resolution imaging and CuAAC labeling.)
3. What are the advantages and disadvantages of different chromophores and how can their photophysical properties be improved?
4. The cyanine dye Cy5 and several of its structural relatives have been reversibly quenched by the phosphine tris(2-carboxyethyl) phosphine (TCEP). Using Cy5 as a model, try to propose the quenching mechanism.
5. Define photoswitchers and photocaged fluorophores. Try to find other types of photocaged fluorophores in the literature.
6. How many types of palladium-mediated cross-coupling reactions have been utilized for biomolecule labeling?
7. Look back at the chemical reactions you have learned from textbooks or other places. Besides the reactions described in this chapter, what kind of reaction might be developed as a bioorthogonal reaction?
8. Since the Cu(I) ions are toxic to live cells, what kind of improvement do you suggest to make it available to living systems?
9. What's the challenge in imaging glycans? How can glycans be labeled with fluorescent probes?

chapter five

Fluorescent proteins for optical microscopy

Pingyong Xu, Mingshu Zhang, and Hao Chang

Contents

5.1 Introduction

Since the first green fluorescent protein (GFP) gene was cloned from the jellyfish *Aequorea victoria* 20 years ago, many GFP-like fluorescent proteins (FPs) have been discovered and applied in the modern life sciences. More and more GFP-like proteins with different color spectra are used for specific labeling and dynamic tracking in living cells. In addition to conventional optical microscopy applications, some FPs can be used for recently developed diffraction-unlimited microscopy techniques. Other FPs with unique properties, such as photoconversion from one color to another, have been developed to meet the specific needs of super-resolution microscopy.

In this chapter, we will first describe the properties of conventional FPs and their related applications in modern optical microscopy. Then, we will focus on recently developed photoactivatable fluorescent proteins (PAFPs) for diffraction-unlimited microscopy. Finally, we provide an outlook on how to develop FPs with specific properties to meet the needs of both traditional and super-resolution microscopy in the future.

5.2 GFP-like proteins for optical microscopy

The most famous gene-fusion tag, GFP, was first purified from the jellyfish *Aequorea victoria* by Osamu Shimomura and his coworkers in 1962 (Shimomura et al. 1962). This wild-type GFP emits bright green fluorescence under UV light. However, for the next 10 years, very little attention was paid to the value of GFP. In 1992, the gene for GFP was cloned by Douglas Prasher and coworkers. In 1994, a breakthrough occurred when the cloned gene for GFP was expressed in *Escherichia coli* and *Caenorhabditis elegans* by Martin Chalfie and colleagues to track the expression of different gene products. This study significantly increased the popularity of GFP as a marker for live-cell imaging in heterologous organisms (Chalfie et al. 1994). Around the same time, Roger Y. Tsien and colleagues solved the mechanism of how GFP emits fluorescence by structural analysis (Ormo et al. 1996) and extended the color palette by developing different variants (Heim et al. 1995). Today, the variants of GFP and its homologs with fluorescence emission spectra ranging from blue to far-red have essentially revolutionized biology and optical microscopy.

In this chapter, we discuss the properties of fluorescent proteins, the classes of fluorescent proteins, and the applications for GFP-like proteins as sensors. The aim of this section is to elucidate the development of GFP-like proteins and facilitate the selection of suitable fluorescent tags for imaging.

5.2.1 Properties of fluorescent proteins

Just as light exhibits a wave-particle duality, the family of GFP-like proteins has two separate basic characters, referred to as the fluorescence-protein duality. As fluorescent dyes, the brightness, excitation and emission spectra, fluorescence lifetime, photostability, and phototoxicity are basic properties affecting their use in various applications. As translated proteins, maturation time and oligomerization also influence the performance of FPs.

5.2.1.1 *Fluorescent protein brightness*

The most essential parameter in evaluating FPs is the brightness of the fluorophore. If you need to examine a weak promoter, detect a fluorescent signal from the expression of a single copy of a protein, or perform single-molecule experiments, you need to choose a "brighter" fluorescent protein. Thus, there has been a demand to develop "brighter" FPs. These brighter tags increase the signal-to-noise ratio for optical microscopy. In general, the brightness of an FP is determined by the quantum yield and molar absorption coefficient at the peak of the absorption band:

$$B_{brightness} = \Phi_{\lambda ex/\lambda em} \times \varepsilon_{\lambda ex} \tag{5.1}$$

For example, TagRFP-T and mCherry have similar molar absorption coefficients at the wavelength of excitation maximum, but the quantum yield of the former is twice that of the latter. Therefore, the brightness of TagRFP-T is two times higher than that of mCherry. The brightness of FPs is an important consideration when detecting weak signals.

5.2.1.2 *Fluorescent protein spectral properties*

The absorption, excitation, and emission spectra of FPs are important parameters for investigators to keep in mind when choosing FPs. Generally speaking, the bandwidth of the excitation and emission peaks determines the number of colors that can be separated by lasers and filters. This is a very important factor for multicolor imaging. In general, four colors (such as blue, green, red/orange, and far-red) can be used to simultaneously image distinct proteins or organelles in one cell. Unlike GFP and mCherry, which have short stoke shifts (the distance between the peak wavelengths of absorption and emission), some fluorescent proteins, such as Sapphire and LSS-mKate1, have very long Stoke shifts, which can extend the effective color palette.

5.2.1.3 *Phototoxicity and photostability*

After an extended period of illumination by lasers and high-intensity arc lamps, FP chromophores tend to produce free radicals that can lead to cell death. This light-induced damage is defined as FP phototoxicity. Red fluorescent proteins in particular, such as mCherry and DsRed, will result in artificially high levels of cell death when overexpressed or imaged for extended periods of time (Strack et al. 2008), presumably due to FP phototoxicity. On the other hand, this phototoxicity can be used as a tool to inactivate target proteins. For example, KillerRed, a derivative of DsRed, can kill HeLa cells when illuminated with a green laser. This method is referred to as chromophore-assisted light inactivation (CALI) and may be a useful tool for killing targeted cells for disease therapy (Bulina et al. 2006).

In live-cell experiments and three-dimensional imaging, the photostability of FPs determines the length of time for which images can be captured. mTagBFP, Emerald GFP, TagRFP-T, mCherry, and mKate2 are all highly photostable proteins suitable for long-term imaging. However, it is important to evaluate the photostability of fluorescent proteins in your own experimental system prior to extensive or long-term imaging because photostability differs depending on illumination conditions, fusion proteins, and expression systems.

5.2.1.4 Maturation time and fluorescence lifetime

The chromophore maturation process reflects the time course over which a fluorescent probe can be detected after translation. Molecular oxygen takes part in this process, and the chromophore will not mature without oxygen. Most fluorescent proteins derived from jellyfish, reef corals, and sea anemones mature well at 37°C. Some fluorescent proteins that mature very quickly, such as sfGFP and Venus, have been developed for special applications, such as inclusion body screening, a fluorescent timer (Khmelinskii et al. 2012) and gene expression detection (Li and Xie 2011).

The fluorescence lifetime is defined as the time that a chromophore stays in its excited state before returning to its ground state. Using the fluorescence lifetime as an analytical tool is not very common in live-cell microscopy. However, it is widely used to detect protein-protein interactions in fluorescence lifetime imaging microscopy, which is considered more efficient and accurate than FRET (fluorescence resonance energy transfer) because the measurements are not based on fluorescence intensity (Chang et al. 2007).

5.2.2 Fluorescent color palette

5.2.2.1 Blue and cyan fluorescent proteins

EBFP, one of the first variants derived from *Aequorea* GFP, has low brightness and poor photostability. In recent years, SBFP2, EBFP2, and mTagBFP have been developed for long-term imaging in live cells. Variants in the cyan spectral region are also rarely used. For many years, scientists were restricted to using ECFP and mCerulean, until the development of a novel monomeric teal-colored variant, mTFP1, which has excellent brightness (three times that of ECFP, twice that of mCerulean, and nearly twice that of EGFP) and greater photostability. This new blue fluorescent protein can be combined with yellow fluorescent proteins for FRET analysis.

5.2.2.2 Green and yellow fluorescent proteins

The most widely used GFP tags are two EGFP variants: monomeric GFP mutant 3 (mGFPmut3, mGFP), which contains the A206K mutation, and Emerald GFP (EmGFP). mGFP was a true monomer used in bacterial imaging (Zacharias et al. 2002). EmGFP, which is more photostable, is widely used in structured illumination microscopy. A novel monomeric yellow-green variant, mNeonGreen, derived from the cephalochordate *Branchiostoma lanceolatum*, has very high brightness and excellent photostability. It is also an excellent FRET acceptor when combined with mTurquoise (Shaner et al. 2013).

Nearly all YFPs are among the brightest fluorescent proteins reported. Enhanced YFP (EYFP), which includes four mutations compared with *Aequorea victoria* GFP (avGFP), is widely used but has a somewhat high pKa value (6.9). Like mGFP, mYFP was generated by introducing the A206K mutation and has been used for FRET applications in plasma membranes (Zacharias et al. 2002). Two novel YFPs, Citrine and Venus, were generated by introducing either the single Q69M mutation or five separate mutations, respectively. Citrine has a lower pKa (5.7) than its precursor EYFP. Venus is a quickly maturing fluorescent protein (approximately 7 min in vivo) and is highly suitable for monitoring gene expression. However, one drawback of Venus is that, unlike in EGFP and EYFP, the A206K mutation does not completely disrupt the tendency of the protein to oligomerize.

5.2.2.3 Orange and red fluorescent proteins

In the past 10 years, much effort has been devoted to developing red fluorescent proteins. However, no red FPs are as suitable as GFPs for use as fusion tags and often form artificial aggregates when used for labeling secretory pathway proteins. DsRed was derived from the sea anemone *Discosoma striata* and has maximum excitation and emission peaks of 558 nm and 583 nm, respectively, once fully matured. However, over the course of maturation, this fluorophore exhibits green fluorescence. Roger Y. Tsien and coworkers developed the Fruits series (Shaner et al. 2004), which includes the well-known proteins mCherry and mOrange2, from DsRed. mCherry has outstanding photostability, a low pKa value, and a fast maturation rate. Its emission peak is red-shifted to nearly 600 nm, and it is consistently monomeric. Thus, mCherry is an excellent partner for EGFP for dual color labeling. TagRFP-T and mOrange2 have excellent photostability. TagRFP-T is a mutation of TagRFP (S158T), which was derived from tetrameric eqFP578 (Shaner et al. 2008). Although it is only 80% as bright as TagRFP, it exhibits the highest photostability of all the red proteins. However, we have found that TagRFP-T is not a true monomer (Han et al. 2014). mOrange2 has higher photostability than mOrange, but retains the ability to be photo-converted to far-red. Other RFPs, such as, mKO1, 2 and mKate1, 2, are not widely used. Table 5.1 is a summary of properties of commonly used FPs.

5.2.3 *Fluorescent proteins used as biosensors*

Numerous biosensors based on fluorescent proteins have been developed to detect intracellular signals. In particular, there are many calcium sensors, including GCaMP1~3 (Nakai et al. 2001; Christian and Spremulli 2009), GECO1.2 (Zhao et al. 2011), RECO1 (Zhao et al. 2011) and FRET-based YC6.1 (Truong et al. 2001). One GFP variant, pHluorin (Miesenbock et al. 1998), is used as a pH sensor and has a pKa of 7.1. Other sensors used to monitor ROS (Hanson et al. 2004), chloride ions (Arosio et al. 2010), voltage (Tsutsui et al. 2008), and protease cleavage (Galperin et al. 2004) have also been developed for cell biology imaging.

5.2.4 *How to choose fluorescent proteins*

When designing experiments using FPs, it is important to consider several parameters. First, the FPs should be efficiently expressed in cells, with little or no phototoxicity. The FPs should be bright enough to separate the signal from the background autofluorescence. In general, FPs can be highly expressed if they have low toxicity. Second, suitable filters and lasers must be selected based on the excitation and emission wavelengths of the FPs. This is quite important when performing multicolor imaging. For example, a 488-nm laser is needed to excite GFP and a 561-nm laser to excite mCherry, and the GFP signal and mCherry signal will need to be imaged sequentially to eliminate GFP signal cross-talk in the red channel. Third, the oligomerization property of the FP should be taken into account when labeling membrane proteins (Zhang et al. 2012) or protein complexes (Landgraf et al. 2012). Fourth, for long-term or three-dimensional imaging, the photostability of the FP is an important consideration. Emerald GFP and TagRFP-T are excellent FPs for live-cell imaging over extended periods of time. Finally, other factors, such as transfection reagents and culture medium, that affect cell health may also affect the localization and expression level of tagged proteins. Thus, it is important to perform preliminary experiments and set up strict positive and negative controls before conducting an imaging experiment.

Fortunately, several resources and reviews exist that provide assistance in choosing suitable fluorescent proteins (http://zeiss-campus.magnet.fsu.edu/articles/probes/index.html; http://zeiss-campus.magnet.fsu.edu/articles/probes/index.html; http://www.olympusconfocal.com/applications/fpcolorpalette.html).

Table 5.1 The Properties of Fluorescent Proteins

Protein	Excitation Peak	Emission Peak	Molar Absorption Coefficient	Quantum Yield	Brightness	pKa	Maturation Half-Time	References
Blue/Cyan Fluorescent Proteins								
EBFP2	383	448	32.0	0.56	18	4.5	25 min	Henderson et al. (2007)
SBFP2	380	446	34.0	0.47	16	5.5	ND	Kremers et al. (2007)
mTagBFP	399	456	52.0	0.63	33	2.7	13 min	Subach et al. (2008)
ECFP/mECFP	433/451	475/503	33.0/30.0	0.41	13/12	4.7	ND	Ai et al. (2006)
mCerulean	433/451	475/503	43.0/37.0	0.67	27/24	4.7	ND	Ai et al. (2006)
mTFP1	462	492	64.0	0.85	54	4.3	ND	Ai et al. (2006)
Green/Yellow Fluorescent Proteins								
EGFP/mGFP	488	507	56.0	0.60	34	6.0	94 min	Shaner et al. (2007)
Emerald GFP	487	509	57.5	0.68	39	6.0	ND	Shaner et al. (2007)
sfGFP	485	510	83.3	0.65	54	5.5	Too fast	Shaner et al. (2007)
mNeonGreen	506	517	116.0	0.80	93	5.7	<10 min	Shaner et al. (2013)
EYFP	514	527	83.0	0.61	51	6.9	6.6 min	Shaner et al. (2007)
Citrine/mCitrine	516	529	77.0	0.76	59	5.7	ND	Shaner et al. (2007)
Venus/mVenus	515	528	92.0	0.57	52	6.0	2.0 min	Shaner et al. (2007)

continued

Table 5.1 (continued) The Properties of Fluorescent Proteins

Protein	Excitation Peak	Emission Peak	Molar Absorption Coefficient	Quantum Yield	Brightness	pKa	Maturation Half-Time	References
			Orange/Red Fluorescent Proteins					
mOrange	548	562	71.0	0.69	49	6.5	2.5 h	Shaner et al. (2008)
mOrange2	549	565	58.0	0.60	35	6.5	4.5 h	Shaner et al. (2008)
mCherry	587	610	72.0	0.22	16	4.5	<15 min	Shaner et al. (2008)
TagRFP-T	555	584	81.0	0.41	33	4.6	100 min	Shaner et al. (2008)
FusionRed	580	608	94.5	0.19	18	4.6	130 min	Shemiakina et al. (2012)
mKate2	588	633	62.5	0.40	25	5.4	20 min	Shcherbo et al. (2009)
mNeptune	599	649	57.5	0.18	10	5.8	35 min	Lin et al. (2009)

5.3 Fluorescent proteins for super-resolution imaging

5.3.1 Photoactivable fluorescent proteins for super-resolution imaging

Due to the diffraction limit, optical resolution can rarely reach the single-molecule level (~10 nm) using a traditional fluorescent microscope. However, the emergence of photoactivatable fluorescent proteins (PAFPs) and new super-resolution technologies based on PAFPs has led to a breakthrough in this regard.

Photoactivable fluorescent proteins shift spectra upon light illumination. The first PAFP, PAGFP, was described by Patterson and Lippincott-Schwartz in 2002. It was developed by directed mutagenesis of GFP, but unlike GFP, it is dark in its native state, and emits green light upon 405-nm laser illumination. In 2006, a German scientist named Eric Betzig and an American cell biologist named Jennifer Lippincott-Schwartz developed photoactivated localization microscopy (PALM), a technique based on the photoactivable property of PAGFP, which can precisely localize a single molecule by fitting the center of a well-separated emission spot, pushing the localization precision up to 2–25 nm (Betzig et al. 2006). At approximately the same time, Ando et al. (2004) developed the first photoconvertible fluorescent protein (PCFP), Kaede, which normally exhibits green fluorescence but can be converted to red fluorescence by activation with UV light. Two years later, the same group described another FP variant that could be photoactivated with 400-nm light and could then be reversibly switched off by irradiation with 490-nm light. This protein was the first reversible protein highlighter, Dronpa, which is also known as a reversibly photoswitchable fluorescent protein (PSFP; Ando et al. 2004). These photoactivatable, photoconvertible, and photoswitchable FPs are quite useful when selectively imaging a particular region or time point within living systems, and because the activation is controlled by light, the imaging contrast ratio can be greatly increased. Most importantly, these initial efforts in developing optically controlled FPs along with the invention of new microscopes led to the era of super-resolution imaging.

Since their initial development, both the quantity and the diversity of PAFPs have increased dramatically. Several new FPs that can be photoactivated or photoconverted to a new wavelength have been discovered (Chudakov et al. 2003, 2004; Wiedenmann et al. 2004; Tsutsui et al. 2005; Gurskaya et al. 2006). PSFPs with different photoswitching rates have been developed (Stiel et al. 2007; Mizuno et al. 2010; Grotjohann et al. 2011; Chang et al. 2012), and significant effort is being dedicated to improve the brightness, photostability, maturation rate, and oligomeric state of these PAFPs. Meanwhile, several new imaging technologies have been invented, including dual color-PALM (Subach et al. 2009), photoquenching

Table 5.2 Commonly Used PAFPs

PAFP Classes	Excitation (nm)	Emission (nm)	Brightness	Oligomeric state	pKa
			PAFP		
PAGFP	504	517	13,750	Monomer	NR
PAmCherry1	564	595	8,280	Monomer	6.3
PATagRFP	564	595	25,080	Monomer	5.3
			PCFP		
PSCFP2	400/490	468/511	8,600/10,810	Monomer	4.3/6.1
mKikGR	505/580	515/591	34,000/18,000	Monomer	6.6/5.2
Dendra2	490/553	507/573	19,250	Monomer	6.6/6.9
mEos2	506/573	519/584	47,000/30,000	Monomer-dimer	5.6/6.4
tdEos	506/569	516/582	59,000/30,000	Tandem dimer	5.7/NR
mClaVGR2	488/566	504/583	15,000/17,000	Monomer	8.0/7.3
PSmOrange	548/634	565/662	5,7783/9,156	Monomer	6.2/5.6
mEos3.1	505/570	513/580	73,400/20,800	Monomer	5.2/6.0
mEos3.2	507/572	516/580	53,300/17,700	Monomer	5.4/5.8
			PSFP		
Dronpa	503	518	80,800	Monomer	5
rsFastlime	496	518	30,100	Monomer	NR
rsEGFP	493	510	16,920	Monomer	NR
Padron	505	522	27,500	Monomer	NR
mTFP0.7	453	488	30,000	Monomer	NR
mGeos-M	503	514	44,000	Monomer	4.5–5.0
			PCFP/PSFP		
mIrisFP	486/546	516/578	44,400/11,440	Monomer	5.7/7.0

FRET (PQ-FRET; Demarco et al. 2006), optical lock-in detection (OLID; Marriottet al. 2008) and reversible saturable optical fluorescence transition (RESOLFT; Hofmann et al. 2005), which have greatly expanded the potential number of applications for PAFPs. Commonly used PAFPs are listed in Table 5.2.

5.3.1.1 Dark-to-bright photoactivators

The first class of PAFPs is dark-to-bright photoactivators. These PAFPs emit little fluorescence in a nonexcited state, but once activated generally by UV light, their fluorescence is intensity-dependent; therefore, they have a large dynamic range. However, due to the lack of initial fluorescence, it is difficult to select a positive cell to begin an imaging experiment. In addition to PAGFP, there are red PAFPs, such as PAmCherry1 (Subach et al. 2009) and PATagRFP (Subach et al. 2010). PAmCherry1 is a derivative of DsRed, and it can be converted from dark to red when excited by a 405-nm laser. Due to its faster maturation rate and higher contrast ratio, PAmCherry1 is often the first choice among red PAFPs. Moreover, by combining PAGFP with PAmCherry1, Subach et al. (2009) successfully performed dual-color PALM in COS-7 cells. As the emission spectra of PAGFP and PAmCherry1 have little overlap, they can be activated simultaneously by a 405-nm laser, then imaged individually at 488 nm and 564 nm. PATagRFP is another PAFP that was derived from eqFP578. Compared to PAmCherry1, PATagRFP is brighter, more photostable, and less sensitive to blue light, and it can also be paired with PAGFP in dual-color PALM (Subach et al. 2010). However, its contrast ratio is nearly 10 times lower than that of PAmCherry1.

5.3.1.2 *Irreversible photoconverters*

PCFPs are the most popular and fastest-growing species among the FPs used in super-resolution imaging. Most PCFPs are green-to-red photoconverters: they emit green light in their native state, and after exposure to UV light, their chromophore electronic conjugate systems extend into the red spectrum (Mizuno et al. 2003; Nienhaus et al. 2005; see Figure 5.1).

EosFP (Wiedenmann et al. 2004) is a tetramer PCFP and was the first PCFP used for PALM. Before photoconversion, its emission maximum is 516 nm, and after photoconversion, its emission maximum shifts to 581 nm. Random mutagenesis led to the generation of two different EosFP dimers and the monomeric mEosFP (Wiedenmann et al. 2004). However, mEosFP cannot mature at 37°C, hindering its use in mammalian cells.

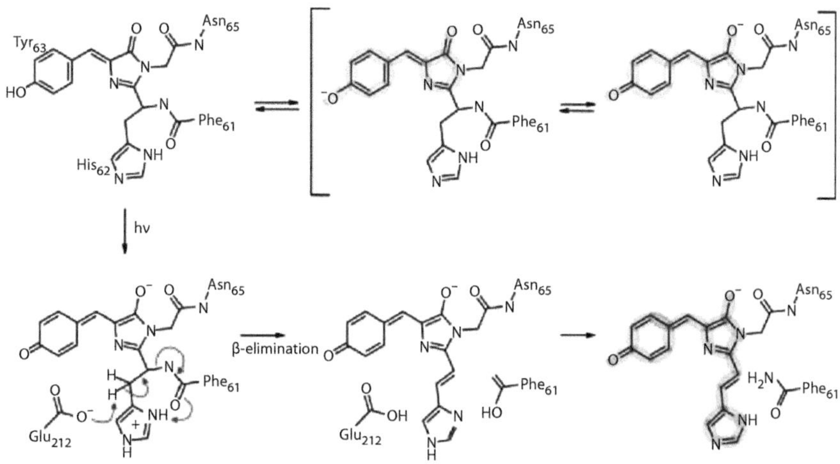

Figure 5.1 Reaction mechanism of green-to-red conversion. (From K. Nienhaus et al. (2005). *Proc. Natl. Acad. Sci. U.S.A.* **102**(26): 9156–9159. With permission.)

To solve this problem, two monomeric EosFP were fused together with a 16-amino-acid linker, yielding tdEosFP (Nienhaus et al. 2006). Although tdEosFP has high brightness and is widely used in super-resolution imaging, its large molecular weight and slow maturation rate might influence the localization of certain proteins (tubulin, histones, gap junctions) (McKinney et al. 2009). McKinney and coworkers reported an improved version of mEosFP, mEos2, which matures well at 37°C. Although mEos2 is not as bright as tdEosFP, it would be a better choice for proteins that mislocalize when labeled with tdEosFP. However, care should be taken when tagging a membrane protein with mEos2 because mEos2 will form oligomers at high concentrations. Recently, Zhang et al. (2012) solved the crystal structure of tetrameric mEos2 and rationally designed improved versions, mEos3.1 and mEos3.2, which are true monomers. Both mEos3.1 and mEos3.2 are brighter than mEos2, have faster maturation rates and exhibit higher photon numbers. Importantly, these two variants exhibit outstanding performance in PALM imaging (Zhang et al. 2012).

In addition to the Eos series, several other PCFPs have been developed by different groups, but their performance in super-resolution imaging remains to be verified. For example, mKikGR, which carries 21 point mutations compared to its parent tetramer KikGR (Habuchi et al. 2008), has an emission spectrum with a 10-nm red shift relative to mEos2, but neither its photostability nor its brightness outperforms mEos2. The monomer Dendra (Gurskaya et al. 2006) was developed from its tetramer precursor by rounds of random and site-directed mutagenesis. The commercial version of Dendra2 (Evrogen) is the first PCFP to be widely used in live-cell tracking imaging experiments. mClavGR2 (Hoi et al. 2010) was derived from the monomeric cyan fluorescent protein mTFP1 after several rounds of mutation and optimization. It is consistently monomeric but has low brightness and poor pH stability (pKa = 8), which may impede its usefulness in further applications. However, it is a useful example of how to develop new FPs from a distant precursor.

5.3.1.3 *Reversible highlighters*

Among the PSFPs, Dronpa is most extensively studied and widely used because of its excellent overall properties. Dronpa will switch "off" under 488-nm laser irradiation, but the fluorescence can be restored by activation with a 405-nm laser, and this cycle can be repeated several hundred times without significant bleaching. It was proposed that the photoswitchable ability of Dronpa is partially based on the transition between the deprotonated (excited state with fluorescence) and protonated (ground state without fluorescence) states (Andresen et al. 2007). Dronpa has two absorption peaks: the main peak at 503 nm represents the deprotonated state and the minor peak at 390 nm represents the natural protonation states. The deprotonated fluorophore emits green fluorescence at 518 nm, with an intensity that is 2.5 times higher than that of EGFP. As Dronpa's application potential in super-resolution imaging has become recognized, more and more derivatives based on Dronpa have been developed. rsFastlime has a very quick photoswitching rate that is nearly 1000 times higher than that of Dronpa (Stiel et al. 2007), which is very useful for reversible saturable optical fluorescence transition (RESOLFT) microscopy (Hofmann et al. 2005; Grotjohann et al. 2011). To broaden the palette of available PSFPs, Andresen and coworkers (Andresen et al. 2008; Stiel et al. 2007) developed the blue light emitter bsDronpa, as well as Padron, which has the opposite optical switching properties of Dronpa. In addition, PSFPs have also been derived from newly discovered fluorescent proteins from sea anemones and corals. KFP1 is a commercially available PSFP whose precursor is the tetrameric kindling fluorescent protein (Chudakov et al. 2003). It emits a red light at approximately 600 nm when excited by a green or yellow (525–580 nm) light, but it will extinguish quickly once the irradiation is withdrawn. In addition, intensive blue light (450–490 nm) can also quench KFP1 quickly, making photoswitching more controllable for super-resolution imaging. mTFP0.7, a mutated form of mTFP1, can act as an optical switch, but has not been used for many applications (Henderson et al. 2007). rsCherry, a derivative of mCherry, can be activated by yellow light and deactivated by blue light. The opposite is true of rsCherryRev, but this protein is not a monomer. Although the overall brightness of these mutants is quite low, at the single-molecule level, their brightness is comparable to mCherry (Stiel et al. 2008).

Grotjohann et al. (2011) recently reported an EGFP mutant, rsEGFP, that can cycle rapidly between on and off states hundreds and thousands of times, further expanding the usefulness of PSFPs in RESOLFT super-resolution imaging.

Chang et al. (2012) developed a series of monomeric green PSFPs with beneficial optical characteristics, such as high photon output per switch, high photostability, and a broad range of switching rates and pH dependence, which make them potentially useful for various applications. One member of this series, mGeos-M, exhibits the highest photon budget and localization precision potential among all green PSFPs. Chang and coworkers propose that mGeos-M may be used as a replacement for Dronpa in applications such as dynamic tracking, dual-color super-resolution imaging, and optical lock-in detection.

5.3.2 Traditional proteins for super-resolution imaging

In addition to super-resolution microscopy techniques designed to take advantage of PAFPs, there are also diffraction-unlimited technologies utilizing both PAFPs and traditional FPs.

Stimulated emission depletion (STED) microscopy is illumination-based super-resolution imaging that uses nonlinear optical approaches to directly modify the point spread function (PSF) and achieve super-resolution. Hein et al. (2008) imaged YFP-labelled ER with subdiffraction resolution by applying STED and demonstrated a 4-fourfold improvement in lateral resolution. Rankin et al. (2011) showed that proteins in cultured cells and even *C. elegens elegans* tagged with GFP can be superresolved by STED. Morozova et al. (2010) developed a far-red fluorescent protein, TagRFP657, which is an efficient marker for stimulated emission depletion STED microscopy. Saturated structured illumination (SSIM) microscopy works by a similar mechanism; therefore, traditional FPs with high photostability can also be used for this technique. For both of these methods, photoactivable FPs are not necessary if the fluorophore can retain its brightness for a sufficient time. However, with a photoactivatable FP, a depletion laser of one million times less lower power can be used (Grotjohann 2011).

Structured illumination (SIM) microscopy uses a shifted and rotated grid pattern of light and can therefore be used with either traditional FPs, such as EGFP or Emerald, with enhanced brightness and photostability, or photoactivatable fluorescent proteins, such as Dronpa (Rego et al. 2012). Using this method, both lateral and axial resolution can be increased by a factor of two.

Recently, a new light-sheet-based super-resolution method for rapid three-dimensional imaging was introduced by Planchon et al. (2011). By combing Bessel beams with structured illumination or two-photon excitation, light sheets that are thinner than 0.5 µm can be produced. In their work, mEmerald and the tandem dimer Tomato were used to perform dual-color live imaging of microtubules.

Spectral precision distance microscopy/spectral position determination microscopy (SPDM) is a high-precision localization method that uses the reversibly bleached states of traditional FPs. Briefly, using a single laser source with a sufficiently high intensity and an embedding medium such as ProLong® Gold (Lemmer et al. 2009) to enhance stochastic "blinking," conventional fluorophores will switch randomly between a very short "on" (fluorescent state) and a long-lived "off" state (reversibly photobleached state—Cremer et al. 2011). This technique has been applied in various ways, to visualize YFP labeling of the plasma membrane structure of human breast cancer cells (Lemmer et al. 2008), the spatial distribution of GFP-tagged P-glycoprotein in the luminal plasma membrane of brain capillary endothelial cells (Huber et al. 2012), and the distribution of two nuclear proteins labeled with autofluorescent GFP and mRFP1 in a dual-color SPDM study (Gunkel et al. 2009).

5.3.3 Requirements for super-resolution imaging

Generally speaking, fluorescent proteins that are suitable for super-resolution imaging should have the following characteristics: high brightness to ensure that sufficient photons can be collected from each molecule; high contrast, which means low background noise and less spontaneous blinking that may contribute to background noise; and consistent monomeric and convenient fusion properties that will guarantee no interference with the localization of the target protein. Furthermore, if single-molecule-based super-resolution imaging methods are used, special issues should be considered. For example, the label intensity should be high enough and the rate of photobleaching (for PSFPs) or photodeactivation (for irreversible PAFPs) and photoactivation should be carefully balanced, so that only a small fraction of the molecules are activated at any arbitrary time while maintaining an activation time that is long enough to collect sufficient photons.

5.4 Strategy for fluorescent protein evolution

Since the discovery and first cloning of the green fluorescent protein (GFP) from jellyfish, there is a continuing effort to develop new FPs with desirable properties. First were the GFP variants, such as S65T and EGFP, with altered excitation and emission spectra, enhanced brightness, and improved pH sensitivity. Next were the PAFPs that can be modulated photophysically. However, many possibilities remain for improved and novel FPs with additional desirable properties. Two strategies are commonly used by researchers to develop new FPs: random mutagenesis followed by laborious screening or structure- or sequence-guided site-directed mutagenesis. It is worth noting that FPs are seldom generated by only one method, and it is necessary to combine these strategies to obtained the final desired FPs.

5.4.1 Random mutagenesis strategy

Theoretically, most FPs with desirable properties can be identified utilizing random mutagenesis. However, as many as five or more rounds of mutation may need to be performed to optimize the initial version. Normally, random mutagenesis would not be the starting point, as site-directed mutagenesis is preferable when possible. For example, when engineering mMaple, McEvoy et al. (2012) first replaced a sequence near the C-terminus of the template protein with the corresponding residues from the close homolog of mTFP1 to reduce the dimerization tendency. They then performed four rounds of protein optimization by screening libraries of many thousands of genetic variants created by random mutagenesis. Screening of the final library led to the identification of mMaple.

One exception is evolution via iterative somatic hypermutation (SHM), in which only random mutagenesis is needed. Roger Y. TsienWang and co-workers (2004) described the generation of a number of monomeric red-shifted mutants by somatic hypermutation (Wang et al. 2004). They utilized somatic hypermutation in mammalian B cells to autonomously diversify and evolve FPs *in situ*, and then screened for the desired phenotype by fluorescence-activated cell sorting (FACS) assays. This method is effective in accumulating reinforced mutations and can be assessed by high-throughput methods; however, few FPs have been generated this way, most likely because B cells have a much lower amplification rate compared to bacteria.

Other than random mutagenesis, semi-random mutagenesis is often a useful tool. When developing mCherry, Shaner et al. (2004) performed semi-random mutagenesis for the first and middle steps. They constructed a library in which Q66 was replaced with random residues, and then they replaced the first seven amino acids in the best candidate (mRFP1.1) with the corresponding residues from enhanced GFP (MVSKGEE), and appended the last seven amino acids of GFP to the C terminus. After additional rounds of random mutagenesis and screening, they performed semi-random mutagenesis again. This time, they randomized position 163 and obtained the pre-version mRFP1.4. After further rounds of directed evolution, they finally obtained mCherry. In another example, Habuchi et al. (2008) first performed site-directed semi-random mutagenesis on KikGR, and then error-prone polymerase chain reactions to introduce additional variations that might affect photoconversion or protein folding ability. The final monomeric version was achieved after 15 cycles of mutagenesis and contains 21 mutations.

5.4.2 *Site-directed mutagenesis strategy*

It is notable that most of the FPs generated by several rounds of random mutagenesis are oligomeric, especially those with high brightness, mainly because fluorescence detection-based assays are generally used for brighter variants. Using site-directed mutagenesis eliminates this problem, simply by choosing sites that are not on the exterior of the FP molecule. Site-directed mutagenesis can be guided by sequence alignment between homologous proteins. To obtain red fluorescent proteins with enhanced photostability while maintaining high brightness, Shaner et al. (2008) directly focused on the amino acids that contribute to mCherry's enhanced photostability and mOrange's higher quantum yield, and after six rounds of directed evolution of mRFP1 they created mApple.

Figure 5.2 **(See color insert.)** Interfaces in tetrameric mEos2.

A more rational design always depends on the x-ray structures of template or homologous FPs. For example, the x-ray structure of an FP is quite useful when engineering monomeric FPs, as candidate residues for mutation are always present at the interface of the oligomerized FP, and can therefore be easily identified by structural analysis. Thus, once the x-ray structure of a tetrameric or dimeric FP is obtained, it is simple to identify the key residues. Sometimes, x-ray structures of homologous proteins are equally helpful.

Our recent study (Zhang et al. 2012) reported the crystal structure of tetrameric mEos2 (a green fluorophore). By carefully examining the residue–residue interactions at two interfaces, we found three key residues that may participate in the oligomerization of mEos2. At the A–B interface, the hydrophobic side chain of Ile102 formed hydrophobic interactions with the side chains of Ile100 and Ile102, and two Tyr121 residues facing each other also formed strong hydrophobic interactions. At the A–C interface, the hydroxyl group of Tyr189 made a water-mediated hydrogen bond with the backbone oxygen atom of His158 (Figure 5.2). From these results, it is reasonable to hypothesize that Ile102 and Tyr121 at the A–B interface and Tyr189 at the A–C interface are critical for the formation of the tetramer. Combined with a sequence alignment between mEos2 and other commonly used PAFPs, we speculated that mutation of Ile102 to asparagine (a similar mutation converts 22G into its monomer counterpart, Dronpa11), Tyr121 to a charged residue such as arginine or lysine, and Tyr189 to alanine would break the β-can–β-can interactions in the tetramer. As expected, the I102N mutant was the most consistently monomeric among all of the mutants and exhibited modest fluorescence. The Tyr121 mutations also reduced oligomerization to varying extents but lowered brightness.

5.5 Conclusion and outlook

Understanding why some FPs have excellent properties for microscopy applications while others display limited performance is vital for facilitating and accelerating the development of new FPs, yet remains a significant challenge. Due to the complexity of photomodulatable mechanisms and their working environment, more structural characterization and computational analysis are needed. In addition, high-throughput library construction methods as well as high-throughput screening assays will be certainly beneficial.

5.5.1 Fluorescent protein evolution for traditional microscopy

Each FP has specific properties that may be applicable for a specific microscopy technique. However, to enhance performance for a particular application, specific properties need to be optimized. Generally speaking, FPs with increased brightness and enhanced photostability are highly desired. These properties, as well as the ability to be fused to a host protein, are basic properties, but very important for three-dimensional or long-term live-cell imaging either in traditional microscopy or super-resolution imaging, especially for FPs in the far-red spectral region. Many of the RFPs currently in use suffer from problems with aggregation and mislocation, even the recently developed FusionRed, which exhibits up to 50% mislocation when fused to some proteins (Shemiakina et al. 2012). Both FusionRed and mCherry are monomeric in in vitro solutions at high concentrations, but they can still form artificial aggregates, especially when used to label membrane proteins. RFPs that are less aggregation-prone and less cytotoxic are in high demand. High-throughput approaches to screen large libraries of RFP mutants and further research regarding the mechanism of RFP aggregation in living cells are required to achieve this goal.

Also in high demand are FPs optimized for specific microscopy and biological applications, such as two-photon microscopy and imaging multiple biosensors in a signaling pathway. High-throughput strategies and carefully designed assays are needed to develop the desired FPs or biosensors.

5.5.2 Fluorescent protein evolution in super-resolution microscopy

To date, many fluorescent protein tags with targeted properties have been developed to satisfy the needs of novel diffraction-unlimited imaging technologies; however, novel proteins with optimized characteristics are still needed for biological applications using super-resolution imaging. First of all, higher resolution is desirable for imaging the detailed structure of cellular organelles or dynamic biological processes. For example, the budding of vesicles from the endoplasmic reticulum, Golgi, or the plasma membrane is a very important biological process for membrane trafficking and cell function. To visualize the budding structure or the dynamic process of budding itself requires very high resolution, on the order of several nanometers. This degree of resolution needs labeling tags that have very high brightness and a high signal-to-noise ratio. In addition, higher labeling density is required to achieve higher resolution, which produces higher background fluorescence due to increased scattering or the fluorescence of PA proteins in the "off" state. Chemical dyes can produce more photons compared to fluorescent proteins, and show a higher signal-to-noise ratio in both single-molecule-based STORM imaging and STED imaging. Therefore, much work is still needed to increase the brightness and signal-to-noise ratio for fluorescent proteins in current use.

To visualize the assembly of a functional organelle or subunit, it is necessary to label two or more proteins and use dual-color or multicolor super-resolution imaging. However, compared to the wide use of green fluorescent proteins, there is an urgent need for new red fluorescent proteins optimized for use in either single-molecule-based PALM/STORM imaging or PSF-modification-based STED/RESOLFT imaging. In PALM/STORM imaging, PCFPs exhibit superior signal-to-noise ratios due to the low background fluorescence. However, this application requires the use of a partner fluorophore with a suitable emission spectrum to be able to differentiate the two signals, and an optimized protocol to detect the different signals from the two FPs. For example, mEos3.1 can be used with PSCFP2 for dual-color PALM imaging. However, it is necessary to obtain the PSCFP2 localization information after all the green state mEos3.1 has been transferred to the red state. On the other hand, PAFPs occupy only one color channel and can be used easily with other fluorescent proteins for dual-color super-resolution imaging. However, there is a lack of high-performance red fluorescent proteins with high photon numbers and labeling density to use as partners with photoactivatable green fluorescent proteins. Moreover, currently used red PAFPs or PSFPs are based on the mCherry/mOrange series, which have poor performance and can perturb the localization of the labeled protein when used as a fusion tag. Thus, there is a need for new red PAFPs or PSFPs to pair with green tags for dual-color PALM/STORM imaging. There is also a significant need for red fluorescent proteins that can be used for dual-color PSF-modification-based and/or illumination pattern-based super-resolution microscopy. For example, red fluorescent proteins with fast switching kinetics would be very valuable for use with rsEGFP for dual-color RESOLFT imaging. Stable red fluorescent proteins are also under development for use in dual-color SIM and nonlinear SIM imaging.

Two cutting-edge techniques in super-resolution imaging are three-dimensional (3D) and live-cell super-resolution microscopy, which allow us to observe the spatial structure and monitor the dynamic movement of labeled proteins at nanometer resolution. Both of these techniques require fluorescent proteins that are highly photostable because multilayer scanning and long-term scanning quench fluorescent proteins very quickly. Both green and red fluorescent photostable proteins are needed for 3D and live-cell super-resolution microscopy, either PAFPs/PSFPs for PALM/STORM imaging, rsEGFP/rsRFP for RESOLFT, or general use GFP/RFP for SIM/nonlinear SIM.

Finally, fluorescent proteins that can be used for both light and electron microscopy would be extremely useful for nanometer-scale protein localization with electron microscopy (EM). However, current EM buffers and fixation procedures quench fluorescent proteins during sample preparation. Fluorescent proteins that can maintain high fluorescence and high resolution in super-resolution imaging after EM are highly desirable for light-electron microscopy.

References

Ai, H. W. et al. (2006). "Directed evolution of a monomeric, bright and photostable version of *Clavularia* cyan fluorescent protein: structural characterization and applications in fluorescence imaging." *Biochem J* **400**(3): 531–540.

Ando, R. et al. (2002). "An optical marker based on the UV-induced green-to-red photoconversion of a fluorescent protein." *Proc Natl Acad Sci USA* **99**(20): 12651–12656.

Ando, R. et al. (2004). "Regulated fast nucleocytoplasmic shuttling observed by reversible protein highlighting." *Science* **306**(5700): 1370–1373.

Andresen, M. et al. (2008). "Photoswitchable fluorescent proteins enable monochromatic multilabel imaging and dual color fluorescence nanoscopy." *Nat Biotechnol* **26**(9): 1035–1040.

Andresen, M. et al. (2007). "Structural basis for reversible photoswitching in Dronpa." *Proc Natl Acad Sci USA* **104**(32): 13005–13009.

Arosio, D. et al. (2010). "Simultaneous intracellular chloride and pH measurements using a GFP-based sensor." *Nat Methods* **7**(7): 516–518.

Betzig, E. et al. (2006). "Imaging intracellular fluorescent proteins at nanometer resolution." *Science* **313**(5793): 1642–1645.

Bulina, M. E. et al. (2006). "A genetically encoded photosensitizer." *Nat Biotechnol* **24**(1): 95–99.

Chalfie, M. et al. (1994). "Green fluorescent protein as a marker for gene expression." *Science* **263**(5148): 802–805.

Chang, C. W. et al. (2007). "Fluorescence lifetime imaging microscopy." *Methods Cell Biol* **81**: 495–524.

Chang, H. et al. (2012). "A unique series of reversibly switchable fluorescent proteins with beneficial properties for various applications." *Proc Natl Acad Sci USA* **109**(12): 4455–4460.

Christian, B. E. and L. L. Spremulli (2009). "Evidence for an active role of IF3mt in the initiation of translation in mammalian mitochondria." *Biochemistry* **48**(15): 3269–3278.

Chudakov, D. M. et al. (2003). "Kindling fluorescent proteins for precise in vivo photolabeling." *Nat Biotechnol* **21**(2): 191–194.

Chudakov, D. M. et al. (2004). "Photoswitchable cyan fluorescent protein for protein tracking." *Nat Biotechnol* **22**(11): 1435–1439.

Cremer, C. et al. (2011). "Superresolution imaging of biological nanostructures by spectral precision distance microscopy." *Biotechnol J* **6**(9): 1037–1051.
Demarco, I. A. et al. (2006). "Monitoring dynamic protein interactions with photoquenching FRET." *Nat Methods* **3**(7): 519–524.
Galperin, E. et al. (2004). "Three-chromophore FRET microscopy to analyze multiprotein interactions in living cells." *Nat Methods* **1**(3): 209–217.
Grotjohann, T. et al. (2011). "Diffraction-unlimited all-optical imaging and writing with a photochromic GFP." *Nature* **478**(7368): 204–208.
Gunkel, M. et al. (2009). "Dual color localization microscopy of cellular nanostructures." *Biotechnol J* **4**(6): 927–938.
Gurskaya, N. G. et al. (2006). "Engineering of a monomeric green-to-red photoactivatable fluorescent protein induced by blue light." *Nat Biotechnol* **24**(4): 461–465.
Habuchi, S. et al. (2008). "mKikGR, a monomeric photoswitchable fluorescent protein." *PLoS One* **3**(12): e3944.
Han, L. et al. (2014). RFP tags for labeling secretory pathway proteins. *Biochem Biophys Res Commun* **447**(3):508–512.
Hanson, G. T. et al. (2004). "Investigating mitochondrial redox potential with redox-sensitive green fluorescent protein indicators." *J Biol Chem* **279**(13): 13044–13053.
Heim, R. et al. (1995). "Improved green fluorescence." *Nature* **373**(6516): 663–664.
Hein, B. et al. (2008). "Stimulated emission depletion (STED) nanoscopy of a fluorescent protein-labeled organelle inside a living cell." *Proc Natl Acad Sci USA* **105**(38): 14271–14276.
Henderson, J. N. et al. (2007). "Structural basis for reversible photobleaching of a green fluorescent protein homologue." *Proc Natl Acad Sci USA* **104**(16): 6672–6677.
Hofmann, M. et al. (2005). "Breaking the diffraction barrier in fluorescence microscopy at low light intensities by using reversibly photoswitchable proteins." *Proc Natl Acad Sci USA* **102**(49): 17565–17569.
Hoi, H. et al. (2010). "A monomeric photoconvertible fluorescent protein for imaging of dynamic protein localization." *J Mol Biol* **401**(5): 776–791.
Huber, O. et al. (2012). "Localization microscopy (SPDM) reveals clustered formations of P-glycoprotein in a human blood-brain barrier model." *PLoS One* **7**(9): e44776.
Khmelinskii, A. et al. (2012). "Tandem fluorescent protein timers for in vivo analysis of protein dynamics." *Nat Biotechnol* **30**(7): 708–714.
Kremers, G. J. et al. (2007). "Improved green and blue fluorescent proteins for expression in bacteria and mammalian cells." *Biochemistry* **46**(12): 3775–3783.
Landgraf, D. et al. (2012). "Segregation of molecules at cell division reveals native protein localization." *Nat Methods* **9**(5): 480–498.
Lemmer, P. et al. (2008). "SPDM: light microscopy with single-molecule resolution at the nanoscale." *Appl Phys B* **93**(1): 1–12.
Lemmer, P. et al. (2009). "Using conventional fluorescent markers for far-field fluorescence localization nanoscopy allows resolution in the 10-nm range." *J Microscopy* **235**(2): 163–171.
Li, G. W. and X. S. Xie (2011). "Central dogma at the single-molecule level in living cells." *Nature* **475**(7356): 308–315.
Lin, M. Z. et al. (2009). "Autofluorescent proteins with excitation in the optical window for intravital imaging in mammals." *Chem Biol* **16**(11): 1169–1179.

Marriott, G. et al. (2008). "Optical lock-in detection imaging microscopy for contrast-enhanced imaging in living cells." *Proc Natl Acad Sci USA* **105**(46): 17789–17794.
McEvoy, A. L. et al. (2012). "mMaple: a photoconvertible fluorescent protein for use in multiple imaging modalities." *PLoS One* 7(12)
McKinney, S. A. et al. (2009). "A bright and photostable photoconvertible fluorescent protein." *Nat Methods* **6**(2): 131–133.
Miesenbock, G. et al. (1998). "Visualizing secretion and synaptic transmission with pH-sensitive green fluorescent proteins." *Nature* **394**(6689): 192–195.
Mizuno, H. et al. (2010). "Higher resolution in localization microscopy by slower switching of a photochromic protein." *Photochem Photobiol Sci* **9**(2): 239–248.
Mizuno, H. et al. (2003). "Photo-induced peptide cleavage in the green-to-red conversion of a fluorescent protein." *Mol Cell* **12**(4): 1051–1058.
Morozova, K. S. et al. (2010). "Far-red fluorescent protein excitable with red lasers for flow cytometry and superresolution STED nanoscopy." *Biophys J* **99**(2): L13–15.
Nakai, J., et al. (2001). "A high signal-to-noise Ca(2+) probe composed of a single green fluorescent protein." *Nat Biotechnol* **19**(2): 137–141.
Nienhaus, G. U. et al. (2006). "Photoconvertible fluorescent protein EosFP: biophysical properties and cell biology applications." *Photochem Photobiol* **82**(2): 351–358.
Nienhaus, K. et al. (2005). "Structural basis for photo-induced protein cleavage and green-to-red conversion of fluorescent protein EosFP." *Proc Natl Acad Sci USA* **102**(26): 9156–9159.
Ormo, M. et al. (1996). "Crystal structure of the *Aequorea victoria* green fluorescent protein." *Science* **273**(5280): 1392–1395.
Patterson, G. H. and J. Lippincott-Schwartz (2002). "A photoactivatable GFP for selective photolabeling of proteins and cells." *Science* **297**(5588): 1873–1877.
Planchon, T. A. et al. (2011). "Rapid three-dimensional isotropic imaging of living cells using Bessel beam plane illumination." *Nat Methods* **8**(5): 417–423.
Prasher, D. C. et al. (1992). "Primary structure of the *Aequorea victoria* green-fluorescent protein." *Gene* **111**(2): 229–233.
Rankin, B. R. et al. (2011). "Nanoscopy in a living multicellular organism expressing GFP." *Biophys J* **100**(12): L63–65.
Rego, E. H. et al. (2012). "Nonlinear structured-illumination microscopy with a photoswitchable protein reveals cellular structures at 50-nm resolution." *Proc Natl Acad Sci USA* **109**(3): E135–143.
Shaner, N. C., et al. (2004). "Improved monomeric red, orange and yellow fluorescent proteins derived from Discosoma sp. red fluorescent protein." *Nat Biotechnol* **22**(12): 1567-1572.
Shaner, N. C. et al. (2013). "A bright monomeric green fluorescent protein derived from *Branchiostoma lanceolatum*." *Nat Methods*
Shaner, N. C. et al. (2008). "Improving the photostability of bright monomeric orange and red fluorescent proteins." *Nat Methods* **5**(6): 545–551.
Shaner, N. C. et al. (2007). "Advances in fluorescent protein technology." *J Cell Sci* **120**(Pt 24): 4247–4260.
Shcherbo, D. et al. (2009). "Far-red fluorescent tags for protein imaging in living tissues." *Biochem J* **418**(3): 567–574.
Shemiakina, II et al. (2012). "A monomeric red fluorescent protein with low cytotoxicity." *Nat Commun* **3**: 1204.

Shimomura, O. et al. (1962). "Extraction, purification and properties of aequorin, a bioluminescent protein from the luminous hydromedusan, *Aequorea*." *J Cell Comp Physiol* **59**: 223–239.
Stiel, A. C. et al. (2008). "Generation of monomeric reversibly switchable red fluorescent proteins for far-field fluorescence nanoscopy." *Biophys J* **95**(6): 2989–2997.
Stiel, A. C. et al. (2007). "1.8 A bright-state structure of the reversibly switchable fluorescent protein Dronpa guides the generation of fast switching variants." *Biochem J* **402**(1): 35–42.
Strack, R. L. et al. (2008). "A noncytotoxic DsRed variant for whole-cell labeling." *Nat Methods* **5**(11): 955–957.
Subach, F. V. et al. (2009). "Photoactivatable mCherry for high-resolution two-color fluorescence microscopy." *Nat Methods* **6**(2): 153–159.
Subach, F. V. et al. (2010). "Bright monomeric photoactivatable red fluorescent protein for two-color super-resolution sptPALM of live cells." *J Am Chem Soc* **132**(18): 6481–6491.
Subach, O. M. et al. (2008). "Conversion of red fluorescent protein into a bright blue probe." *Chem Biol* **15**(10): 1116–1124.
Truong, K. et al. (2001). "FRET-based in vivo Ca^{2+} imaging by a new calmodulin-GFP fusion molecule." *Nat Struct Biol* **8**(12): 1069–1073.
Tsutsui, H. et al. (2008). "Improving membrane voltage measurements using FRET with new fluorescent proteins." *Nat Methods* **5**(8): 683–685.
Tsutsui, H. et al. (2005). "Semi-rational engineering of a coral fluorescent protein into an efficient highlighter." *EMBO Rep* **6**(3): 233–238.
Wang, L. et al. (2004). "Evolution of new nonantibody proteins via iterative somatic hypermutation." *Proc Natl Acad Sci USA* **101**(48): 16745–16749.
Wiedenmann, J. et al. (2004). "EosFP, a fluorescent marker protein with UV-inducible green-to-red fluorescence conversion." *Proc Natl Acad Sci USA* **101**(45): 15905–15910.
Zacharias, D. A. et al. (2002). "Partitioning of lipid-modified monomeric GFPs into membrane microdomains of live cells." *Science* **296**(5569): 913–916.
Zhang, M. et al. (2012). "Rational design of true monomeric and bright photoactivatable fluorescent proteins." *Nat Methods* **9**(7): 727–729.
Zhao, Y. et al. (2011). "An expanded palette of genetically encoded Ca(2)(+) indicators." *Science* **333**(6051): 1888–1891.

Acknowledgments

We acknowledge support by the National Key Basic Research Support Foundation of China (NKBRSFC, grants 2010CB912303 and 2013CB910103), the National Scientific Foundation of China (grants 31170818 and 31270910), and the project from Chinese Academy of Sciences (KSCX2-EW-Q-11).

Problems

1. What are the advantage and disadvantage of using FPs for fluorescent imaging?
2. How was the first FP discovered?

3. Do FPs share similar three-dimensional structures?
4. How many types of FPs can be used in super-resolution imaging? What are their differences?
5. What are the key properties of FPs that can be used in point spread function modification super-resolution imaging?
6. What are the key properties of FPs that can be used in single molecular localization super-resolution imaging?
7. In which ways do scientists develop new FPs?
8. What is the trend in developing FPs?

chapter six

Single-molecule imaging with quantum dots

Mohammad U. Zahid and Andrew M. Smith

Contents

6.1 Introduction

Quantum dots (QDs) are a class of fluorescent nanocrystal with unique optical and electronic properties (Smith et al. 2008; Smith and Nie 2009a). Originally investigated as components for light-emitting devices, solar cells, and catalysts in the early 1980s, these particles were introduced in 1998 as optical tags for bioimaging and biological detection (Chan and Nie 1998; Bruchez et al. 1998). Over the ensuing 15 years, their most significant contribution to biomedicine has been in the field of single-molecule fluorescence microscopy, for which they have filled a major need for bright fluorescent probes with long-term photostability. These particles are now widely used for imaging the dynamics of individual molecules and the interactions between molecules in complex biological environments.

In this chapter, we first describe the structure and optical properties of QDs and discuss engineering strategies to enhance their properties for bioimaging applications. We then explore recent biological questions that have been answered through the implementation of QDs as single-molecule emitters, and finally we summarize early-stage attempts to incorporate these particles into advanced optical imaging modalities of nanoscopy and orientation-specific imaging.

6.2 Quantum dot structure and photophysics

6.2.1 Quantum dot structure

Figure 6.1 depicts a prototypical QD used in single-molecule imaging. The core is a nanocrystal composed of the semiconductor cadmium selenide (CdSe) with a diameter of 2 to 6 nm, surrounded by an insulating crystalline shell of cadmium sulfide (CdS) and/or zinc sulfide (ZnS). Fluorescence and light absorption primarily take place in the core; the shell serves to enhance and protect the optical and electronic properties of the QD (Dabbousi et al. 1997). The crystalline surface facets are coated with organic ligands and/or polymers that stabilize the nanocrystal as a colloidal suspension in biological media and prevent nonspecific adhesion to proteins and cellular structures. These outer layers can be specifically tagged with biomolecules such as peptides, nucleic acids, and small-molecule ligands. The most commonly used tagging strategy is based on QDs covalently conjugated to streptavidin, a modular protein adaptor

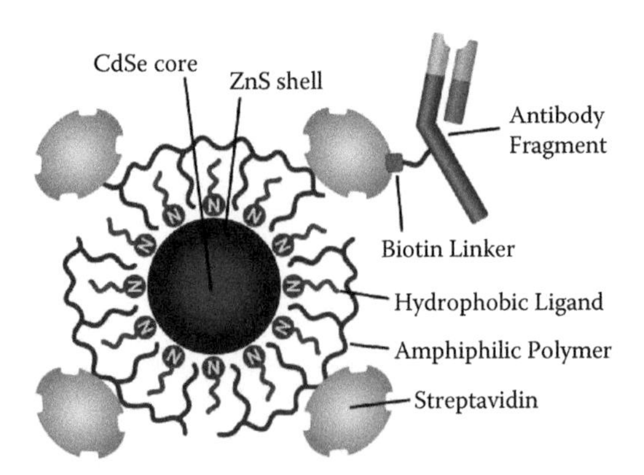

Figure 6.1 Structure of a prototypical streptavidin-coated QD. See the text for details.

that binds with high affinity to the small molecule biotin, which is readily linked to various biomolecules such as antibodies (see Figure 6.1 Wu et al. 2002; Goldman et al. 2002).

6.2.2 Photophysical properties

The most distinctive feature of a QD is its size-tunable fluorescence color (see Figure 6.2a and b). If the QD length dimensions are near to or smaller than a critical threshold (the Bohr exciton diameter), then the QD is within the "quantum confinement regime" (Smith and Nie 2009a). In this size range, the wavelength of fluorescence emission is determined by the nanocrystal size. In the process of fluorescence, a ground state QD first absorbs an incident photon which excites an electron to a higher energy electronic state. This leaves behind an empty electronic orbital that behaves like a positively charged particle analogous to the negatively charged electron. When the electron and hole recombine, their energy is converted into a fluorescent photon. The conversion efficiency, or quantum yield (QY),

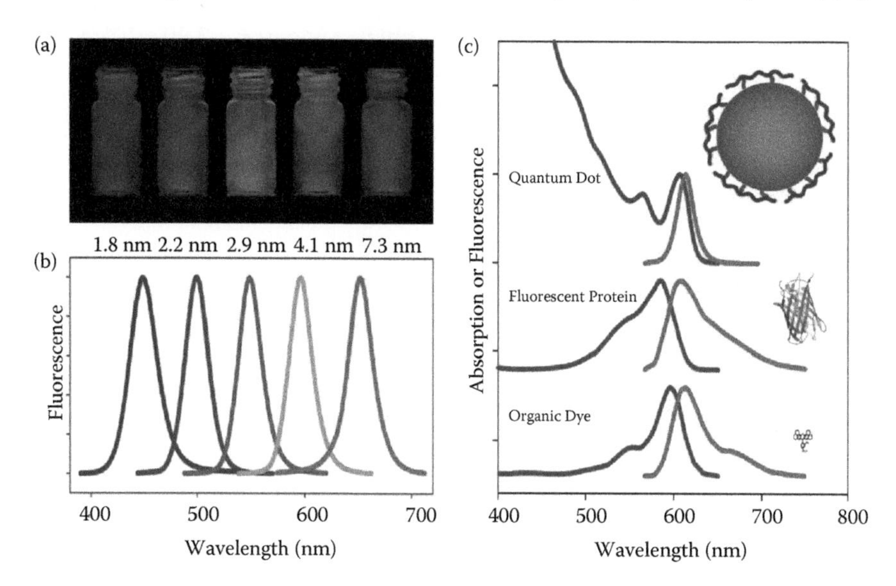

Figure 6.2 Optical properties of QDs. (a) Vials containing five sizes of QDs composed of CdSe dispersed in solution are illuminated with an ultraviolet lamp. (b) Fluorescence spectra of QDs depicted in (a). (c) Absorption spectra (dark shade) and fluorescence spectra (light shade) of QDs are compared with those of a fluorescent protein (mCherry) and an organic dye (Texas Red), showing the broad absorption spectra of QDs and narrow and symmetric fluorescence spectra of QDs. The relative sizes of these fluorescent labels are depicted next to their optical spectra. (From Andrew M. Smith, Mary M. Wen, and Shuming Nie. 2010. *The Biochemist* 32 (3) (June): 12. With permission.)

is the number of photons emitted divided by the number of photons absorbed, and this value is largely a factor of the quality and thickness of the inorganic shell.

The electron-hole pair (e^-h^+) is collectively called the exciton, which, to a first approximation, behaves like a quantum mechanical "particle in a box." The energy (E) of excitonic light absorption and emission is

$$E = \frac{h^2}{8m_r r^2} + E_g.$$

where h is the Planck constant (~6.626 × 10^{-34} J), r is the radius of a spherical QD, E_g is the semiconductor bandgap energy (E_g = 1.76 eV or 2.82 × 10^{-19} J for CdSe), and m_r is the reduced mass of the charge carriers ($m_r^{-1} = m_e^{-1} + m_h^{-1}$), a function of the effective electron mass m_e and effective hole mass m_h, which are characteristic properties of each material (for CdSe, $m_e = 0.119\ m_o$ and $m_h = 0.570\ m_o$, where m_o is the free electron mass, ~9.11 × 10^{-31} kg). The wavelength (λ) of light emission is inversely proportional to the exciton energy ($\lambda = hc/E$ where c is the speed of light in a vacuum, ~3.00 × 10^8 m s^{-1}). Thus, the wavelength of fluorescence depends on both size and the nanocrystal composition, with larger sizes yielding longer wavelengths. This "effective mass approximation" becomes increasingly less accurate at smaller sizes; more sophisticated approximations can improve accuracy but at a significantly increased computational cost.

6.2.3 *Comparison with other fluorophores*

Four key attributes of QDs are particularly advantageous for single-molecule imaging applications relative to other fluorescent reporters such as organic dyes, fluorescent proteins, and fluorescent beads. First, their fluorescence brightness is 10 to 100 times greater than that of organic dyes and proteins, which results in a higher signal-to-noise ratio with decreased excitation power requirements (Chan and Nie 1998). This effect arises from the crystalline nature of these materials, in which hundreds to thousands of bonding electrons collectively oscillate to generate massive extinction coefficients, compared with just tens of electrons in organic dyes and proteins. Second, their emission stability is 100 to 1000 times greater than that of other fluorophores, which enables long-term tracking of single molecules without signal decay (Chan and Nie 1998; Bruchez et al. 1998; Wu et al. 2002). This attribute derives from the insulating shell, as the QD can withstand oxidation or decomposition of many bonds without detriment to the sensitive core bonds in which absorption and emission take place. This is in contrast to organic dyes and proteins, for which breaking a single bond will lead to often irreversible fluorescence quenching. These first two attributes are also characteristic of some fluorescent beads; however; QDs provide this in a much more compact form (beads are generally an order of magnitude larger; see Section 6.3).

Third, QD fluorescence emission bands are narrow and symmetric and their excitation bands are broad (see Figure 2), which allows for the simultaneous excitation of many fluorescence colors with little color crosstalk compared to conventional fluorophores. Fourth, their color is readily tuned over a broad spectrum by adjusting either the particle size or composition, spanning wavelengths of 300 to 5000 nm, far beyond what is possible with conventional fluorophores. Collectively, these attributes are responsible for the capacity to image and track multiple distinct colors of individual particles for long durations in complex oxidizing environments. QDs can also be synthesized with a wide range of sizes, shapes, compositions, and composite structures, which allows for precise tuning of certain optical parameters (e.g., fluorescence lifetime and polarization) that cannot be predictably altered in conventional fluorophores. This tunability is possible due to the well-understood physical laws governing QDs derived from decades of study of semiconductor physics (Smith and Nie 2009a), and this tunability has been an enabling feature for their implementation in advanced microscopy techniques (see Sections 6.5 and 6.6).

6.3 New QD technology

Two major needs have driven QD engineering in recent years: the need for more compact, precise probes, and the need for continuous, nonblinking light emission (Smith and Nie 2009b). Here we will describe recent technological advances to overcome these problems, which are summarized in Figure 6.3.

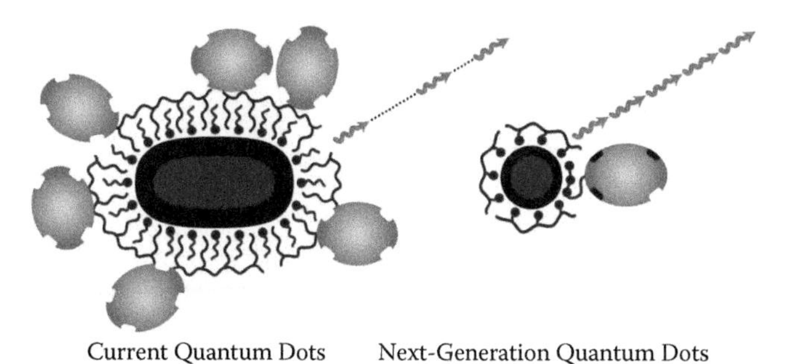

Figure 6.3 Schematic depiction of current QDs used for single-molecule imaging and the next generation of QDs currently in development. New QDs are compact, with a thin, robust coating bound to a molecular adaptor with a single binding site (here depicted as streptavidin with a single binding site) and continuous (nonblinking) fluorescence emission.

6.3.1 *Size minimization*

Current QDs are macromolecules with a hydrodynamic size near 15 to 35 nm, which is in an intermediate size range between large fluorescent beads (~50 to 500 nm) and small organic dyes (1 nm) and proteins (~4 nm). In some instances the rather bulky size of QDs limits access to crowded cellular environments (Smith, Wen, and Nie 2010; Groc et al. 2007) and limits the efficiency of energy transfer (Hohng and Ha 2004), which has driven efforts toward size minimization. The QD hydrodynamic size arises from two distinct structural domains. First the nanocrystalline core/shell component is usually ~5 to 8 nm in diameter with a quasi-spherical or rod-like shape. QDs can be readily synthesized as small as ~2 nm; however, significant advantages are gained from larger structures: more atoms give rise to increased brightness, and thicker crystalline shells improve stability. Researchers have recently balanced these offsets by preparing core/shell materials that are just large enough for bright and stable emission for specific imaging modalities (e.g., epifluorescence and confocal microscopy). QDs as small as ~3.5 nm in diameter with suitable brightness and stability for single-molecule epifluorescence imaging have recently been reported (Smith and Nie 2012), and aqueous QDs with diameters less than 5 nm have been produced with exceptionally high QY (~80%) (O. Chen et al. 2013).

The hydrodynamic size is also a function of the organic coating, which has traditionally been based on densely packed micelle-like hydrophobic bilayers that are highly stable (see Figure 6.3; Wu et al. 2002; Dubertret et al. 2002). Monolayer coatings, in contrast, can be very thin, but they are typically composed of ligands such as mercaptoacetic acid, for which the thiol group adsorbs to the QD through a weak interaction that readily dissociates in aqueous media (Aldana, Wang, and Peng 2001). To overcome this problem, new monolayer coatings have been designed based on families of linear polymers or peptides containing multiple thiols, amines, or imidazoles that bind to the nanocrystal surface through multivalent interactions (Smith and Nie 2008; W. Liu et al. 2009; Xu et al. 2012) and substantially enhance adsorption strength. These new coatings yield nanoparticles with hydrodynamic diameters as small as 5 to 10 nm that are stable for years at room temperature. To prevent fouling or adsorption to proteins and cellular structures these coatings are usually ligated to polyethylene glycol (PEG), a flexible polymer with neutral charge that does not substantially interact with biomolecules. However, PEG intrinsically expands the total particle size by 5 to 20 nm (Choi et al. 2009), so PEG replacements based on hydroxyl groups or zwitterions that are both net-neutral and compact are being explored (Kairdolf et al. 2008; W. Liu et al. 2007).

6.3.2 Precise bioconjugation strategies

Conventional conjugation strategies have relied on amide bond formation between an activated carboxylic acid (e.g. succinimidyl ester) and an amine or thioether formation from a maleimide and thiol (Xing et al. 2007). These methods are problematic because competing hydrolysis reactions reduce efficiency, and the large number of chemically equivalent functional groups on biomolecules leads to random attachment orientation on the QD surface. To overcome this problem, biomolecules can be modified with metal chelating groups (e.g., hexa-his peptides) to bind rapidly, specifically, and efficiently to the metal atoms on the QD surface (Goldman et al. 2005). While this technique has been widely used to generate QD probes bound to specific proteins and nucleic acids with precise geometric orientation, the requirement for surface facet exposure is generally a detriment as it reduces the long-term stability of the probes. It is desirable to instead modify only the coating layer. Recently "click" chemistry reactions have emerged that are highly resistant to hydrolysis and significantly increase reaction efficiency between QDs bearing azide groups and biomolecules bearing alkynes (P. Zhang et al. 2012; Schieber et al. 2012). Additionally, new engineered protein technologies are being used for bioconjugation: proteins of interest are fused to enzymes that have been modified to ligate to specific chemical functional groups that are non-native to cells or present at only very low levels (Jing and Cornish 2011). Examples include chloroalkane ligands for Halo-Tag proteins (So, Yao, and Rao 2008), benzylguanine groups for SNAP-Tag proteins (Petershans, Wedlich, and Fruk 2011), and lipoic acid derivatives for lipoic acid ligase (Uttamapinant et al. 2010). These couplings are precise in orientation and also orthogonal relative to one another, such that multiple tagged proteins can be targeted simultaneously without cross-reactivity (D. S. Liu et al. 2012).

QDs and other macromolecules are chemically polyvalent, in that they display hundreds to thousands of chemically equivalent functional groups. Therefore, when they are conjugated to other molecules, the resulting population of conjugates is heterogeneous in the number of molecules bound per QD. For single-molecule imaging applications it is critical to achieve a 1:1 ratio so that targets are not cross-linked together and so that probe binding is quantitatively dictated by the concentration of the target (Saint-Michel et al. 2009; Courty et al. 2006). Researchers often underfunctionalize QDs such that only a small percentage of QDs are bound to a target protein (~10% of the population), and thus the fraction of dual-functionalized QDs is small (<0.6% by Poisson statistics—Courty et al. 2006; Lidke et al. 2004). Sometimes it is possible to purify specific conjugation ratios (Gerion et al. 2002; Fu et al. 2004); QDs conjugated to an engineered monovalent streptavidin protein (native streptavidin is a tetramer with four binding sites for biotin) can be separated using gel electrophoresis to generate conjugates that bind to precisely one biotinylated biomolecule (Howarth et al. 2008). Further developments in monovalent conjugation schemes are expected to have a broad impact in the near term.

6.3.3 Nonblinking quantum dots

Like nearly all known fluorophores, the light emission from individual QDs turns "on" and "off" rapidly and randomly—a phenomenon known as blinking. For single-molecule imaging applications this can be either an asset or a liability. Blinking is a simple means to clearly distinguish individual molecules from aggregates or clusters: individual QDs exhibit only "on" or "off" behavior, whereas clusters exhibit intermediate brightness states or continuous emission. Some forms of nanoscopy can exploit these random fluctuations to improve image resolution (see Section 6.5); however, blinking is largely a detrimental attribute for long-duration single-molecule tracking because when the signal is temporarily lost it is unclear if the QD turned off or if it left the plane of focus, and in a dense field of emitters reconnecting the correct trajectory when the emitter turns back "on" is statistically nontrivial.

There is strong evidence that the dominant blinking mechanism for most QDs (i.e. CdSe/ZnS QDs) results from ionization trapping of electrons or holes at crystalline defects on the particle surface: when the electron or hole is trapped or ejected, it cannot recombine with the opposite charge carrier, resulting in the "off" fluorescence state (Galland et al. 2011). When this charge carrier returns to the nanocrystal core, emission returns. Blinking can be reduced and sometimes virtually eliminated through overgrowth of a thick insulating shell (Y. Chen et al. 2008; Mahler et al. 2008). However, this induces a commensurate increase in nanocrystal size. Recently, researchers have found that QDs with intermediate shell thickness can exhibit properties approaching those of thick-shell QDs (O. Chen et al. 2013). Engineering a new generation of QDs that are nonblinking as well as small with precise conjugation will continue to be an ongoing challenge over the next decade.

6.4 Novel biological phenomena

In 2003 and 2004, two seminal reports described the first use of QDs as probes for single molecules in living cells and demonstrated the major advantages of QDs over other available fluorophores (Lidke et al. 2004; Dahan et al. 2003). In this section, we describe how QDs have become critical tools in the past decade for evaluating the dynamics and behavior of biological molecules and for revealing the organization and molecular heterogeneity of living cells and tissues.

6.4.1 Single-molecule dynamics

When QDs are bound to membrane proteins, motor proteins, or other biological molecules, patterns emerge from their spatial trajectories that can be used to deduce biomolecular behavior. In single-molecule trajectory analysis, the position of a single emitter in an image is first pinpointed in two dimensions (x, y) by calculating the centroid of the point spread function (PSF), and its displacement is calculated between sequential image frames. Trends in molecular behavior can be found from the mean of the squared displacement (MSD) for a time increment t:

$$\mathrm{MSD}(t) = \left[x(t+t_0) - x(t_0)\right]^2 + \left[y(t+t_0) - y(t_0)\right]^2.$$

where x and y denote the position of the particle of interest at a given time, t_0 is the initial time point and the angled brackets denote a combined average over all increments of time t. When MSD values are plotted as a function of t, a linear plot indicates random Brownian motion, while a plot with a decreasing slope indicates confined motion and a plot with an increasing slope indicates motion with a preferred directional orientation (see Figure 6.4).

For Brownian motion in two dimensions, the MSD plot yields the *diffusion coefficient* (D) by the following equation:

$$\lim_{t \to 0} \mathrm{MSD}(t) = 4Dt.$$

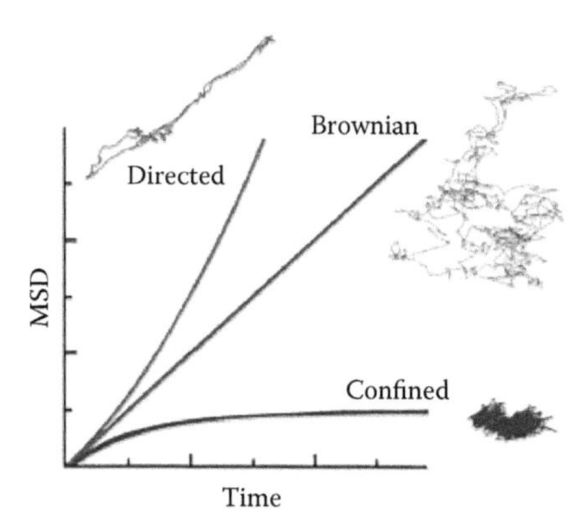

Figure 6.4 A plot of MSD values as a function of time increment that shows the differences between Brownian motion, directed motion, and confined motion. (From F. Pinaud et al. 2010. *Nature Methods* 7 (4): 275–285. With permission.)

The trajectory data can also be used to calculate an *instantaneous diffusion coefficient* (D_{inst}), by averaging only over a few time points to determine if the molecule diffuses into regions with different local environments with characteristic viscosities. QDs are particularly effective for D_{inst} calculations because their brightness offers a higher signal-to-noise ratio, which allows for high accuracy reconstruction of the PSF and long duration tracking. In contrast, organic dyes can only be tracked for a few seconds before irreversible photobleaching eliminates the fluorescent signal, and they require much higher excitation intensity for detection, which can be detrimental to cellular health.

6.4.2 *Imaging and tracking of membrane proteins*

The plasma membrane is a heterogeneous, dynamically changing structure that regulates the exchange of energy and matter between the cell and the surrounding medium, integrates the cell's mechanical components with the extracellular matrix, and senses environmental change. Membrane proteins that mediate these effects exhibit complex behaviors that remain poorly understood. Membrane protein-bound QDs have recently played a key role in elucidating membrane protein function, especially for detecting molecular confinement in membrane subdomains (Pinaud and Dahan 2011; Crane et al. 2008; Crane and Verkman 2008), and the field of neuronal biology has been one of the greatest beneficiaries of this work. Using QD-tagged receptors for neurotransmitters, receptor behaviors have been shown to differ when they are within the neuronal synapse compared to when they diffuse out of the synapse, and long-term imaging has allowed observation of the entire endocytosis/exocytosis receptor recycling process (Groc et al. 2007; Dahan et al. 2003). These observations have revealed distinct receptor behaviors in the presence of neuromodulatory drugs (Mikasova et al. 2008; Porras et al. 2012). QD studies of plasma membrane proteins have connected short-term single-molecule dynamics with long-term molecular behavior that previously had not been possible. For example, QD-tagged receptors for the neurotransmitter gamma-aminobutyric acid (GABA) were shown to redistribute in the presence of an extracellular gradient of GABA to amplify gradient sensing for nerve growth cone signaling (Bouzigues et al. 2007). The retrograde transport of vesicles containing individual dimers of nerve growth factor (NGF) has been observed across the entire length of axons to the soma (Cui et al. 2007). Growth factor receptor dynamics have also been explored in a variety of other cellular systems, and the multicolor imaging capability of QDs has played a key role in understanding the dimerization of the epidermal growth factor receptor (EGFR) and its transport in the cell after binding to its cognate receptor EGF (see Figure 6.5; Kawashima et al. 2010; Lidke et al. 2005).

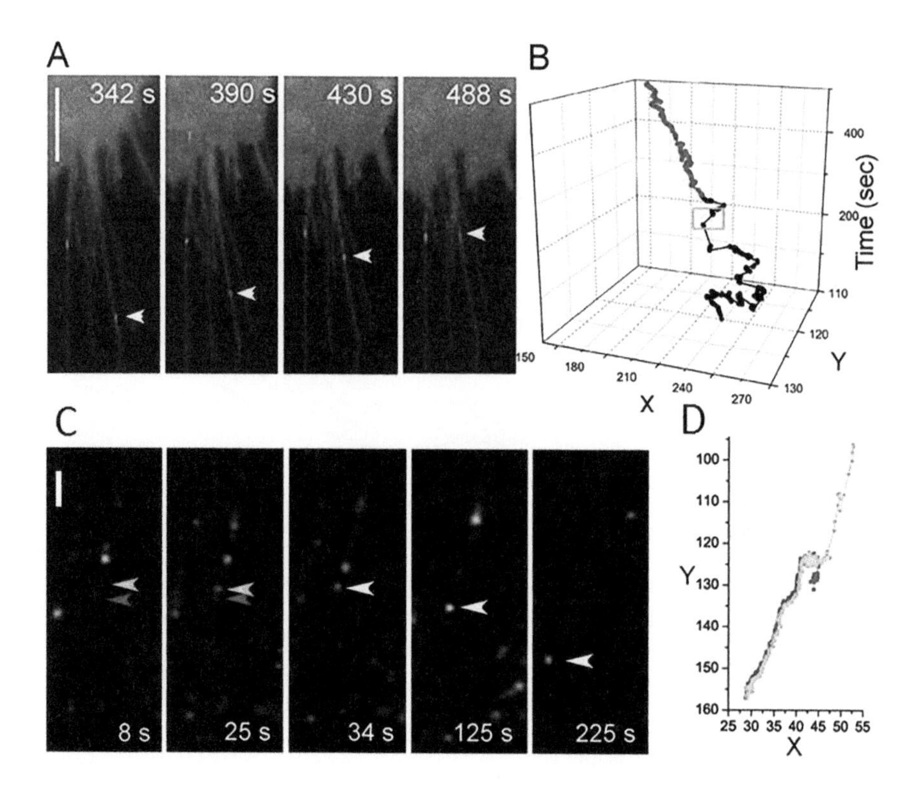

Figure 6.5 Receptor oligomerization and retrograde transport of the epidermal growth factor receptor (EGFR). (a) Selected frames of a GFP-expressing cell from a time series after binding EGF-QD (indicated by arrowhead) followed by addition of free EGF (50 ng/ml) at 300 s. Bar, 5 μm. Images are contrast enhanced. (b) Trajectory of the indicated monomer EGF-QD–EGFR complex (a, arrowhead) on a filopodium that exhibits random diffusion (dark shade) until the addition of unlabeled EGF (box), after which the complex undergoes active retrograde transport (light shade). (c) Time series of two different colored EGF-QD–EGFR complexes (EGF-QD525 tracked by light shade arrowhead and EGF-QD605 tracked by dark shade arrowhead) on a single filopodium of an A431 cell showing merging (single arrowhead at 34 s) followed by transport to the cell body (bottom left). Bar, 5 μm. Images are contrast enhanced. (d) Trajectory of the two receptors before and after dimerization. (From D. S. Lidke et al. 2005. *Science Signaling* 170 (4): 619. With permission.)

6.4.3 Imaging single molecules in living tissue

Single-molecule dynamical studies have also been extended to tissues in living animals with the use of confocal microscopy and surgically implanted imaging windows. Hideo Higuchi and colleagues have used QD-antibody conjugates to target receptors of breast tumor cells following intravenous injection in live anesthetized mice (Gonda et al. 2010; Tada

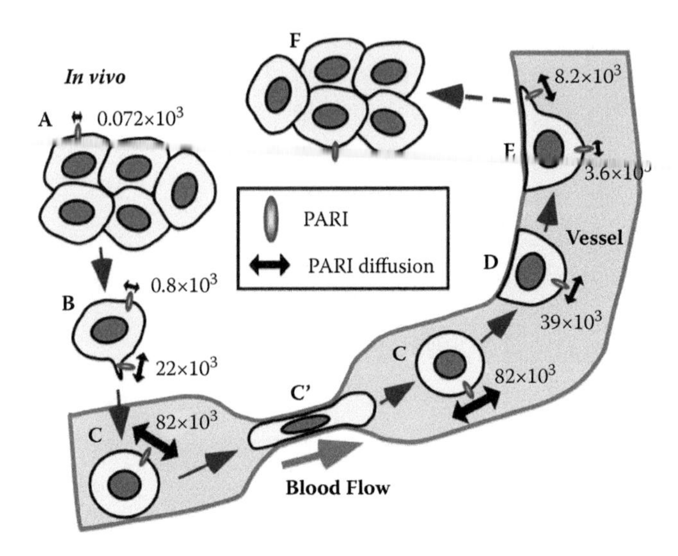

Figure 6.6 Dynamics of the membrane receptor PAR1 on tumor cells in vivo during metastasis. (A) Cells far from blood vessels; (B) cell near blood vessels; (C) cell in the bloodstream; (C′) cells in narrow blood vessels; (D) cells adhering to the inner vascular surface without directional migration; (E) cells migrating directionally on the surface; (F) cells after extravasation. Numerical values show the diffusion coefficient in nm^2/s. (From K. Gonda et al. 2010. *Journal of Biological Chemistry* 285 (4): 2750–2757. With permission.)

et al. 2007). Using either an antibody against the HER2 oncogene product or the protease-activated receptor 1 (PAR1), confocal imaging revealed diffusion dynamics of both receptors on tumor cells in vivo. PAR1 exhibited distinct diffusion regimes on tumor cells during four different stages of tumor cell metastasis, reaching a maximum mobility when the cancer cells were in free in the blood vessels (see Figure 6.6). As both HER2 and PAR1 are important mediators of breast cancer pathophysiology, these studies provide important clues in tumor progression.

6.4.4 *Enzyme imaging*

Motor proteins are enzymes that convert chemical energy from ATP into mechanical motion to power directional translocation along a substrate. The cytoskeletal motor proteins myosin, dynein, and kinesin traverse cytoskeletal tracks to perform a variety of cellular functions, such as intracellular cargo delivery and cellular division. Specific modes of motion of these proteins have been established in vitro using purified proteins tagged with QDs, but it is a major challenge to deliver QDs into live cells for in vivo verification. Intracellular delivery mechanisms include microinjection, electroporation, osmotic rupture of endosomes, or cationic lipid

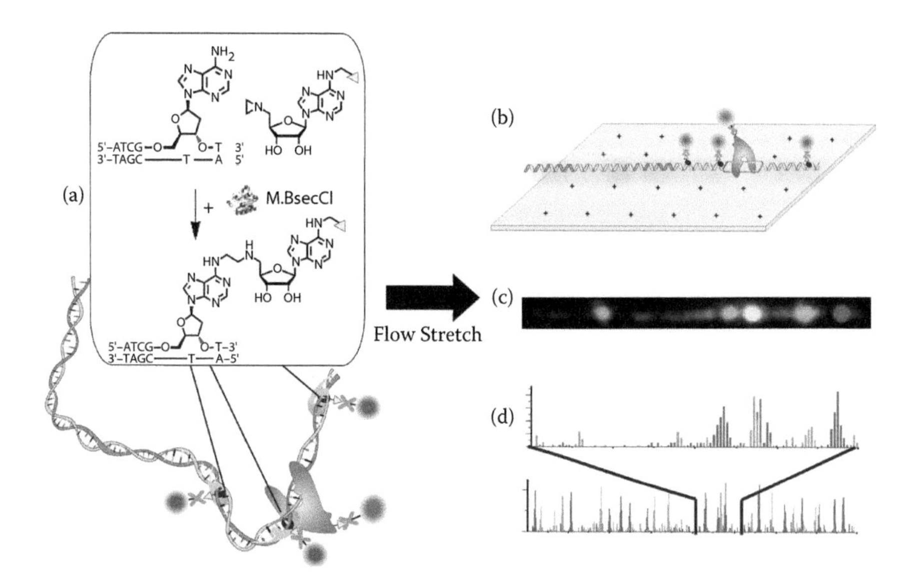

Figure 6.7 **(See color insert.)** (a) Schematic representation of RNA polymerase labeled with green QDs and bound to T7 bacteriophage DNA. The DNA has been enzymatically labeled with biotin (yellow triangle) at rare sequences (5′-ATCGAT-3′), which can be labeled with streptavidin-QDs to serve as reference points on the genome. (b) RNA polymerase-bound T7 bacteriophage genome flow stretched over a polylysine surface. (c) Image of flow-stretched, dye-stained T7 bacteriophage DNA (white) with QD-labeled RNA polymerase (green) and sequence-specific labels with red QDs as sequence references. Overlapping red and green signals are shown in yellow. (d) Conceptual representation of a genome-wide map of promoters (green) and reference sites (red). (From S. Kim et al. 2012. *Angewandte Chemie* 124 (15): 3638–3641. With permission.)

transfection, each of which has certain limitations (Smith, Wen, and Nie 2010). However studies have confirmed that QD-motor protein conjugates exhibit similar behaviors for both kinesin (Courty et al. 2006) and myosin Va (Nelson et al. 2009) inside and outside of cells. Nucleic acid motors are a second class of enzymes that include polymerases, helicases, topoisomerases, and other enzymes involved in the maintenance, repair, and replication of genetic material. The multicolor imaging capacity of QD probes has been used to image and map the binding of these proteins to DNA with nanometer-scale resolution, which allows precise mapping of enzyme binding locations, such as promoters, across entire genomes (see Figure 6.7; Ebenstein et al. 2009).

QDs have also been instrumental in determining how enzymes that are not powered by ATP efficiently scan through entire genomes to find their target sequence. Simple calculations show that three-dimensional diffusion of enzymes and random attachment to DNA alone would not

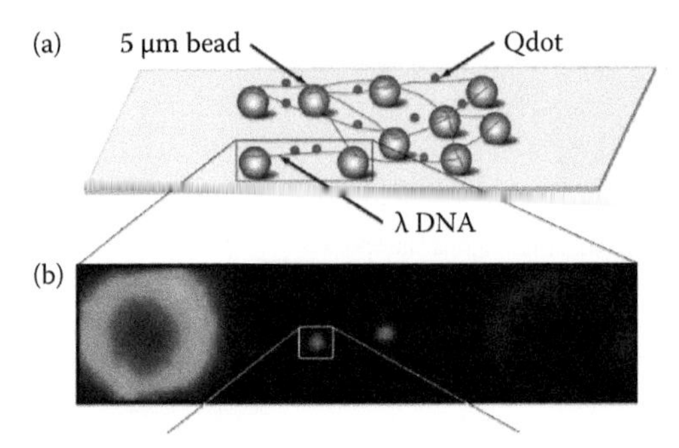

Figure 6.8 **(See color insert.)** Single-molecule N-glycosylase imaging. (a) DNA "tightropes" extended between stationary 5-μm silica beads suspend DNA off of the substrate. (b) Image of faint DNA strung between beads (green) with bound QD-labeled N-glycosylase (red). (From A. R. Dunn et al. 2011. *Nucleic Acids Research* 39 (17): 7487–7498. With permission.)

provide the efficiency necessary for genomic maintenance and repair. Single-QD imaging has revealed that enzymes such as EcoRV, UvrA, UvrB, and N-glycosylases randomly attach to DNA but then slide along the DNA while rotating with the helix to scan for their targets (see Figure 6.8; Dikić et al. 2012; Dunn et al. 2011; Kad et al. 2010). This one-dimensional search process powered by thermal fluctuations is much more efficient than a random-attachment search.

6.5 *Super-resolution imaging applications*

The resolution of traditional light microscopy is limited by the diffraction limit to approximately half of the wavelength of light, or about 200 to 300 nm. Single-molecule imaging applications of QDs discussed so far have achieved a spatial resolution much smaller than the optical diffraction limit because the majority of the emitters have been spatially isolated, as the targets have only been sparsely tagged. However, this methodology, termed fluorescence imaging with one nanometer accuracy (FIONA), is not effective if the particles are separated by a distance smaller than the optical diffraction limit. Recent developments in super-resolution microscopy have resulted in the optical resolution of objects as small as tens of nanometers for densely packed emitters by using fluorescent probes that can be optically switched between "on" and "off" states. Spatially deterministic optical modulation through the use of STED microscopy and spatially stochastic modulation (e.g., PALM and STORM) are the subjects of Chapters 1 and 2, respectively. Unlike photoswitchable dyes, there is no generalizable mechanism to modulate QD fluorescence, but they have still been successfully integrated into both of the aforementioned microscopy types. For stochastic imaging applications, blinking provides random

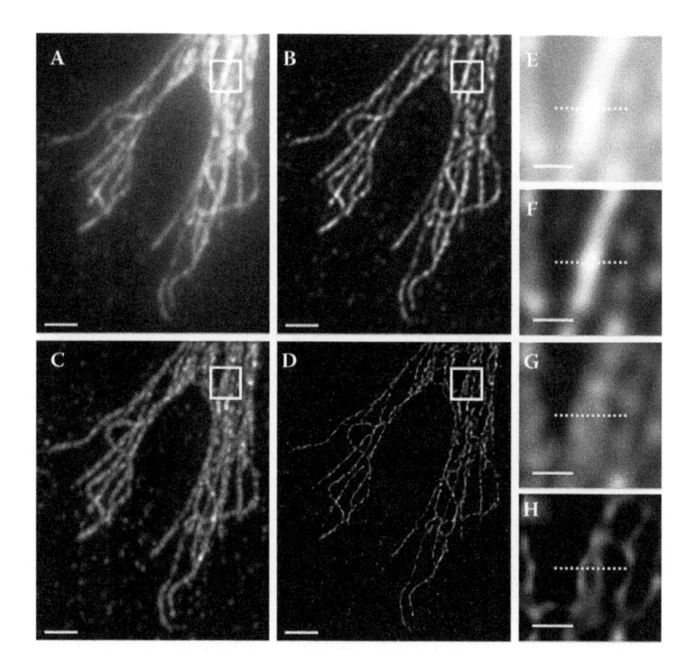

Figure 6.9 SOFI images of QD-625-labeled fibroblasts. (A) Original image generated by time averaging all frames of the acquired movie (3,000 frames, 100 ms per frame). (B) The image in *A* deconvolved. (C) Second-order SOFI image. (D) The image in *C* deconvolved. (E–H) Magnified views of the boxed regions in *A–D*. (Scale bars: *A–D*, 2 μm; *E–H*, 500 nm.) (From T. Dertinger et al. 2009. *Proceedings of the National Academy of Sciences U.S.A.* 106 (52): 22287–22292. With permission.)

on-and-off fluctuations, which has been used as a mechanism to achieve a fivefold improvement in resolution of alpha-tubulin microtubules in fibroblasts through super-resolution optical fluctuation imaging (SOFI—see Figure 6.9; Dertinger et al. 2009).

Using the SOFI technique, enhanced resolution is attained simply by waiting for QDs to randomly turn off so that more discrete fluorescence points can be imaged. This can be quite time-consuming and requires extensive data analysis, so more recent applications of QDs in super-resolution microscopy have utilized some of the uniquely tunable photophysical characteristics of these particles to improve resolution and data collection time. Watanabe and coworkers (2010) created "blinking enhanced" QDs with very thin ZnS shells, which spend a longer fraction of time in the "off" state. The researchers also improved calculation speed by tracking the variance of emission rather than cumulant of emission, a technique they called variance imaging for superresolution (VISion), which was shown to allow 90-nm spatial and 80-ms temporal resolution in live-cell vesicle imaging.

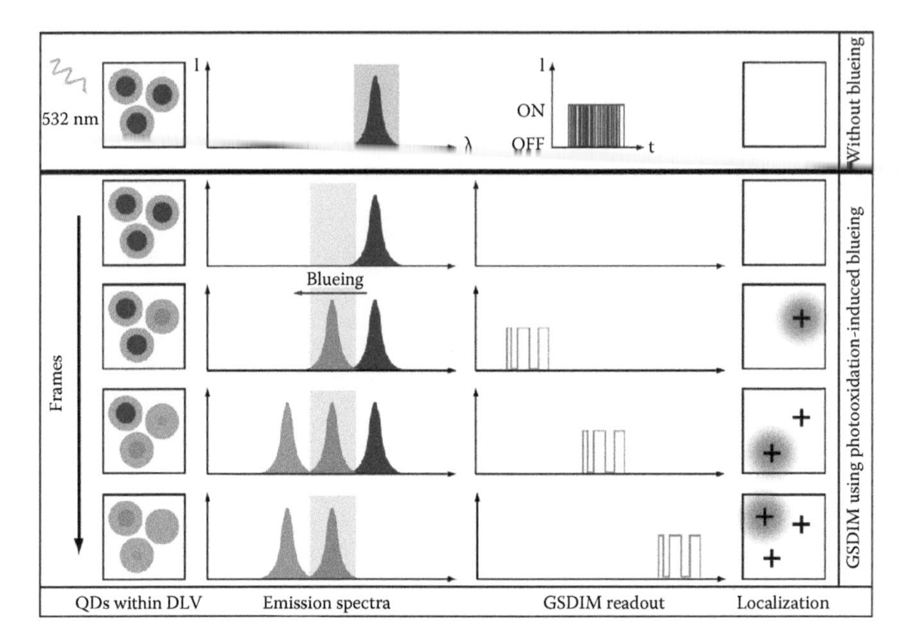

Figure 6.10 Blue shifting-enabled GSDIM. Without blueing, the signal from QDs overlaps within a diffraction limited volume (DLV) at a given time point due to rather short off periods. Disparate time periods between illumination and onset of blueing separate the fluorescence traces. Successive spectral translation of subsets of QDs into the blue-shifted detection window yields a diffraction unlimited image benefiting from the huge number of photons gained from a dot before bleaching or further blueing. (From P. Hoyer et al. 2010. *Nano Letters* 11 (1): 245–250. With permission.)

While QD emission modulation is still not yet consistently controllable, QDs can be irreversibly photoswitched through photooxidation. As QDs are photooxidized, their size decreases, which shifts the fluorescence wavelength to the blue. This optical shifting mechanism was incorporated into the established ground state depletion microscopy followed by individual molecule return (GSDIM) technique (Hoyer et al. 2010). In this case, the pseudo-ground state QDs are considered to be those emitting in a window that is slightly photooxidized compared to the native QDs. While GSDIM operates by forcing an entire field of emitters into the "off" state and then watching the stochastic return of single emitters to the "on" state, QD-GSDIM is based on watching QDs shift over the photooxidation spectral window. As QD blueing is a stochastic process, distinct QDs shift into the window discretely, revealing individual QDs from dense populations (see Figure 6.10), resulting in ~12-nm spots.

Deterministic super-resolution methods based on STED microscopy have the greatest potential for high spatial resolution with high temporal resolution, but these methods require the capacity to optically deplete the emitting state through stimulated emission. Conventional QDs have been resistant to this application because their excitation spectra overlap with their depletion spectra, which prevents efficient depletion. Recently Irvine and colleagues (2008) developed a novel QD based on ZnSe QDs doped with Mn^{2+} ions. The key feature of these QDs is that the ion dopant is responsible for emission, and this dopant is at a much lower energy than the absorption onset. Thus, the emitting state can be efficiently and reversibly depleted with a pulsed laser without exciting the particles. This method, termed reversible saturable optical fluorescence transitions (RESOLFT), has been demonstrated to resolve structures separated by 85 nm (Irvine et al. 2008).

6.6 *Z-axis microscopy of QDs*

6.6.1 *Imaging QD orientation through defocused microscopy*

QDs can be synthesized with a variety of crystal structures with cubic or hexagonal symmetry. Those with hexagonal symmetry are of particular interest because their light emission is preferentially polarized along a unique axis of the crystal, which gives rise to polarized light emission and asymmetric PSFs, with increasing asymmetry for QDs with more asymmetric shapes. This PSF asymmetry is greatest when the QDs are out of the optical focal plane of the image, which increases the relative intensity of the outer lobes of the PSF (see Figure 6.11). By focusing approximately 0.5 to 0.9 μm away from the plane containing QDs, the orientation can be determined, and this has been used to track the orientation and location of glycine receptors in HeLa cells (Brokmann et al. 2005), as well as the orientation of myosin V with 10° accuracy at a time resolution of 33 ms

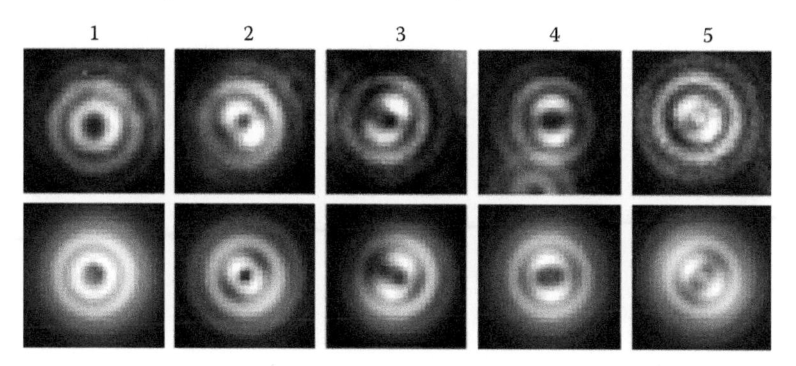

Figure 6.11 Defocused images of single CdSe/ZnS quantum dots at a focus position of f = 0.9 μm (top row) as well as calculated images. (From R. Schuster et al. 2005. *Chemical Physics Letters* 413 (4): 280–283. With permission.)

(Ohmachi et al. 2012). Many nuances of this methodology have not yet been worked out and the ideal QD aspect ratio/size combination is not yet known. Also the PSF image may show an artifact of apparent rotation after "off" blinking states, which is possibly due to warping of the exciton by surface defects, an effect that can be reduced by surface passivation with an inorganic shell or polymer (Li et al. 2010).

6.6.2 *Imaging QDs in three dimensions*

The studies reviewed so far have been limited to resolving structures within a single x-y plane, but many new microscopy techniques have been developed to resolve structures in all three spatial dimensions. Resolving structures in the z-axis is a challenge due to greater interference from out-of-focus light, depth-dependent attenuation of excitation and emitted light, as well intrinsically lower resolution in wide-field imaging modes in this dimension. Z-axis resolution can be substantially improved with confocal line-scanning microscopy techniques and multiphoton excitation, and depth penetration can be improved with multiphoton excitation with longer-wavelength light. QDs have especially large multiphoton cross-sections, often 100 to 1000 times that of conventional organic dyes. By exploiting this immense absorbing capacity, Ruobing Zhang and colleagues (2011) developed a holographic imaging procedure using two-photon excitation of QDs, where they achieved 2- to 3-nm accuracy in all three dimensions. Holographic excitation allowed simultaneous excitation of 80 diffraction limited spots—in contrast to conventional confocal imaging techniques with a single excitation point. This increases imaging speeds by a factor of 80. Using this technique, they imaged the motion of myosin V motors, the LamB receptor distribution on *Escherichia coli* cells, and the endocytosis of EGF receptors in breast cancer cells (R. Zhang et al. 2011). Another study integrated the asymmetric absorption profile of quantum rods and the superior depth penetration of two-photon excitation in order to combine increased depth discrimination and orientation tracking in one imaging platform (Rothenberg et al. 2004).

Another approach to imaging in the z-axis is through the simultaneous collection of emitted light from multiple focal planes using multiple cameras (Ram et al. 2008, 2012). This eliminates the time needed to separately image multiple focal planes and maximizes the "photon budget" of a given sample to reduce photodamage. This technique was used to continuously track a transferrin-tagged QD in three dimensions as it attached to a cell, underwent endocytosis, and transferred between adjacent cells (see Figure 6.12; Ram et al. 2008). QDs are exceptional markers for multifocal-plane microscopy techniques because their intense brightness provides a large number of photons for unambiguous identification of molecules as they move between planes.

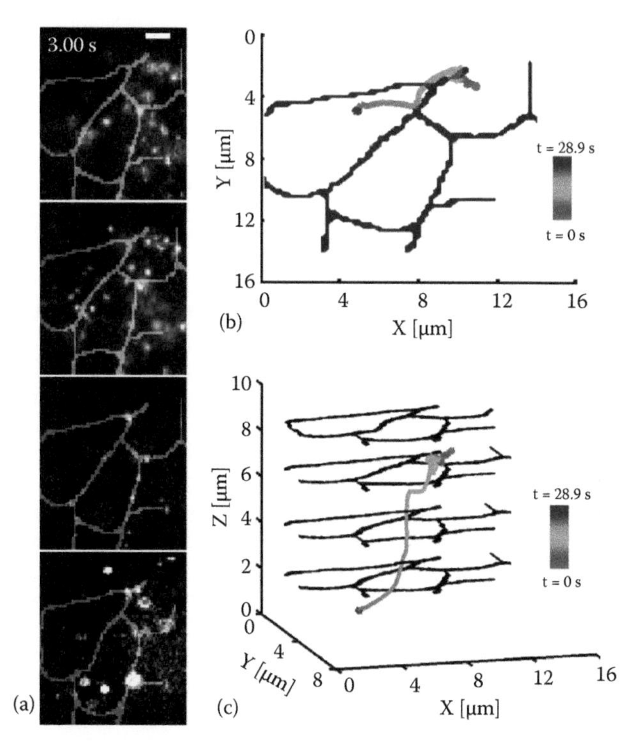

Figure 6.12 **(See color insert.)** Multifocal imaging of live cells taking up transferrin proteins bound to QDs. In panel *A*, four focal planes are shown for a single time point in which cell plasma membranes are marked green and QDs are shown in gray scale. Scale bar = 5 µm. Panels *B* and *C* show the X-Y projection and the full three-dimensional trajectory, respectively, of the transferrin-QD molecule highlighted with a red arrow in panel *A*. The trajectories are color-coded to indicate time. The Transferrin-QD molecule is initially seen inside one cell and then moves toward the lateral plasma membrane and undergoes exocytosis. The molecule is then immediately endocytosed by the adjacent cell. (From S. Ram et al. 2012. *Biophysical Journal* 103 (7): 1594–1603. With permission.)

6.7 Conclusions

QDs have had an important impact in the past decade on our understanding of complex processes intrinsic to biology. However, the use of QDs in biological applications is still in its infancy, as upcoming advances will drastically enhance the capabilities of QDs due to their capacity for optical and structural engineering. The near future will see the development of QDs with sizes as small as fluorescent proteins, precise valency for just one biomolecular target, and negligible nonspecific interactions. With the simultaneous advancement of super-resolution imaging techniques, QDs are poised to enable dynamic tracking of specific molecules at resolutions previously only achievable with electron microscopy techniques.

References

Aldana, Jose, Y. Andrew Wang, and Xiaogang Peng. 2001. "Photochemical Instability of CdSe Nanocrystals Coated by Hydrophilic Thiols." *Journal of the American Chemical Society* 123 (36): 8844–8850.

Bouzigues, Cédric, Mathieu Morel, Antoine Triller, and Maxime Dahan. 2007. "Asymmetric Redistribution of GABA Receptors During GABA Gradient Sensing by Nerve Growth Cones Analyzed by Single Quantum Dot Imaging." *Proceedings of the National Academy of Sciences U.S.A.* 104 (27): 11251–11256.

Brokmann, Xavier, Marie-Virgine Ehrensperger, Jean-Pierre Hermier, Antoine Triller, and Maxime Dahan. 2005. "Orientational Imaging and Tracking of Single CdSe Nanocrystals by Defocused Microscopy." *Chemical Physics Letters* 406 (1): 210–214.

Bruchez, Marcel, Mario Moronne, Peter Gin, Shimon Weiss, and A. Paul Alivisatos. 1998. "Semiconductor Nanocrystals as Fluorescent Biological Labels." *Science* 281 (5385): 2013–2016.

Chan, Warren C. W., and Shuming Nie. 1998. "Quantum Dot Bioconjugates for Ultrasensitive Nonisotopic Detection." *Science* 281 (5385): 2016–2018.

Chen, Ou, Jing Zhao, Vikash P. Chauhan, Jian Cui, Cliff Wong, Daniel K. Harris, He Wei, et al. 2013. "Compact High-Quality CdSe—CdS Core—Shell Nanocrystals with Narrow Emission Linewidths and Suppressed Blinking." *Nature Materials* 12 (5): 445–451.

Chen, Yongfen, Javier Vela, Han Htoon, Joanna L. Casson, Donald J. Werder, David A. Bussian, Victor I. Klimov, and Jennifer A. Hollingsworth. 2008. "'Giant Multishell CdSe Nanocrystal Quantum Dots with Suppressed Blinking." *Journal of the American Chemical Society* 130 (15): 5026–5027.

Choi, Hak Soo, Binil Itty Ipe, Preeti Misra, Jeong Heon Lee, Moungi G. Bawendi, and John V. Frangioni. 2009. "Tissue-and Organ-Selective Biodistribution of NIR Fluorescent Quantum Dots." *Nano Letters* 9 (6): 2354–2359.

Courty, Sébastien, Camilla Luccardini, Yohanns Bellaiche, Giovanni Cappello, and Maxime Dahan. 2006. "Tracking Individual Kinesin Motors in Living Cells Using Single Quantum-Dot Imaging." *Nano Letters* 6 (7): 1491–1495.

Crane, Jonathan M., Alfred N. Van Hoek, William R. Skach, and A. S. Verkman. 2008. "Aquaporin-4 Dynamics in Orthogonal Arrays in Live Cells Visualized by Quantum Dot Single Particle Tracking." *Molecular Biology of the Cell* 19 (8): 3369–3378.

Crane, Jonathan M. and A. S. Verkman. 2008. "Long-Range Nonanomalous Diffusion of Quantum Dot-Labeled Aquaporin-1 Water Channels in the Cell Plasma Membrane." *Biophysical Journal* 94 (2): 702–713.

Cui, Bianxiao, Chengbiao Wu, Liang Chen, Alfredo Ramirez, Elaine L. Bearer, Wei-Ping Li, William C. Mobley, and Steven Chu. 2007. "One at a Time, Live Tracking of NGF Axonal Transport Using Quantum Dots." *Proceedings of the National Academy of Sciences U.S.A.* 104 (34): 13666–13671.

Dabbousi, B. O., J. Rodriguez-Viejo, Frederic V. Mikulec, J. R. Heine, Hedi Mattoussi, R. Ober, K. F. Jensen, and M. G. Bawendi. 1997. "(CdSe) ZnS Core-Shell Quantum Dots: Synthesis and Characterization of a Size Series of Highly Luminescent Nanocrystallites." *Journal of Physical Chemistry B* 101 (46): 9463–9475.

Dahan, Maxime, Sabine Levi, Camilla Luccardini, Philippe Rostaing, Beatrice Riveau, and Antoine Triller. 2003. "Diffusion Dynamics of Glycine Receptors Revealed by Single-Quantum Dot Tracking." *Science Signalling* 302 (5644): 442.

Dertinger, T., R. Colyer, G. Iyer, S. Weiss, and J. Enderlein. 2009. "Fast, Background-Free, 3D Super-Resolution Optical Fluctuation Imaging (SOFI)." *Proceedings of the National Academy of Sciences U.S.A.* 106 (52): 22287–22292.

Dikić, Jasmina, Carolin Menges, Samuel Clarke, Michael Kokkinidis, Alfred Pingoud, Wolfgang Wende, and Pierre Desbiolles. 2012. "The Rotation-Coupled Sliding of EcoRV." *Nucleic Acids Research* 40 (9): 4064–4070.

Dubertret, Benoit, Paris Skourides, David J. Norris, Vincent Noireaux, Ali H. Brivanlou, and Albert Libchaber. 2002. "In Vivo Imaging of Quantum Dots Encapsulated in Phospholipid Micelles." *Science* 298 (5599): 1759–1762.

Dunn, A. R., N. M. Kad, S. R. Nelson, D. M. Warshaw, and S. S. Wallace. 2011. "Single Qdot-Labeled Glycosylase Molecules Use a Wedge Amino Acid to Probe for Lesions While Scanning Along DNA." *Nucleic Acids Research* 39 (17): 7487–7498.

Ebenstein, Yuval, Natalie Gassman, Soohong Kim, Josh Antelman, Younggyu Kim, Sam Ho, Robin Samuel, Xavier Michalet, and Shimon Weiss. 2009. "Lighting up Individual DNA Binding Proteins with Quantum Dots." *Nano Letters* 9 (4): 1598–1603.

Fu, Aihua, Christine M. Micheel, Jennifer Cha, Hauyee Chang, Haw Yang, and A. Paul Alivisatos. 2004. "Discrete Nanostructures of Quantum dots/Au with DNA." *Journal of the American Chemical Society* 126 (35): 10832–10833.

Galland, Christophe, Yagnaseni Ghosh, Andrea Steinbrück, Milan Sykora, Jennifer A. Hollingsworth, Victor I. Klimov, and Han Htoon. 2011. "Two Types of Luminescence Blinking Revealed by Spectroelectrochemistry of Single Quantum Dots." *Nature* 479 (7372): 203–207.

Gerion, Daniele, Wolfgang J. Parak, Shara C. Williams, Daniela Zanchet, Christine M. Micheel, and A. Paul Alivisatos. 2002. "Sorting Fluorescent Nanocrystals with DNA." *Journal of the American Chemical Society* 124 (24): 7070–7074.

Goldman, Ellen R., Eric D. Balighian, Hedi Mattoussi, M. Kenneth Kuno, J. Matthew Mauro, Phan T. Tran, and George P. Anderson. 2002. "Avidin: a Natural Bridge for Quantum Dot-Antibody Conjugates." *Journal of the American Chemical Society* 124 (22): 6378–6382.

Goldman, Ellen R., Igor L. Medintz, Andrew Hayhurst, George P. Anderson, J. Matthew Mauro, Brent L. Iverson, George Georgiou, and Hedi Mattoussi. 2005. "Self-Assembled Luminescent CdSe—ZnS Quantum Dot Bioconjugates Prepared Using Engineered Poly-Histidine Terminated Proteins." *Analytica Chimica Acta* 534 (1): 63–67.

Gonda, K., T. M. Watanabe, N. Ohuchi, and H. Higuchi. 2010. "In Vivo Nano-Imaging of Membrane Dynamics in Metastatic Tumor Cells Using Quantum Dots." *Journal of Biological Chemistry* 285 (4): 2750–2757.

Groc, Laurent, Mathieu Lafourcade, Martin Heine, Marianne Renner, Victor Racine, Jean-Baptiste Sibarita, Brahim Lounis, Daniel Choquet, and Laurent Cognet. 2007. "Surface Trafficking of Neurotransmitter Receptor: Comparison Between Single-Molecule/Quantum Dot Strategies." *Journal of Neuroscience* 27 (46): 12433–12437.

Hohng, Sungchul, and Taekjip Ha. 2004. "Near-Complete Suppression of Quantum Dot Blinking in Ambient Conditions." *Journal of the American Chemical Society* 126 (5): 1324–1325.

Howarth, Mark, Wenhao Liu, Sujiet Puthenveetil, Yi Zheng, Lisa F. Marshall, Michael M. Schmidt, K. Dane Wittrup, Moungi G. Bawendi, and Alice Y. Ting. 2008. "Monovalent, Reduced-Size Quantum Dots for Imaging Receptors on Living Cells." *Nature Methods* 5 (5): 397–399.

Hoyer, Patrick, Thorsten Staudt, Johann Engelhardt, and Stefan W Hell. 2010. "Quantum Dot Blueing and Blinking Enables Fluorescence Nanoscopy." *Nano Letters* 11 (1): 245–250.

Irvine, Scott E., Thorsten Staudt, Eva Rittweger, Johann Engelhardt, and Stefan W. Hell. 2008. "Direct Light-Driven Modulation of Luminescence from Mn-Doped ZnSe Quantum Dots." *Angewandte Chemie* 120 (14): 2725–2728.

Jing, Chaoran, and Virginia W. Cornish. 2011. "Chemical Tags for Labeling Proteins Inside Living Cells." *Accounts of Chemical Research* 44 (9): 784–792.

Kad, Neil M., Hong Wang, Guy G. Kennedy, David M. Warshaw, and Bennett Van Houten. 2010. "Collaborative Dynamic DNA Scanning by Nucleotide Excision Repair Proteins Investigated by Single-Molecule Imaging of Quantum-Dot-Labeled Proteins." *Molecular Cell* 37 (5): 702–713.

Kairdolf, Brad A., Michael C. Mancini, Andrew M. Smith, and Shuming Nie. 2008. "Minimizing Nonspecific Cellular Binding of Quantum Dots with Hydroxyl-Derivatized Surface Coatings." *Analytical Chemistry* 80 (8): 3029–3034.

Kawashima, Nagako, Kenichi Nakayama, Kohji Itoh, Tamitake Itoh, Mitsuru Ishikawa, and Vasudevanpillai Biju. 2010. "Reversible Dimerization of EGFR Revealed by Single-Molecule Fluorescence Imaging Using Quantum Dots." *Chemistry – A European Journal* 16 (4): 1186–1192.

Kim, Soohong, Anna Gottfried, Ron R. Lin, Thomas Dertinger, Andrew S. Kim, Sangyoon Chung, Ryan A. Colyer, Elmar Weinhold, Shimon Weiss, and Yuval Ebenstein. 2012. "Enzymatically Incorporated Genomic Tags for Optical Mapping of DNA-Binding Proteins." *Angewandte Chemie* 124 (15): 3638–3641.

Li, Qiang, Xiao-Jun Chen, Yi Xu, Sheng Lan, Hai-Ying Liu, Qiao-Feng Dai, and Li-Jun Wu. 2010. "Photoluminescence Properties of the CdSe Quantum Dots Accompanied with Rotation of the Defocused Wide-Field Fluorescence Images." *Journal of Physical Chemistry C* 114 (32): 13427–13432.

Lidke, Diane S., Keith A. Lidke, Bernd Rieger, Thomas M. Jovin, and Donna J. Arndt-Jovin. 2005. "Reaching Out for Signals: Filopodia Sense EGF and Respond by Directed Retrograde Transport of Activated Receptors." *Science Signaling* 170 (4): 619.

Lidke, Diane S., Peter Nagy, Rainer Heintzmann, Donna J. Arndt-Jovin, Janine N. Post, Hernan E. Grecco, Elizabeth A. Jares-Erijman, and Thomas M. Jovin. 2004. "Quantum Dot Ligands Provide New Insights into erbB/HER Receptor-Mediated Signal Transduction." *Nature Biotechnology* 22 (2): 198–203.

Liu, Daniel S., William S. Phipps, Ken H. Loh, Mark Howarth, and Alice Y. Ting. 2012. "Quantum Dot Targeting with Lipoic Acid Ligase and HaloTag for Single-Molecule Imaging on Living Cells." *ACS Nano* 6 (12): 11080–11087.

Liu, Wenhao, Hak Soo Choi, John P. Zimmer, Eiichi Tanaka, John V. Frangioni, and Moungi Bawendi. 2007. "Compact Cysteine-Coated CdSe (ZnCdS) Quantum Dots for in vivo Applications." *Journal of the American Chemical Society* 129 (47): 14530–14531.

Liu, Wenhao, Andrew B. Greytak, Jungmin Lee, Cliff R. Wong, Jongnam Park, Lisa F. Marshall, Wen Jiang, et al. 2009. "Compact Biocompatible Quantum Dots via RAFT-Mediated Synthesis of Imidazole-Based Random Copolymer Ligand." *Journal of the American Chemical Society* 132 (2): 472–483.

Mahler, Benoit, Piernicola Spinicelli, Stephanie Buil, Xavier Quelin, Jean-Pierre Hermier, and Benoit Dubertret. 2008. "Towards Non-Blinking Colloidal Quantum Dots." *Nature Materials* 7 (8): 659–664.

Mikasova, Lenka, Laurent Groc, Daniel Choquet, and Olivier J. Manzoni. 2008. "Altered Surface Trafficking of Presynaptic Cannabinoid Type 1 Receptor in and Out Synaptic Terminals Parallels Receptor Desensitization." *Proceedings of the National Academy of Sciences U.S.A.* 105 (47): 18596–18601.

Nelson, Shane R., M. Yusuf Ali, Kathleen M. Trybus, and David M. Warshaw. 2009. "Random Walk of Processive, Quantum Dot-Labeled Myosin Va Molecules Within the Actin Cortex of COS-7 Cells." *Biophysical Journal* 97 (2): 509.

Ohmachi, Masashi, Yasunori Komori, Atsuko H Iwane, Fumihiko Fujii, Takashi Jin, and Toshio Yanagida. 2012. "Fluorescence Microscopy for Simultaneous Observation of 3D Orientation and Movement and Its Application to Quantum Rod-Tagged Myosin V." *Proceedings of the National Academy of Sciences U.S.A.* 109 (14): 5294–5298.

Petershans, Andre, Doris Wedlich, and Ljiljana Fruk. 2011. "Bioconjugation of CdSe/ZnS Nanoparticles with SNAP Tagged Proteins." *Chemical Communications* 47 (38): 10671–10673.

Pinaud, Fabien, Samuel Clarke, Assa Sittner, and Maxime Dahan. 2010. "Probing Cellular Events, One Quantum Dot at a Time." *Nature Methods* 7 (4): 275–285.

Pinaud, Fabien, and Maxime Dahan. 2011. "Targeting and Imaging Single Biomolecules in Living Cells by Complementation-Activated Light Microscopy with Split-Fluorescent Proteins." *Proceedings of the National Academy of Sciences U.S.A.* 108 (24): E201–E210.

Porras, Gregory, Amandine Berthet, Benjamin Dehay, Qin Li, Laurent Ladepeche, Elisabeth Normand, Sandra Dovero, et al. 2012. "PSD-95 Expression Controls l-DOPA Dyskinesia through Dopamine D1 Receptor Trafficking." *Journal of Clinical Investigation* 122 (11): 3977.

Ram, Sripad, Dongyoung Kim, Raimund J. Ober, and E. Sally Ward. 2012. "3D Single Molecule Tracking with Multifocal Plane Microscopy Reveals Rapid Intercellular Transferrin Transport at Epithelial Cell Barriers." *Biophysical Journal* 103 (7): 1594–1603.

Ram, Sripad, Prashant Prabhat, Jerry Chao, E. Sally Ward, and Raimund J. Ober. 2008. "High Accuracy 3D Quantum Dot Tracking with Multifocal Plane Microscopy for the Study of Fast Intracellular Dynamics in Live Cells." *Biophysical Journal* 95 (12): 6025–6043.

Rothenberg, Eli, Yuval Ebenstein, Miri Kazes, and Uri Banin. 2004. "Two-Photon Fluorescence Microscopy of Single Semiconductor Quantum Rods: Direct Observation of Highly Polarized Nonlinear Absorption Dipole." *Journal of Physical Chemistry B* 108 (9): 2797–2800.

Saint-Michel, Edouard, Grégory Giannone, Daniel Choquet, and Olivier Thoumine. 2009. "Neurexin/Neuroligin Interaction Kinetics Characterized by Counting Single Cell-Surface Attached Quantum Dots." *Biophysical Journal* 97 (2): 480–489.

Schieber, Christine, Alessandra Bestetti, Jet Phey Lim, Anneke D. Ryan, Tich-Lam Nguyen, Robert Eldridge, Anthony R. White, et al. 2012. "Conjugation of Transferrin to Azide-Modified CdSe/ZnS Core—Shell Quantum Dots Using Cyclooctyne Click Chemistry." *Angewandte Chemie International Edition* 51 (42): 10523–10527.

Schuster, Roman, Michael Barth, Achim Gruber, and Frank Cichos. 2005. "Defocused Wide Field Fluorescence Imaging of Single CdSe/ZnS Quantum Dots." *Chemical Physics Letters* 413 (4): 280–283.

Smith, Andrew M., Hongwei Duan, Aaron M. Mohs, and Shuming Nie. 2008. "Bioconjugated Quantum Dots for in vivo Molecular and Cellular Imaging." *Advanced Drug Delivery Reviews* 60 (11) (August 17): 1226–40.

Smith, Andrew M. and Shuming Nie. 2008. "Minimizing the Hydrodynamic Size of Quantum Dots with Multifunctional Multidentate Polymer Ligands." *Journal of the American Chemical Society* 130 (34): 11278–11279.

———. 2009a. "Semiconductor Nanocrystals: Structure, Properties, and Band Gap Engineering." *Accounts of Chemical Research* 43 (2): 190–200.

———. 2009b. "Next-Generation Quantum Dots." *Nature Biotechnology* 27 (8) (August): 732–733.

———. 2012. "Compact Quantum Dots for Single-Molecule Imaging." *Journal of Visualized Experiments : JoVE* (68): e4236.

Smith, Andrew M., Mary M. Wen, and Shuming Nie. 2010. "Imaging Dynamic Cellular Events with Quantum Dots the Bright Future." *Biochemist* 32 (3) (June): 12.

So, Min-kyung, Hequan Yao, and Jianghong Rao. 2008. "HaloTag Protein-Mediated Specific Labeling of Living Cells with Quantum Dots." *Biochemical and Biophysical Research Communications* 374 (3): 419–423.

Tada, Hiroshi, Hideo Higuchi, Tomonobu M. Wanatabe, and Noriaki Ohuchi. 2007. "In Vivo Real-Time Tracking of Single Quantum Dots Conjugated with Monoclonal Anti-HER2 Antibody in Tumors of Mice." *Cancer Research* 67 (3): 1138–1144.

Uttamapinant, Chayasith, Katharine A. White, Hemanta Baruah, Samuel Thompson, Marta Fernández-Suárez, Sujiet Puthenveetil, and Alice Y. Ting. 2010. "A Fluorophore Ligase for Site-Specific Protein Labeling Inside Living Cells." *Proceedings of the National Academy of Sciences U.S.A.* 107 (24): 10914–10919.

Watanabe, Tomonobu M., Shingo Fukui, Takashi Jin, Fumihiko Fujii, and Toshio Yanagida. 2010. "Real-Time Nanoscopy by Using Blinking Enhanced Quantum Dots." *Biophysical Journal* 99 (7): L50–L52.

Wu, Xingyong, Hongjian Liu, Jianquan Liu, Kari N. Haley, Joseph A. Treadway, J. Peter Larson, Nianfeng Ge, Frank Peale, and Marcel P. Bruchez. 2002. "Immunofluorescent Labeling of Cancer Marker Her2 and Other Cellular Targets with Semiconductor Quantum Dots." *Nature Biotechnology* 21 (1): 41–46.

Xing, Yun, Qaiser Chaudry, Christopher Shen, Koon Yin Kong, Haiyen E. Zhau, Leland W. Chung, John A. Petros, et al. 2007. "Bioconjugated Quantum Dots for Multiplexed and Quantitative Immunohistochemistry." *Nature Protocols* 2 (5): 1152–1165.
Xu, Jianmin, Piotr Ruchala, Yuval Ebenstain, J. Jack Li, and Shimon Weiss. 2012. "Stable, Compact, Bright Biofunctional Quantum Dots with Improved Peptide Coating." *Journal of Physical Chemistry B* 116 (36): 11370–11378.
Zhang, Pengfei, Shuhui Liu, Duyang Gao, Dehong Hu, Ping Gong, Zonghai Sheng, Jizhe Deng, Yifan Ma, and Lintao Cai. 2012. "Click-Functionalized Compact Quantum Dots Protected by Multidentate-Imidazole Ligands: Conjugation-Ready Nanotags for Living-Virus Labeling and Imaging." *Journal of the American Chemical Society* 134 (20): 8388–8391.
Zhang, Ruobing, Eli Rothenberg, Gilbert Fruhwirth, Paul D. Simonson, Fangfu Ye, Ido Golding, Tony Ng, Ward Lopes, and Paul R. Selvin. 2011. "Two-Photon 3D FIONA of Individual Quantum Dots in an Aqueous Environment." *Nano Letters* 11 (10): 4074–4078.

Problems

1. Plot the wavelength (in nanometers) of excitonic light absorption and emission versus nanoparticle radius (in nanometers) for a CdSe QD for radii between 1 and 10 nm.
2. (a) What four key properties of QDs set them apart from traditional fluorophores? (b) Given that larger-scale imaging systems can use excitation powers that are orders of magnitude smaller per QD than the requirements for single-molecule-scale systems, which of these four properties would scale up in a beneficial way for imaging at larger length scales?
3. What are the key differences between current QDs and next-generation QDs?
4. Given the following analytic solution for an MSD plot, $MSD(t) = 1 - e^{-0.5t}$ [μm^2] where time increment t is in units of seconds, (a) does this MSD plot indicate directed motion, confined motion, or Brownian diffusion? (b) Calculate the diffusion coefficient.
5. Describe how two biological fields have benefited from recent advances in QD technology.
6. In Figure 6.10, an adaptation of GSDIM with QDs is demonstrated. Would an imaging process like this allow long-term tracking of individual particles in living cells?

chapter seven

Fluorescence detection and lifetime imaging with stimulated emission

Po-Yen Lin, Jianhong Ge, Cuifang Kuang, and Fu-Jen Kao

Contents

7.1 Introduction

The need to visualize objects that cannot be observed by the naked eye motivates the development of microscopy. During the past few decades, numerous improvements have been made in the field. In the 1930s, the first electron microscope was invented. It provided extremely high-resolution images (Erni et al. 2009). In 1981, a scanning tunneling microscope yielded images with atomic resolution (Binning et al. 1993). These imaging platforms have demonstrated superior resolving power for identifying objects on the atomic scale. However, the imaging conditions of these microscopes are unsuitable for the observation of biological samples. For example, the sample environment in an electron microscope typically involves a vacuum that prevents its application to a living sample. The optical microscope, however, provides a flexible sample environment and is an essential tool in both biology and medical research (Miyawaki et al. 2003; Tsien 2003; Periasamy and Clegg 2009).

Fluorescence microscopy has been extensively applied in biology and medicine owing to its high sensitivity and specificity. As a noninvasive imaging platform, it is used not only to visualize the morphology of biological samples, but also to elucidate the dynamics of biomolecules. The basic concept of fluorescence microscopy is the use of a light source with high energy to excite fluorescent molecules into an excited state. These excited molecules then fall to their ground state, emitting a photon of lower energy. If a suitable set of filters is chosen, these emitted photons can be extracted from the incident light, improving the image contrast. The use of fluorescence in imaging has led to the identification of cells and cellular organelles with a high specificity in transparent biomaterial. In fact, the fluorescence microscope can be used to visualize single molecules (Xie and Dunn 1994).

Fluorescence emission is a spontaneous process in which emitted photons are distributed among all solid angles (4 Pi). Accordingly, fluorescence is typically detected using high numerical aperture (NA) optics to provide a high spatial resolution and to ensure the effective collection of the spontaneously emitted photons. Although high NA optics allows efficient collection by covering a large solid angle, the working distance and the depth of focus are limited accordingly. This limited working distance prevents long-distance fluorescence detection and shrinks the observation environment.

In the following, the use of stimulated emission to detect fluorescent molecules is discussed. Stimulated emission (SE) detection has been demonstrated to be an effective method for detecting fluorescent or nonfluorescent molecules (Min et al. 2009). The underlying principle of SE detection is to force an electronic transition of an excited fluorophore to stimulate emission rather than a radiative or nonradiative decay process, by increasing the intensity of the incident stimulation beam to an extent that can be measured using the lock-in detection method. Hence, the SE signal depends on the number of molecules in the excited state (concentration), the pulse separation between the excitation and stimulated beams (fluorescence lifetime), and the polarization states of the incident laser beam (molecular orientation). It also has an overall quadratic dependence on power, enabling three-dimensional optical sectioning in imaging (Min et al. 2009). Additionally, the spatial coherence of stimulated fluorescence emission is maintained, resulting in emission in a narrow cone in the forward direction, as presented in Figure 7.1. This characteristic can be exploited in the imaging of fluorescent molecules with long working distances (Dellwig et al. 2010).

The de-excitation aspect of stimulated emission allows the fluorescence lifetime of a target molecule to be probed by adjusting the time delay between excitation and stimulation pulses (Dong et al. 1995). Fluorescence lifetime imaging microscopy (FLIM) is an effective tool for obtaining critical information on molecular dynamics and it is used extensively in the

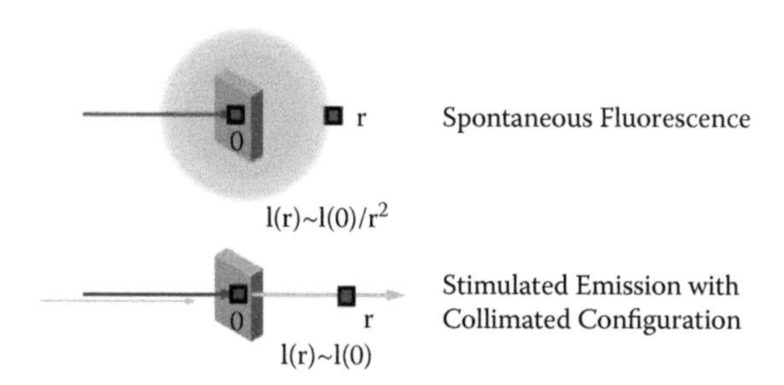

Figure 7.1 Stimulated emission detection diagram. Spatial coherence of stimulated emission results in a narrow emission cone.

determination of pertinent cellular parameters and tissue characteristics, including ion concentration (Hille et al. 2009) and the pH of the environment (Hanson et al. 2002). It is also used, for example, in the identification of carious dental tissue (Lin et al. 2011). FLIM measurements are generally made by detecting spontaneously emitted photons using a method that operates in the time or frequency domain (Sun et al. 2011). In this chapter, fluorescence lifetime measurements are made by detecting a stimulated emission. The temporal resolution of this method (pump-probe scheme) is typically determined by the pulse width of the laser. The fluorescence lifetime images that are based on stimulated emission detection are obtained using a scanning probe laser beam, which dwells on each pixel before moving to the next, or they are obtained as entire images, which together constitute a series of time slice images. The basic fluorescence and stimulated emission processes are introduced below.

7.1.1 Characteristics of fluorescence

Fluorescence occurs when a system (such as an atom or molecule) is excited to a higher energy level by the absorption of a photon, and then spontaneously returns to a lower energy level with the emission of a photon. Basic principles of fluorescence can be elucidated using the Jablonski diagram, which was named after the Polish physicist Aleksander Jablonski (Lakowicz 2009).

The diagram in Figure 7.2 presents the electronic states of a molecule and the transitions between them. The electronic states are arranged vertically in order of increasing energy. Following the absorption of a photon, many processes occur with varying probabilities, but relaxation to the lowest vibrational energy level of the first excited state is the most likely process. This process is known as internal conversion or vibrational relaxation and typically occurs in a picosecond or less. An excited molecule exists in the lowest excited singlet state (S_1) for periods of the order of nanoseconds (for most organic dyes) before finally relaxing to the ground

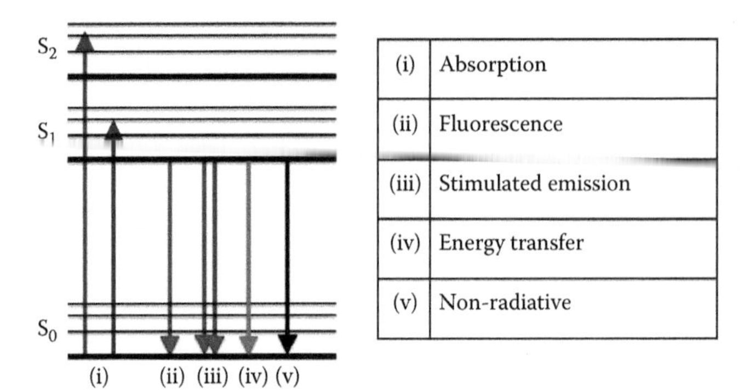

Figure 7.2 Simplified Jablonski energy diagram. The several possible energy relaxation pathways for excited fluorophores include spontaneous emission (fluorescence), stimulated emission, nonradiative process, and energy transfer to other fluorophores.

state (S_0). If relaxation from S_1 to S_0 states is accompanied by the emission of a photon, then the process is formally known as fluorescence or radiative decay. In a process that differs from fluorescent decay, energy relaxation from S_1 to S_0 can also occur nonradiatively via vibrations or collisions. The ratio of radiative transition rate k_r to the total decay rate $(k_r + k_{nr})$ is called the quantum efficiency,

$$Q = \frac{k_r}{k_r + k_{nr}}, \tag{7.1}$$

where k_{nr} is the nonradiative decay rate. For most organic dyes used as fluorescent labels, the quantum efficiency generally exceeds 50%; for example, for fluorescein it is 90%. In contrast, many endogenous fluorophores in cells/tissue have low quantum efficiency owing to the dominant nonradiative decay. Such nonrediative decay includes dynamic collisional quenching, resonance energy transfer, internal conversion, and intersystem crossing. Therefore, a change in the rate associated with any pathway will influence both the fluorescence quantum yield and the excited state lifetime. For example, the fluorescence lifetime of some dyes has been used as a sensitive metric of the concentration of oxygen in solutions, for example, pyrenes, lanthanides, or phosphorescent dyes. Fluorophores in the excited state interact with oxygen molecules, quenching their fluorescence and reducing the fluorescence lifetime in a concentration-dependent manner, as described by the Stern-Volmer relationship. The population in the excited state (S_1) decays according to

$$\frac{dS_1}{dt} = (k_r + k_{nr})S_1. \tag{7.2}$$

This phenomenon is an exponential decay of the excited state population, $S_1(t) = S_{1ini} \exp(-t/\tau)$. In fluorescence measurement, the intensity is proportional to the population of the excited state S_1, so the fluorescence intensity is given by

$$I(t) = I_0 e^{(-t/\tau)} \tag{7.3}$$

where I_0 is the intensity at time zero. The lifetime τ is the reciprocal of the total decay rate, $\tau = (k_r + k_{nr})^{-1}$. Generally, the reciprocal of the lifetime is the sum of the rates of depopulation of the excited state. The fluorescence lifetime is an intrinsic property of a fluorophore, and it provides information on its local environment, because k_{nr} varies with its environment.

7.1.2 Stimulated emission

In a two-level system, the interaction of light with matter is described by Einstein's model of absorption, spontaneous emission, and stimulated emission. Using Einstein's coefficient, the lifetime, known as the intrinsic lifetime, is expressed as

$$A_{10} = \frac{8\pi h}{\lambda^3} B_{01} = \frac{1}{\tau_n}$$

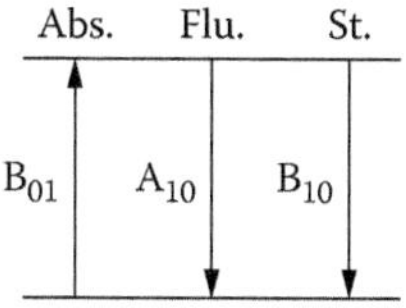

Abs. Flu. St.

B_{01} A_{10} B_{10}

where τ_n denotes the lifetime without any nonradiative decay; λ is the wavelength of absorption/emission; h is Planck's constant, and A_{10} and B_{01} are Einstein's coefficients of fluorescence and absorption, respectively. Stimulated emission is the process by which an excited atom or molecule interacts with incident light of the correct wavelength, and then emits a photon of the same wavelength, phase, and polarization as those of the incident light. Stimulated emission provides an energy relaxation pathway via which excited molecules can return to their ground state by spontaneous emission (fluorescence process). Therefore, another laser beam can be used to cause the excited fluorophores to enter a nonfluorescent state (dark state) by stimulated emission. STED microscopy is based on this mechanism (Hell 2002). In the STED process, stimulated emission can also alter the fluorescence lifetime of fluorophores. S_1 is a solution to the following equations (Vicidomini et al. 2011).

$$\begin{aligned} \frac{dS_0}{dt} &= -k_{exc}S_0 + (k_r + k_{\mathrm{STED}})S_1 \\ \frac{dS_1}{dt} &= +k_{exc}S_0 - (k_r + k_{\mathrm{STED}})S_1 \end{aligned} \tag{7.4}$$

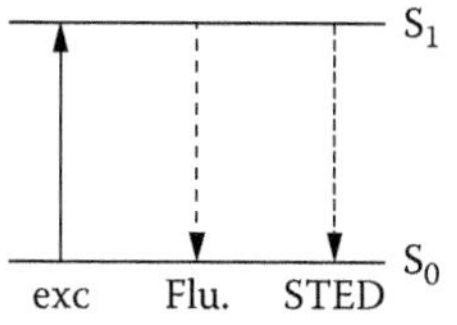

S_1 S_0

exc Flu. STED

where $k_r = 1/\tau$ is the spontaneous emission rate, which is determined by the excited-state lifetime τ in the absence of STED light; $k_{STED} = \sigma_{STED} I_{STED}$ is the stimulated emission rate, which is determined by the simulated emission cross-section σ_{STED} and the intensity of the STED beam I_{STED}, and $k_{exc} = \sigma_{abs} I_{exc}$ is the excitation rate, which is a function of the absorption cross-section and the intensity of the excitation beam. Therefore, in the presence of a STED beam, the excited-state lifetime $\tau = 1/(k_r + k_{STED})$ is shortened as the intensity of the STED beam increases. This feature of STED has also been used to improve the resolution of STED microscopic images under a low-intensity STED beam (Vicidomini et al. 2011). Another difference between spontaneous and stimulated emission is that the spontaneous process results in emission in the form of a dipole radiation pattern, whereas the stimulated process leads to emission in a narrow cone in the forward and backward directions (Boyd 2003). Therefore, the spatial coherence of stimulated emission can be exploited in imaging with a long working distance, including fluorescence and stimulated Raman scattering. In this context, stimulated emission has been used to improve the effectiveness of time-resolved microscopy, such as long working distance imaging.

7.1.3 Pump-probe (dual-beam) technique

Fluorescence lifetime measurement techniques can be classified as time-domain (TD) and frequency domain (FD) techniques (Gratton et al. 2003; Buranachai et al. 2008). FLIM, based on these two methods, is routinely used for monitoring molecular dynamics in living cells, involving the detection of protein-protein interactions. Along with the TD and FD methods, the pump-probe technique can also be used for making time-resolved measurements (Hamilton et al. 1986; Dong et al. 1995). In this technique, two very short laser pulses are used as the pump beam in succession to excite fluorophores, and a probe beam is then used to monitor the excited states of the molecules. The temporal evolution of the molecular dynamics can be studied by varying the pulse delay between pump and probe pulses. The temporal resolution of this method is limited by the pulse width of the laser, and usually of the order of picoseconds to femtoseconds, making it superior to that achieved using conventional time-correlated single photon counting (TCSPC) modules. Various pump-probe techniques exist, but the main idea of each is to use separated pulses to study the ultrafast dynamics. However, the disadvantage of the pump-probe approach is the need for high-power pulsed laser sources that would cause photobleaching and photodamage. To solve these problems, the probe beam can be tuned to the proper wavelength to induce stimulated emission, resulting in the amplification of the probe beam and fluorescence quenching. The measurement of the gain of the probe beam and the loss of fluorescence as functions of the pulse delay can reveal the excited state dynamics of fluorophores. However, the setups for measuring these two signals differ considerably.

For fluorescence quenching, the probe beam, also called the STED beam, is used to quench spontaneous emission by stimulated emission (Gryczynski et al. 1997). STED pulses must arrive before the emission of the excited fluorophore. Accordingly, fluorescence quenching depends on the separation between the excitation and STED pulses. The effect of various time delays t_d on the extent of fluorescence quenching should reveal the decay dynamics of the fluorophore. The changes in intensity are measured as $(I_0 - I)/I_0$, where I_0 and I are the intensities in the absence and presence of the probe beam (Gryczynski et al. 1997), and

$$I_0 - I \big/ I_0 = q\exp(-t_d/\tau) \tag{7.5}$$

This method is still based on fluorescence measurement and depends on a highly sensitive detector, such as photomultiplier tube (PMT) or avalanche phototide (APD). The measurement of the gain of the probe beam must be made differently owing to the high background that is generated by the probe beam. According to Einstein's coefficient, the cross-section for stimulated emission is comparable to the absorption cross-section. Interaction with the excited fluorophores increases the number of photons in the probe (stimulated) beam. The gain of the stimulated beam is given by

$$dI_s/I_s = N_2\sigma/A \tag{7.6}$$

where N_2 is the number of excited fluorophores; σ is the stimulated emission cross-section (~10^{-16} cm^2); and A is the waist area of the beam (~10^{-9} cm^2). For a single molecule, the gain of the stimulated beam is approximately 10^{-7}. This small signal is difficult to measure directly because it is buried in the noise of the stimulated beam. To solve this noise problem in the measurement of the stimulated photon, the lock-in technique is used to detect the intensity change of the stimulated beam that is caused by the stimulated photons. In lock-in detection, the intensity of the excitation beam is modulated sinusoidally with a fixed frequency ω, resulting in the modulation of the stimulated beam at the same frequency, since the population in excited state N_2 depends on the intensity of the excitation beam.

Accordingly, the gain of the stimulated beam can be extracted using a lock-in amplifier. The population of the excitation state can be determined by adjusting the delay between the excitation and stimulated pulses. Additionally, stimulation emission can be used for imaging nonfluorescent fluorophores by the competition between stimulated emission and nonradiative radiation (Min et al. 2009). In this chapter, the spatial coherence of stimulated emission is exploited in imaging with a long working distance.

7.2 Setup for stimulated emission imaging with long working distance

Figure 7.3 schematically depicts fluorescence lifetime detection with stimulated emission. A pulsed diode laser (Picoquant, Berlin, Germany) with a wavelength of 635 nm is the excitation source, which is synchronized with the mode-locked Ti:sapphire laser (Mira900, Coherent, USA) through a fast photodiode (TDA 200, Picoquant, Berlin, Germany). The Ti:sapphire laser, as the stimulation beam, is operated at 740 nm with a repetition rate of 76 MHz. The selected wavelengths are highly consistent with the absorption and emission spectra of the ATTO 647N dye (ATTO-TEC, Germany). In particular, the wavelength of the stimulation beam is set to match the red-shift region of the ATTO 647N emission spectrum, to prevent re-excitation of the ATTO 647N.

Based on the fluorescence lifetime of the targeted fluorophores, various time delay schemes may be used. For species with very short fluorescence lifetimes (of the order of picoseconds or shorter), the use of a precise mechanical translation stage for optical path adjustment is preferred, as in the conventional pump-probe technique. For fluorophores with a longer fluorescence lifetime (of the order of nanoseconds or longer), the electronic adjustment of the time delay is preferred. The delay can be electronically controlled by adjusting the length of the cable that connects

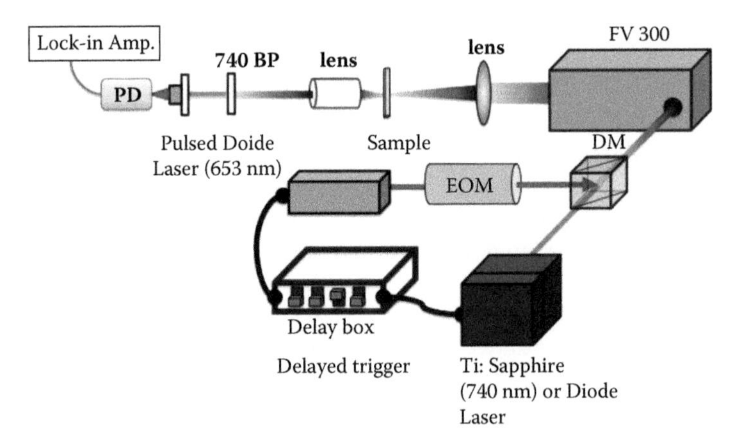

Figure 7.3 **(See color insert.)** Setup for low-NA stimulated emission imaging. EOM, electro-optical modulator; BP, band pass filter; DM, dichroic mirror; PD, photodiode.

the fast photodiode to the trigger of the pulsed diode laser driver, providing a delay step of approximately 0.12 ns. The transition time for the commonly used 50-ohm cable is approximately 5 ns per meter. Changing the length of the cable may be a time-consuming method for adjusting the pulse delay, but it offers a stable and cost-effective means of measuring fluorescence lifetime. Additionally, a programmable electronic delay improves the acquisition time. In such a configuration, the temporal resolution is limited by timing jittering (~100 ps) rather than pulse width (van Munster and Gadella 2005). Notably, the mechanical and electronic control of the pulse delay can be combined.

An all-semiconductor laser system for SE imaging has been demonstrated (Ge et al. 2013). A pair of synchronized gain switched diode lasers with wavelengths of 635 nm and 700 nm serves as excitation and stimulation light sources, respectively. The fully electronic configuration provides highly flexible and versatile FLIM imaging. Additionally, the variable repetition rate allows measurements to be made of fluorescent samples with a wide range of lifetimes. In the SE detection experiment, the critical alignment step ensures the spatial and temporal overlap of the excitation and stimulation pulses. Spatial alignment is easily achieved by observing the overlapping of the focus spots of the two beams using a CCD camera. To ensure temporal overlap, a TCSPC module is used to determine the arrival times of the pulses by measuring an instrument response function (IRF).

For imaging, the excitation and stimulation beams were coupled collinearly using a dichroic mirror (720SP, Semrock) and fed into a commercially available laser scanning unit (FV300, Olympus). Two objectives with low NA (10×, NA 0.3) were separately used to focus and collect the transmitted stimulation signals. The collecting objective was placed slightly out of focus to minimize the collection of spontaneous fluorescence. The transmission beam that carried the stimulated emission photons was detected using a biased Si detector (DET36A, Thorlab) with a large area of detection to compensate for the motion of the beam that was caused by the scanning of the galvo-mirror. The apertureless detection scheme prevented the formation of any artifact in the pump-probe measurement, such as by the thermal lens effect (Lu et al. 2010). Bandpass filters (747/33, Semrock) and a neutral-density filter were used in combination to transmit the stimulation beam only, and to reduce the intensity of the stimulation beam to prevent saturation of the photodiode. Notably, direct measurement of stimulated emission photons was difficult because of the very high background SE laser intensity. Therefore, to detect the stimulated emission signal, lock-in detection was used to recover the stimulated emission photons that were carried in the stimulation beam. To this end, the intensity of the excitation beam was modulated using an electro-optical modulator (EOM, ConOptics) at 30 kHz, which was driven by a function generator and provided a reference frequency for a lock-in amplifier (SR830, Stanford Research USA). The time constants of the lock-in amplifier were set to 1 ms and 300 μs to capture the images and measure the lifetime, respectively. The authenticity of the signal is confirmed by blocking the excitation and/or the stimulation beams, as well as changing the relative time delay between the two pulses. In particular, the time delay examination is more critical to confirming the authenticity of the SE signal, because the signal that results from the fluorescence or thermal lens effect yields the same feature as obtained in the blocking test. The output of the lock-in amplifier was fed into the A/D converter of the laser scanning system (FV300) and Fluoview software was used to reconstruct the image. Notably, subharmonically synchronized laser pulses were used for SE detection. This method reduces the effect of laser intensity noise by means of synchronizing two pulse lasers with subharmonic repetition rates. For example, Yb-fiber laser pulses at a repetition rate of 38 MHz have been synchronized to Ti:sapphire laser pulses at a repetition rate of 76 MHz and used in SRS microscopy with the shot noise sensitivity (Ozeki et al. 2010).

The above methods that involve a photodiode suppress the laser intensity noise to improve sensitivity. Alternatively, a balance detector can be used to measure the SE photons (Rittweger et al. 2007). A balance detector measures the difference between the intensities of two laser beams. To do so, the stimulation beam is divided into two beams, the probe beam and the reference beam. Once stimulated emission occurs, the gain of the stimulation beam, which is the difference between the intensities of the probe and reference beams, can be measured using a balance detector.

7.3 Characteristics of the stimulated emission signal

A stimulated emission signal is produced when the excitation beam interacts with molecules. The gain of the stimulated beam intensity is expressed as

$$\Delta I_s = N_2 \sigma_{st} I_s / A \propto N_1 \, I_{exc} \, I_s, \tag{7.7}$$

where N_1 and N_2 are the numbers of fluorophores in ground and excited states, respectively; σ_{st} is the stimulated emission cross-section (~10^{-16} cm^2), and A is the waist area of the beam (~10^{-9} cm^2).

Figure 7.4 plots the dependence of the stimulated emission signal on the power of the excitation and stimulation beams. In Figure 7.4a, the stimulated emission signal increases linearly with the excitation power. This linear dependence reveals that the nonsaturation condition was satisfied at the excitation intensity used (~10^4 W/cm^2). The nonsaturation condition also prevented the photobleaching effect. The saturation intensity

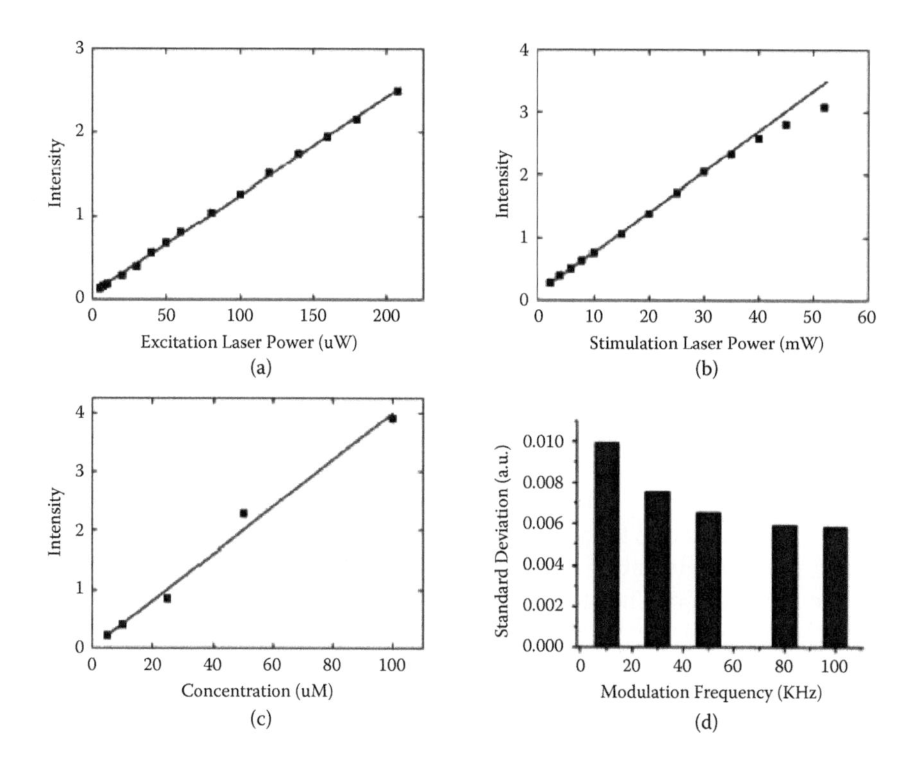

Figure 7.4 Dependence of stimulated emission signal on (a) power of excitation beam; (b) power of stimulation beam; (c) concentration. The sensitivity limit for concentration in this experiment is approximately a few micromoles. (d) The standard deviation of the signal as a function of modulation frequency on excitation.

of organic dyes is approximately ~MW/cm^2, which greatly exceeds that used here. The ATTO 647N dye has been extensively used in stimulated emission depletion (STED) microscopy and exhibits excellent photostability under highly intense illumination. Figure 7.4b plots the dependence of SE signal on stimulation power. When the stimulation beam irradiates excited ATTO 647N molecules, stimulated emission or up-conversion occurs and contributes to the signal. However, stimulated emission has been demonstrated to be the dominant fluorescence-quenching process for ATTO 647N dyes (Rittweger et al. 2007). Therefore, the effect of up-conversion is insignificant. Initially, the signal depends linearly on power. When the power of the stimulation beam exceeds 30 mW (0.6 GW/cm^2), stimulated emission begins to saturate, because of depletion of the excited state. The fluorescence depletion efficiency is defined as a function of the intensity of the stimulation beam according to $\eta(r) = \exp[-\sigma I_{ST}(r)T_{ST}]$, where T_{ST} is the duration of the stimulation pulse, and σ and $I_{ST}(r)$ are the cross-section of the stimulated emission and the intensity of the stimulation pulse, respectively (Westphal and Hell 2005). Therefore, the stimulated emission signal as a function of intensity is $1 - \eta(r)$. The saturation condition observed provides, in principle, diffraction-unlimited imaging, as revealed by saturation excitation microscopy (Fujita et al. 2007). Figure 7.4c also plots the dependence of the stimulated emission signal on the dye concentration. The mean powers of excitation and stimulation are 100 μW and 20 mW, respectively. The linear relationship supports quantitative analysis, as reported elsewhere in relation to stimulated Raman scattering (SRS) microscopy (Freudiger et al. 2008). Notably, in the setup for stimulated emission imaging with a long working distance, the image resolution is traded off for increased working distance of the imaging system.

The sensitivity of SE measurement with lock-in detection can be improved by increasing the modulation frequency. The noise in the laser intensity is typically present at low frequencies (from DC to kilohertz), and is the so-called $1/f$ noise. As the modulation frequency increases, the noise is suppressed to the shot noise level. Figure 7.4 plots the dependence of the standard deviation (or fluctuation) of the intensity of the signals on the modulation. The standard deviation of the signal declines as the modulation frequency increases. In the proposed system, the noise reduction is strongest at a frequency of 80~100 kHz.

7.4 *Time-resolved imaging using stimulated emission*

The time-resolved images shown in Figure 7.5 were captured using various time delays in steps of 0.5 ns. Since the stimulated emission signal is proportional to the population of excited molecules, the fluorescence decay curves were obtained by reading out the same pixels across the time delay images.

Figure 7.5a plots the intensity of the stimulated signal as a function of pulse separation. The gain of the stimulated beam is consistent with the quenching of fluorescence. In the light quenching experiment, the intensity of fluorescence as a function of relative delay is given by Equation (7.5). Therefore, the stimulated emission signal as a function of the delay between the excitation and stimulation pulses is $\Delta I_s \propto \exp(t_d/\tau)$, where t_d and τ are the pulse separation and fluorescence lifetime of the fluorophore, respectively. Fitting the curve using the exponential decay function yields a fluorescence lifetime of 3.0 ns. For comparison, the fluorescence lifetime of ATTO 647 N has been reported to be 1.8 ns or 3.4 ns (Kolmakov et al. 2010; Bückers et al. 2011). Such a long lifetime of a few nanoseconds is difficult to measure using the conventional pump-probe configuration, which involves the use of a mechanical moving stage to adjust the delay between pulses. Accordingly, the electronic delay provides a highly

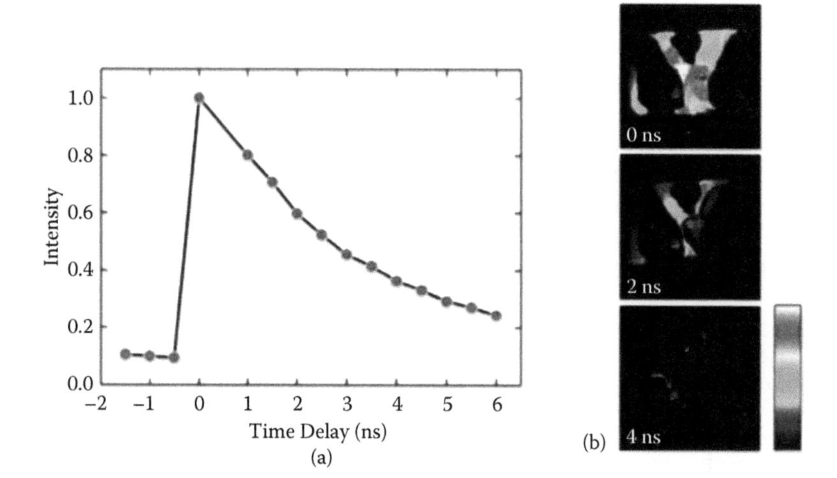

Figure 7.5 **(See color insert.)** Time-resolved fluorescence images. (a) Stimulated signal versus time delay between excitation and stimulation pulses. (b) Stimulated emission images of ATTO 647 N sample injected into Y-shape microfluidic sample with various interpulse delays. Scanned size: 600 μm × 600 μm.

effective method of probing the changing population of the excited state. Figure 7.5b plots the stimulated emission images at various time delays. The ATTO 647 N sample was dissolved in gel solution and injected into a Y-shaped microfluidic channel for imaging. The image acquisition time is approximately 10 minutes, and could be improved by increasing the modulation frequency. The maximum-intensity image was obtained at zero time delay, as verified by TCSPC measurement. The image intensity declined as the time delay increased from 0 ns to 6 ns. The stimulated emission signals depended on the population of the excited state following the arrival of the stimulation pulse, and this dependence was used to distinguish specimens with various fluorescence lifetimes. Additionally, the low-NA objective that was used supported stimulated emission imaging with a long working distance.

7.5 Polarization dependence of stimulated emission

The efficiency of stimulated emission depends on many factors. First, the wavelength of the stimulation beam must be sufficiently red-shifted from the absorption spectrum to prevent the excitation of fluorescence from unexcited molecules. A second consideration is that the temporal separation between the excitation and stimulation pulses must be short enough to minimize the effects of orientational averaging within the distribution of excited states. Additionally, the stimulated emission process is optimized for parallel excitation and stimulation field polarization, which is determined by the angular distribution of molecular transition dipole moments ($\sim \cos^2 \theta$). When a fluorophore is illuminated under polarized light, the molecule whose absorption transition moment is aligned parallel to the incident electric field exhibits the highest probability of absorption. Therefore, the stimulated emission signal depends on the time-dependent angular distribution of the transition moments of the excited molecules. In the experiment on time-resolved anisotropy, the anisotropy is defined as

$$r(t) = \frac{I_{\parallel}(t) - GI_{\perp}(t)}{I_{\parallel}(t) + G2I_{\perp}(t)}, \tag{7.8}$$

where $I_{\parallel}(t)$ and $I_{\parallel}(t)$ are the degrees of fluorescence intensity decay parallel and perpendicular to the polarization of the excitation, respectively. G takes into account the sensitivities of the detection system for detecting the two components. When a single detector system is used, the G factor must always be determined because the rotation of the emission polarizer can change the position of the focused image of the fluorescence, affecting the effective sensitivity. In the polarization-stimulated emission imaging system described here, no polarizer is used in the detection part. The parallel and vertical components are probed by changing the polarization of the stimulation beam. Figure 7.6 plots the time-resolved anisotropy measurements of ATTO 647 N. Since the stimulated emission depends on the population in the excited state, the polarization-resolved SE signal reveals the angular distribution of the transition moments of the excited molecules. The low-NA optics used also prevents the depolarization effect that would otherwise occur in high-NA optics (Török et al. 1998). Fluorescence

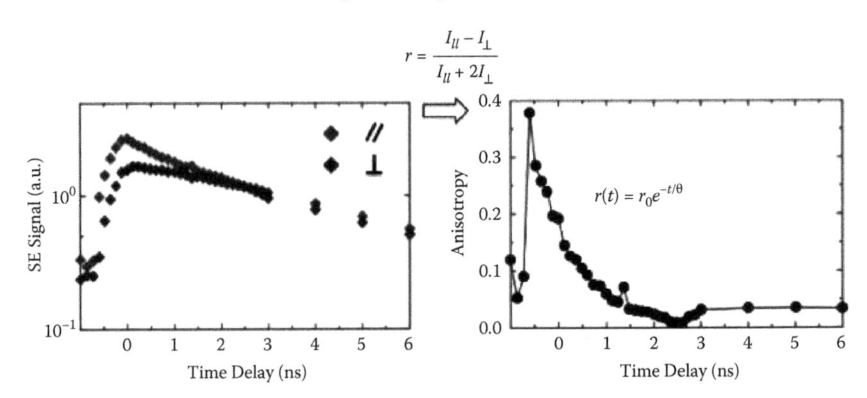

Figure 7.6 Parallel (red) and perpendicular (blue) polarization-stimulated emission signal as a function of time delay. The calculated anisotropy decay curve is plotted on the right.

Table 7.1 Comparison of Fluorescence and Stimulated Emission Imaging Methods

	Fluorescence imaging	Stimulated emission imaging
Characteristic	Incoherent Isotropic emission	Coherent Forward emission
Working distance	Limited (~millimeter)	Long (~ meter)
Lifetime imaging	Yes	Yes
Cost	Expensive electronics	Cost-effective
Temporal resolution	Moderate (~100 ps)	High (~100 fs or limited by pulse width) Or electronics jittering limited
Anisotropy	Yes	Yes
Dark fluorophores	No	Yes
Extendibility		Extendible to SRS (chemical imaging is possible) and other coherent optical processes

anisotropy has been widely used to measure the binding constants and kinetics of reactions that change the rotational time of molecules (Levitt et al. 2011). The rotational diffusion of fluorophores dominates the depolarization of the fluorescence. Depolarization by the rotational diffusion of spherical rotors satisfies the Perrin equation:

$$\frac{r_0}{r} = 1 + \frac{\tau}{\theta} = 1 + 6D\tau \tag{7.9}$$

where τ is the fluorescence lifetime; θ is the rotational correlation time, and D is the rotational diffusion coefficient. The binding of fluorophore to a molecule significantly changes the rotational correlation time. Therefore, the degree of binding can be calculated from the variation of the anisotropies of the partially bound, free, and fully bound (large excess of protein) states, measured by titrating the binding partners.

Table 7.1 compares the fluorescence and simulated emission imaging methods. Though stimulated emission imaging has been demonstrated to be versatile in imaging and so useable for a diverse range of purposes, single-molecule sensitivity for stimulated emission imaging has not yet been achieved.

7.6 Summary

The feasibility of using an electronic trigger control in pump-probe-based stimulated emission imaging is demonstrated. In this method, the change in the time delay between excitation and stimulation pulses is exploited to probe the fluorescence lifetime of ATTO 647 N. The approach also prevents optical distortion when the pulse delay is adjusted with a long optical path displacement. An imaging system that uses electronic synchronization and gain-switched diode lasers as light sources can be very compact and robust. Stimulated emission supports imaging with a long working distance owing to its inherent spatial coherence. Stimulated emission imaging can be realized using a low-NA optical imaging system that can be operated in a wide range of environments. Depletion of the excited states is responsible for saturation of the stimulated emission signal at high stimulation intensity. This nonlinearity indicates a means of improving the resolution in a stimulated emission imaging system.

One interesting application of stimulated emission is the characterization of the dark fluorophores in living tissue, such as hemoglobin and cytochromes. Fast nonradiative decay dominates the energy relaxation of these dark fluorophores, and so the use of fluorescence detection methods is very difficult. SE detection can lighten these dark fluorophores and make them visible (Min et al. 2009). The application of SE imaging with an all-semiconductor laser configuration enables a low-cost, high-throughput characterization apparatus to be established. Moreover, SE detection can also be applied to optical coherence tomography (OCT) by incorporating the making of interferometric measurements. This approach, SE detection, with improved sensitivity and a simplified setup will be potentially effective in biomedical research and may provide unprecedented opportunities.

References

Bückers, J., D. Wildanger, et al. (2011). "Simultaneous multi-lifetime multi-color STED imaging for colocalization analyses." *Optics Express* 19(4): 3130–3143.

Binning, G., H. Rohrer, et al. (1993). Surface studies by scanning tunneling microscopy. H. Neddermyer, ed., 31–35 *Scanning Tunneling Microscopy*, Springer: 31–35.

Boyd, R. W. (2003). *Nonlinear Optics*. New York: Academic Press.

Buranachai, C., D. Kamiyama, et al. (2008). "Rapid frequency-domain FLIM spinning disk confocal microscope: lifetime resolution, image improvement and wavelet analysis." *Journal of Fluorescence* 18(5): 929–942.

Dellwig, T., M. R. Foreman, et al. (2010). "Coherent long-distance signal detection using stimulated emission: a feasibility study." *Chinese Journal of Physics* 48(6): 873–884.

Dong, C., P. So, et al. (1995). "Fluorescence lifetime imaging by asynchronous pump-probe microscopy." *Biophysical Journal* 69(6): 2234–2242.

Erni, R., M. D. Rossell, et al. (2009). "Atomic-resolution imaging with a sub-50-pm electron probe." *Physical Review Letters* 102(9): 096101.

Freudiger, C. W., W. Min, et al. (2008). "Label-free biomedical imaging with high sensitivity by stimulated Raman scattering microscopy." *Science* 322(5909): 1857–1861.

Fujita, K., M. Kobayashi, et al. (2007). "High-resolution confocal microscopy by saturated excitation of fluorescence." *Physical Review Letters* 99(22): 228105.

Ge, J., Y. Wang, et al. (2013). "Fluorescence lifetime imaging with pulsed diode laser enabled stimulated emission." SPIE BiOS, International Society for Optics and Photonics.

Gratton, E., S. Breusegem, et al. (2003). "Fluorescence lifetime imaging for the two-photon microscope: time-domain and frequency-domain methods." *Journal of Biomedical Optics* 8(3): 381–390.

Gryczynski, I., S. W. Hell, et al. (1997). "Light quenching of pyridine2 fluorescence with time-delayed pulses." *Biophysical Chemistry* 66(1): 13–24.

Hamilton, C. E., J. L. Kinsey, et al. (1986). "Stimulated emission pumping: new methods in spectroscopy and molecular dynamics." *Annual Review of Physical Chemistry* 37(1): 493–524.

Hanson, K. M., M. J. Behne, et al. (2002). "Two-photon fluorescence lifetime imaging of the skin stratum corneum pH gradient." *Biophysical Journal* 83(3): 1682–1690.

Hell, S. (2002). "Increasing the resolution of far-field fluorescence light microscopy by point-spread-function engineering." *Topics in Fluorescence Spectroscopy* 5: 361–426.

Hille, C., M. Lahn, et al. (2009). "Two-photon fluorescence lifetime imaging of intracellular chloride in cockroach salivary glands." *Photochemical & Photobiological Sciences* 8(3): 319–327.

Kolmakov, K., V. N. Belov, et al. (2010). "Red-emitting rhodamine dyes for fluorescence microscopy and nanoscopy." *Chemistry-A European Journal* 16(1): 158–166.

Lakowicz, J. R. (2009). *Principles of Fluorescence Spectroscopy*. New York: Springer.

Levitt, J. A., P. H. Chung, et al. (2011). "Fluorescence anisotropy of molecular rotors." *ChemPhysChem* 12(3): 662–672.

Lin, P.-Y., H.-C. Lyu, et al. (2011). "Imaging carious dental tissues with multiphoton fluorescence lifetime imaging microscopy." *Biomedical Optics Express* 2(1): 149.

Lu, S., W. Min, et al. (2010). "Label-free imaging of heme proteins with two-photon excited photothermal lens microscopy." *Applied Physics Letters* 96(11): 113701–113703.

Min, W., S. Lu, et al. (2009). "Imaging chromophores with undetectable fluorescence by stimulated emission microscopy." *Nature* 461(7267): 1105–1109.

Miyawaki, A., A. Sawano, et al. (2003). "Lighting up cells: labelling proteins with fluorophores." *Nature Cell Biology* 5: S1–S7.

Ozeki, Y., Y. Kitagawa, et al. (2010). "Stimulated Raman scattering microscope with shot noise limited sensitivity using subharmonically synchronized laser pulses." *Optics Express* 18(13): 13708–13719.

Periasamy, A., and Robert M. Clegg (2009). "FLIM applications in the biomedical sciences." In *FLIM Microscopy in Biology and Medicine*, edited by A. Periasamy and Robert M. Clegg, 385. Boca Raton, FL: CRC Press.

Rittweger, E., B. Rankin, et al. (2007). "Fluorescence depletion mechanisms in super-resolving STED microscopy." *Chemical Physics Letters* 442(4): 483–487.

Sun, Y., R. N. Day, et al. (2011). "Investigating protein-protein interactions in living cells using fluorescence lifetime imaging microscopy." *Nature Protocols* 6(9): 1324–1340.

Török, P., P. Higdon, et al. (1998). "On the general properties of polarised light conventional and confocal microscopes." *Optics Communications* 148(4-6): 300–315.

Tsien, R. Y. (2003). "Opinion: imagining imaging's future." *Nature Reviews Molecular Cell Biology* 4: SS16–SS21.

van Munster, E. B. and T. W. Gadella (2005). "Fluorescence lifetime imaging microscopy (FLIM)." in *Microscopy Techniques*, edited by J. Rietdorf, 143–175. New York: Springer.

Vicidomini, G., G. Moneron, et al. (2011). "Sharper low-power STED nanoscopy by time gating." *Nature Methods* 8(7): 571–573.

Westphal, V. and S. W. Hell (2005). "Nanoscale resolution in the focal plane of an optical microscope." *Physical Review Letters* 94(14): 143903.

Xie, X. S. and R. C. Dunn (1994). "Probing single molecule dynamics." *Science* 265(5170): 361–364.

Problems

1. Derive the gain of the stimulated beam as the following relationship:

$$dI_s/I_s = N_2\sigma/A,$$

   where N_2 is the number of excited fluorophores; σ is the stimulated emission cross-section; and A is the waist area of the beam. Also evaluate the gain of the stimulated emission beam at the single-molecule level.
2. Compare the fluorescence and stimulated emission detection methods.
3. Explain the saturation of a stimulated emission signal as increasing the intensity of a stimulated emission beam.
4. Describe how the pump-probe technique can be used to measure fluorescence lifetime.
5. What are the factors that affect the efficiency of stimulated emission?

chapter eight

Fiber optic microscopy

Jin U. Kang and Xuan Liu

Contents

8.1 Introduction

This chapter discusses the principle of fiber optic microscopy using both single-mode fiber and fiber bundle. In general, optical fibers used in fiber optic microscopy are primarily used for remote illuminating light from the source to the imaging site and collected imaging light from the imaging site to the detector. Optical fibers are highly flexible and thin. These properties make them ideal for use in in situ and endoscopic imaging (Kimura and Wilson 1991; Gmitro and Aziz 1993); however, these benefits come with consequences: the small core fiber diameter makes the light coupling into fiber difficult and this limited fiber aperture acts as a spatial filter. This chapter will discuss some of these issues.

Fiber optics have played a major role in the development of modern communication networks (Agrawal 2010). Now, many fiber optic devices and techniques developed for the fiber communications industries are helping advance other areas of science. Specifically in microscopy, optical

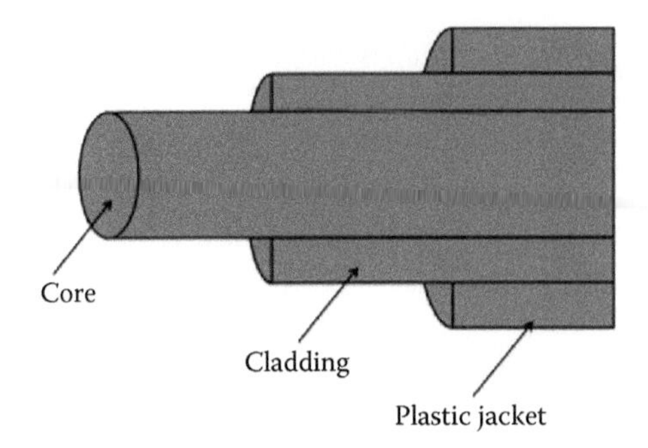

Figure 8.1 Cross-section of an optical fiber.

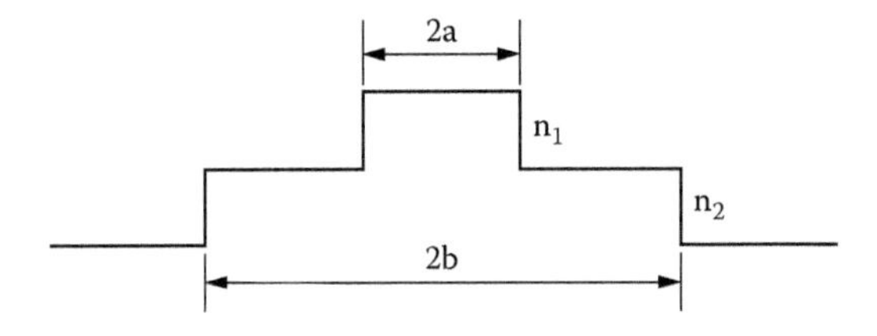

Figure 8.2 Index profile of a step-index fiber.

fiber endoscopes and other imaging devices have become compact, inexpensive, and high-performing over the past 10 years.

Most optical fibers are made of glass, because SiO_2 (silica) glass exhibits extremely low absorption and scattering losses and is easy to fabricate. The basic geometry of optical fiber is shown in Figure 8.1. To efficiently confine and guide the light, the core of the fiber has to have a higher refractive index than that of the cladding, as shown in Figure 8.2. Different dopants can be used to modify the refractive index of silica: P_2O_5 and GeO_2 can be used to increase the refractive index of SiO_2 glass while B_2O_3 and F can be used to reduce the index of SiO_2 glass (Yariv 1997; Azadeh 2009; Agrawal 2010).

In addition to single-core fibers shown in Figures 8.1 and 8.2, flexible fiber bundles having multiple cores are being widely used in endoscopic imaging. A fiber bundle can have multiple (more than 10,000) cores surrounded by a common cladding, as shown in Figure 8.3 (Han et al. 2010). The use of a fiber bundle for microscopy allows direct image transfer from the distal end of the fiber to its proximal end. This allows us to miniaturize the imaging probes as well as to image moving or live biological samples noninvasively (Flusberg et al. 2008; Han et al. 2010; Liu et al. 2011).

Fiber is highly effective in confining the light inside the core, which is usually several micrometers in diameter. Therefore, when used in microscopy for illumination and signal collection, the fiber core can serve as a

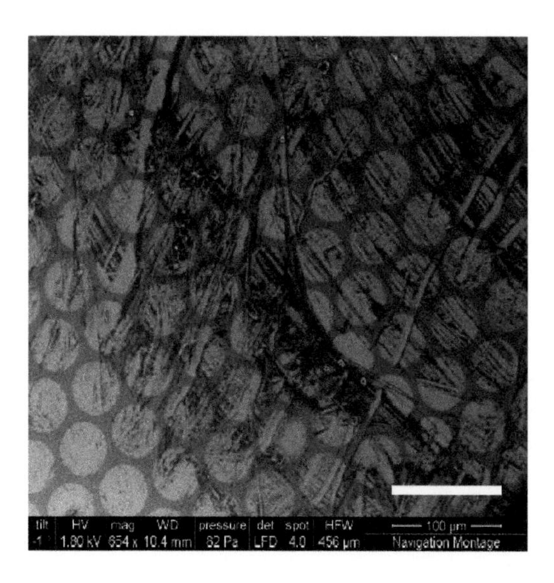

Figure 8.3 Surface image of a fiber bundle end taken by a scanning electron microscope showing circular cores (scale bar: 100 μm).

near-perfect pinhole that can serve as a spatial filter, which is an essential element of confocal microscopes. Compact, lightweight, thin, and flexible optical fibers are perfectly adaptable to an endoscope to reach imaging sites where bulk optics cannot access. Optical fibers are also immune to electromagnetic interference; therefore, microscopes based on fiber optics can be used in conjunction with MRI (magnetic resonance imaging), CT (computed tomography), and other imaging modalities. Finally, it should be noted that optical fiber devices need less maintenance than bulk optic devices, which makes the operation easy and easier to maintain.

8.2 Basic fiber optic microscopy configurations

Both single-mode fiber and fiber bundle-based fiber optic microscopes are essentially scanning microscopes. In the case of single-mode fiber microscopes, each image pixel is obtained sequentially; in the case of fiber bundle microscopes all imaging pixels are obtained in parallel. Figure 8.4 illustrates basic concepts of the two types of microscope.

8.2.1 Fiber optic microscopy based on single-mode fiber (SMF)

Figure 8.5 shows the basic setup for a microscope based on SMF (Kim 2007). The output of a light source, which is usually fiber coupled, is sent to a fiber optic circulator. The light comes out of port 2 and is subsequently collimated by a collimating lens. The light is then focused onto a target

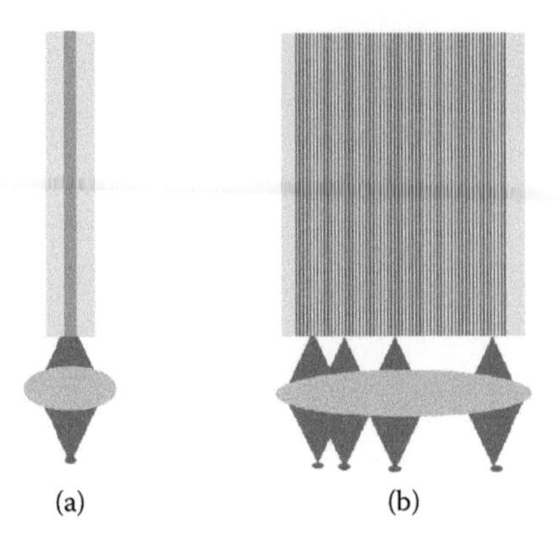

Figure 8.4 Fiber imager concept based on (a) single-mode fiber; (b) fiber bundle.

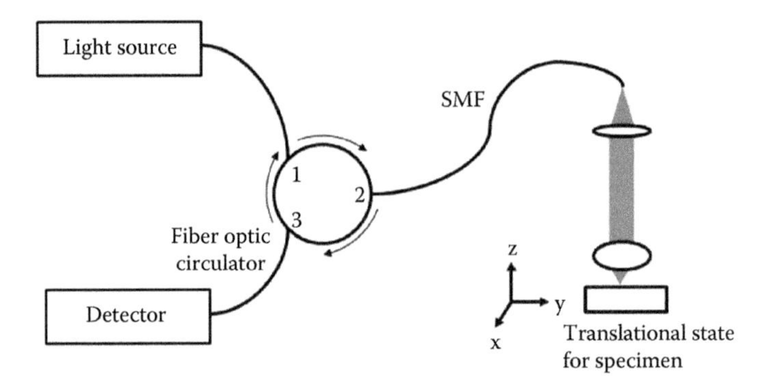

Figure 8.5 Schematic of fiber optic microscopy based on single-mode fiber.

using an objective lens. The reflected light from the target is then coupled back to the fiber. The reflected light comes out of port 3 of the circulator and then gets detected. Various mechanical scanning methods can be used to form two- or even three-dimensional images using this basic setup. One simple method uses translational stages to scan the specimen, as shown in Figure 8.5. The setup is easy to implement and highly stable since the collimating and focusing optics do not move.

As discussed in the Introduction, the core of the fiber acts as a near-perfect pinhole. Therefore, fiber optic microscopy is inherently performing confocal microscopy that blocks out-of-focus beams. This is illustrated in Figure 8.6, which simplified the imaging optics as a thin lens. The illuminating spot on the target represents the fiber core image. The backscattered light from the target can be detected only if it is from the in-focus plane and overlaps with the modal shape of the fiber core. The lights coming from out-of-focus planes are effectively rejected by the fiber (Kimura

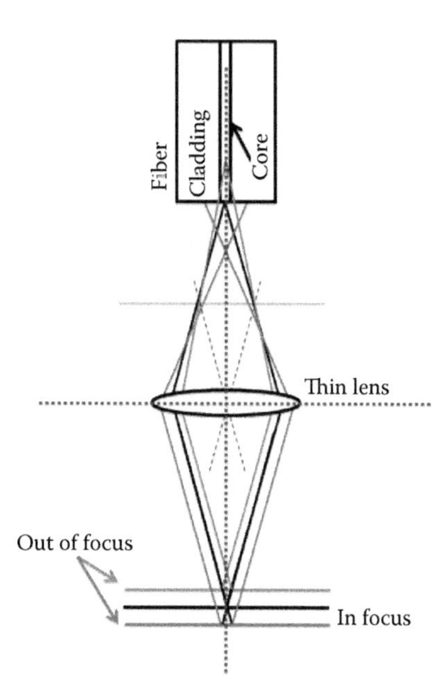

Figure 8.6 Confocal gating of a fiber microscope based on SMF.

and Wilson 1991; Delaney et al. 1994; Izatt et al. 1994; Tearney et al. 1998; Bird and Gu 2003; Fu and Gu 2007; Kim 2007).

Using the Rayleigh criterion for confocal microscopy (Vo-Dinh 2003), the axial (r_z) and lateral ($r_{x,y}$) resolution of a fiber optic confocal microscope can be calculated using Equations (8.1) and (8.2) in which λ, n, and NA represent wavelength, refractive index of the medium, and the objective numerical aperture (Born and Wolf 1999; Jonkman and Stelzer 2002; Wang and Wu 2012).

$$r_z = \frac{1.4\lambda n}{NA^2} \tag{8.1}$$

$$r_{x,y} = \frac{0.4\lambda}{NA} \tag{8.2}$$

According to Equations (8.1) and (8.2), axial and lateral resolution can be improved by using a stronger lens with a higher NA.

8.2.2 *Wide-field fiber bundle microscopy*

Two-dimensional images can be obtained from a fiber bundle microscope using wide-field illumination and a two-dimensional image sensor

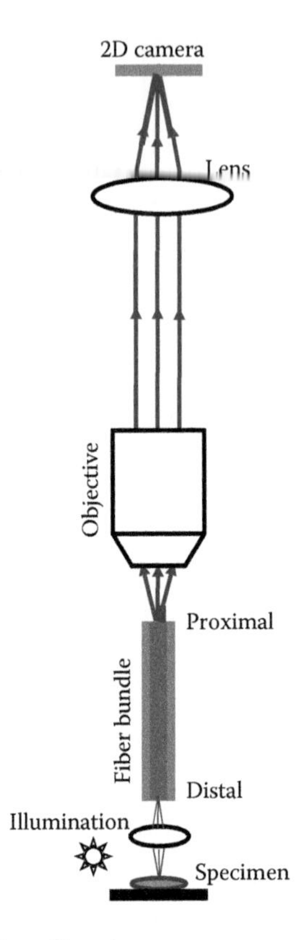

Figure 8.7 Wide-field fiber bundle microscope.

(CCD or CMOS camera), as shown in Figure 8.7. For such wide-field illumination, it is desirable to use an extended light source with a low coherence, such as a thermal light source or light-emitting diode (LED; Muldoon et al. 2007; Han et al. 2010; Sun et al. 2010; Liu et al. 2011). If a spatially coherent light source such as a laser is used to illuminate the fiber bundle, the cross-coupling between adjacent cores in the fiber bundle can introduce strong cross-talk and undesired interference patterns (Oh et al. 2006; Chen et al. 2008). A wide-field microscope operating in this mode cannot differentiate photons from different axial planes and does not exhibit a confocal effect; therefore, the lateral resolution can be calculated based on a conventional microscope using Equation (8.3) (Delaney et al. 1994; Born and Wolf 1999).

$$r_{x,y} = \frac{0.6\lambda}{NA} \tag{8.3}$$

Figure 8.7 illustrates its configuration. The target is imaged at the distal facet of the fiber bundle. The image is then sampled by the fiber cores, which serve as an image conduit to transport the image to the proximal facet of the fiber bundle. To preserve the diffraction-limited lateral resolution shown as Equation (8.3), the image of the target on the fiber bundle facet has to be sampled by the fiber cores at a frequency greater than twice its highest spatial frequency; otherwise, the lateral resolution is limited simply by the core size and pitch of the fiber bundle. Moreover, when the proximal facet of the fiber bundle is imaged to the camera sensor plane, the two-dimensional pixel array of the camera has to sample the proximal facet of the fiber bundle at least twice the frequency of the highest spatial frequency of the fiber bundle core; otherwise, the aliasing artifact can smear the image.

8.2.3 Laser-scanned fiber bundle microscopy

Two-dimensional images can also be obtained from a fiber bundle microscope by scanning a point light source with a mechanical scanner such as a galvanometer, as shown in Figure 8.8 (Gmitro and Aziz 1993; Liang et al.

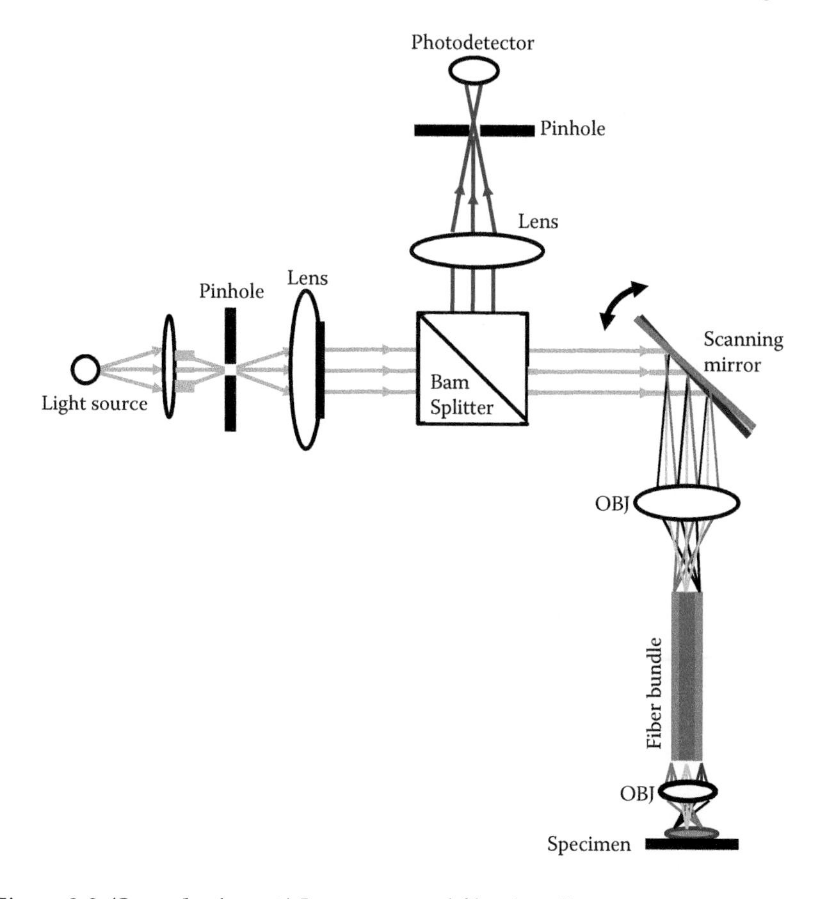

Figure 8.8 **(See color insert.)** Laser-scanned fiber bundle microscopy.

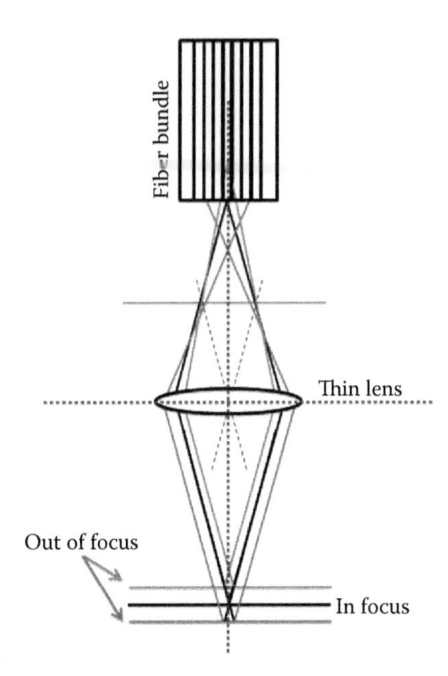

Figure 8.9 Illustration of confocal effect consequence of laser-scanned field fiber bundle microscopy.

2001; Flusberg et al. 2005, 2008; Xie et al. 2005; Han et al. 2009, 2010). Since the image is obtained pixel by pixel, cross-talk between adjacent cores in the fiber bundle is no longer a problem and a laser source with high spatial coherence can be used to achieve high coupling efficiency.

The confocal effect of the laser-scanning fiber bundle microscope is schematically presented in Figure 8.9, which shows that photons from the in-focus plane couple into a single core of the fiber bundle, while photons from the out-of-focus plane couple into multiple cores of the fiber bundle. For both cases, the signals are guided by the fiber core (or cores) from the distal end to the proximal end. The small pinhole in front of the detector that is in conjugate with the probing light spot is used to effectively reject the signal from the out-of-focus planes guided by the multiple fiber cores; therefore, the resolution of the laser-scanning field fiber optic microscopy based on fiber bundle can be estimated using Equations (10.1) and (10.2).

8.3 Advancement of fiber optic microscopy

8.3.1 Fiber optic confocal microscopy using an eye-safe near-infrared fiber laser

As stated in the previous sections, optical fiber is an ideal pinhole having a very small- (a few microns) diameter aperture and a highly efficient light

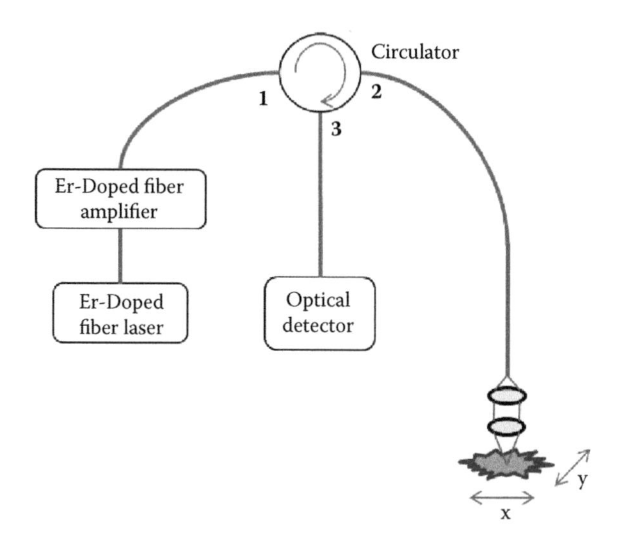

Figure 8.10 Schematic of a 1.55-μm near-infrared fiber optic confocal microscope.

waveguide; thus, use of an optical fiber in confocal configuration negates the need for a pinhole (Kim 2007; Liu et al. 2011). A simple and highly effective near-infrared fiber optic confocal microscope can be built using all fiber optic components developed by the communications industry. An example of such a setup is shown in Figure 8.10; a 1.55-μm light source is used for the confocal microscope. This wavelength is safe for human eyes. Fiber optic components developed for optical communications for this wavelength band are widely available and can be directly used to build the microscope. Acquired three-dimensional data (x, y, intensity) can be combined to an intensity map using standard imaging software. Both the translational stage and optical power meter are controlled by a personal computer and the intensity at each position is matched to each x-y coordinate.

To characterize the resolution of a microscopic imaging system, a United States Air Force (USAF) test target with the pattern shown in Figure 8.11(a) can be used. Figure 8.11(b) and (c) (120 × 120 pixels, 0.5 μm step size) show images of the test target obtained using a fiber optic confocal microscope system with 20× and 60× objective lenses, respectively. Figure 8.11(c) demonstrates a resolution of 1.1 μm, which is higher than the resolution of Figure 8.11(b), due to the use of a larger NA objective lens.

Figure 8.12(a) shows the conventional microscopic image of a semiconductor substrate coated with photoresistor pattern (SAMPLE-1). Figure 8.12(b) and (c) show fiber optic confocal microscopic image of SAMPLE-1 containing 100 (x-direction) × 100 (y-direction) pixels with 6 μm step size, and an image of SAMPLE-1 containing 200 (x-direction) × 200

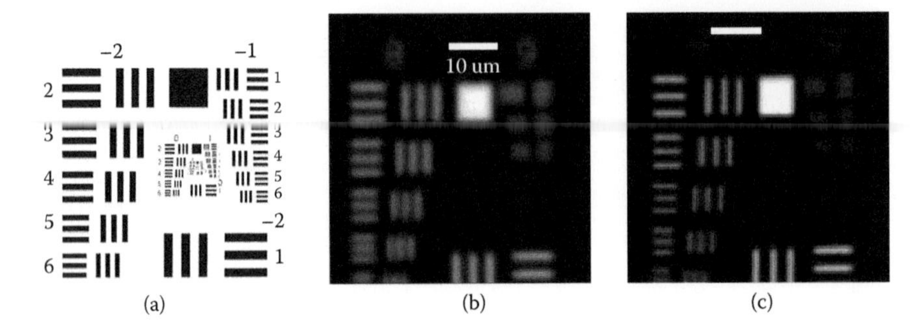

Figure 8.11 (a) Pattern of United States Air Force (USAF) test target; fiber optic confocal microscope image of USAF test target obtained with 20× objective lens (b) and 60× objective lens (c).

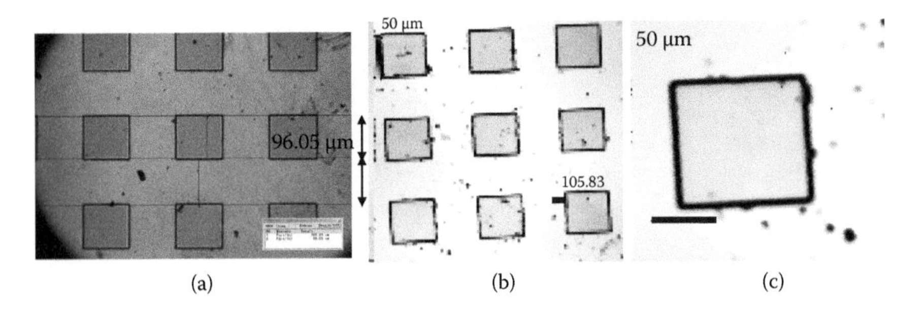

Figure 8.12 (a) Optical microscope image of SAMPLE-1 with 450× objective lens; (b) images of SAMPLE-1 taken by fiber optic confocal microscope for (a) $N = 100$ and $d = 6$ μm; (c) $N = 200$ and $d = 1$ μm (N, number of pixels in x- or y-direction; d, step size).

(y-direction) pixels with 1 μm step size. Figure 8.13(a) shows the conventional microscopic image of a collagen film (SAMPLE-2) made of solid type III calfskin collagen. Figure 8.13(b) and (c) are images of SAMPLE-2 obtained using a fiber optic confocal microscopic system. The number of sampling points is 100 for Figure 8.13(b) in both x- and y-directions with a step size of 15 μm; the number of sampling points is 100 for Figure 8.13(c) in both x- and y-directions with a step size of 1 μm.

8.3.2 *Dark field illuminated fiber bundle*

A separate fiber is commonly used to illuminate the specimen in a wide-field fiber bundle imager; however, in a reflectance fiber bundle microscope, it is often desired to use only one fiber bundle for illumination and image acquisition. The consequence of such a setup is that the specular

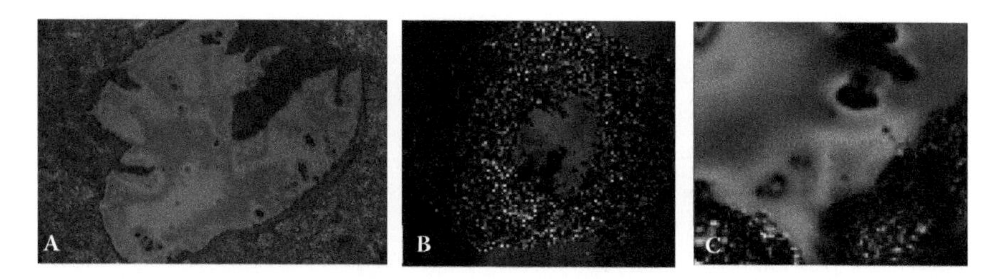

Figure 8.13 (a) Optical microscope image of SAMPLE-1 with 450× objective lens; (b) images of SAMPLE-1 taken by fiber optic confocal microscope for (a) $N = 100$ and $d = 15$ μm; (c) $N = 100$ and $d = 1$ μm (N, number of pixels in x- or y-direction; d, step size).

reflectance from the end facets of the fiber bundle can be much larger than the backscattered light from the sample. To suppress the specular reflection from the facet of the fiber bundle, researchers have proposed various methods, including the use of index-matching gel, cross polarization detection, and dark field illumination (Han et al. 2010; Liu et al. 2011). Compared to other methods, the dark field illumination technique can effectively suppress specular reflection using a simpler configuration and lower cost (Liu et al. 2011; Villiger, Pache, and Lasser 2011). The configuration for a dark field illuminated reflectance fiber bundle endoscopic microscope is illustrated in Figure 8.14(a). A circular optical stop (an opaque disk) and a round aperture are inserted into the illumination and detection arm, respectively. Due to the circular optical stop and, therefore, the annular illumination, as shown in Figure 8.14(b), the light specular reflection from the fiber bundle end would follow the path shown by purple lines in Figure 8.14(c). Adjusting the aperture size of the iris can completely block the specular light from reaching the camera. On the other hand, signal photons backscattered by the sample follow a different path, as shown in Figure 8.14(d). This is because the geometry of the light beam exiting the fiber bundle is determined by the fiber modes supported. Unlike traveling in free space, light incident into the fiber bundle will be coupled into the guided modes determined by the physical properties of the fiber bundle. Therefore, photons traveling in guided mode in a fiber core will "forget" the illumination configuration. The output beam will form a cone with a diverging angle determined by the NA of the fiber bundle. Therefore, although the iris causes some signal loss, the optical field that passes the iris still contains a relatively unattenuated light signal from the specimen and is imaged by the camera. Figure 8.15 (a) and (b) show the image of a resolution target (1963A resolution target, Edmund Optics) obtained from a fiber bundle microscope with and without dark field illumination. Clearly, bars in Figure 8.15(a)

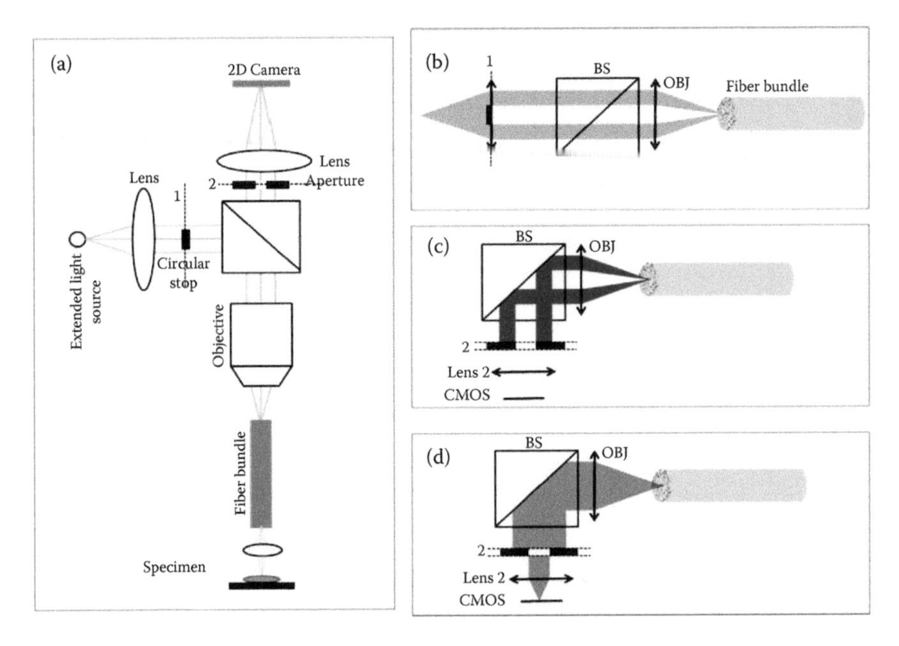

Figure 8.14 **(See color insert.)** (a) Configuration of dark field illuminated reflectance fiber bundle microscope; (b) optical path of illumination light; (c) optical path of specular reflection from the fiber bundle end; (d) optical path of signal light back scattered by the specimen and guided by the fiber bundle.

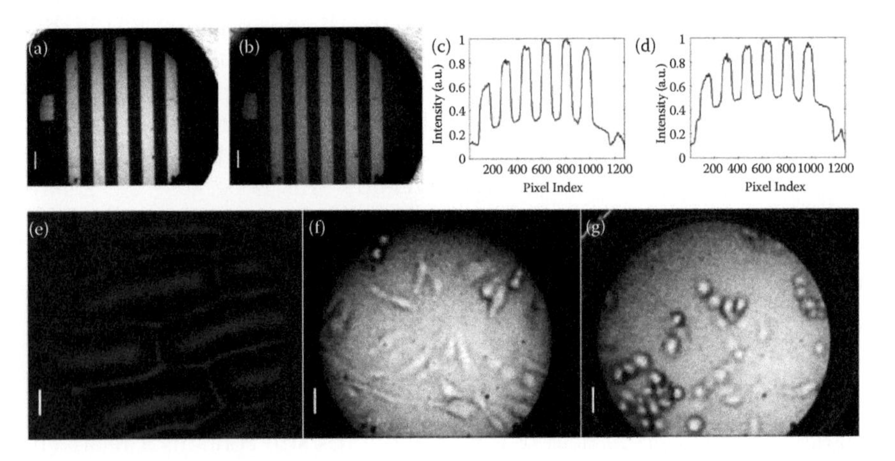

Figure 8.15 Reflectance images of NBS 1963A resolution target (a) with dark field illumination; (b) without dark field illumination. (c) Image intensity along the 512th row of the image in Figure 8.10a; (d) image intensity along the 512th row of the image in Figure 8.10 (b). (e) Onion skin cells; (f) KAT-18 cells; (g) FTC-133 cells. Scale bars represent 50 μm.

exhibit much higher visibility than in Figure 15(b). The signal amplitudes along the central (512th) row of the images are shown in Figure 8.15(c) and (d). Clearly, Figure 15(c) shows higher contrast between the high reflective and low reflective part of the resolution target. Intrinsic contrast images from label-free cell samples (onion cells, thyroid cancer cell lines, KAT-18, and FTC-133) obtained from the dark field illuminated fiber bundle endoscope are shown in Figure 8.15 (e)–(g), indicating the high sensitivity of the imaging system.

8.3.3 *Pixelation effect removal from fiber bundle probe-based optical coherence tomography imaging*

A distinct artifact in images obtained from the fiber bundle is the pixelation effect due to the characteristic arrangement of fiber core arrays. The pixelation effect can be removed by applying filters in the spatial or frequency domain (Flusberg et al. 2005; Han et al. 2010; Han and Yoon 2011). A simple method combining histogram equalization and Gaussian filtering can effectively remove the structural artifact of the fiber bundle and enhance the image quality with minimum blurring of the object's image features. Histogram equalization is a simple and effective way to adjust contrast using the histogram of the image, especially for low-contrast gray images with limited scale levels. It is based on a point transformation of gray levels in the input image so that intensities of the resulting image can be better distributed by effectively spreading out the most frequent pixel values. A Gaussian smoothing filter is a form of weighted average with different coefficients multiplied to the pixels having a peak at the center and tapering down as the Euclidean distance from the center pixel increases. The Gaussian filter can accommodate the circularity of the fiber structure and its corresponding symmetrical dimension by controlling the filter parameters.

The effectiveness of the depixelation method is demonstrated in Figure 8.16. Figure 8.16(a) shows an image of the USAF target obtained from a laser-scanned fiber bundle microscope. The fiber bundle used for this study is a 76-mm-long rigid coherent fiber bundle (Edmund Optics Inc.) having a numerical aperture of 0.53, individual core spacing of 50 μm (total number of fibers: 3,000), and refractive indices for core and cladding, 1.49 and 1.58.

The image in Figure 8.16(a) shows pixelation artifact and low contrast. Figure 8.16(b) is the histogram equalized image which shows improved contrast. When a Gaussian filter is applied to Figure 8.16(a), a depixelated image can be obtained, as shown in Figure 8.16(c). When a Gaussian filter is applied to the image after histogram equalization (Figure 8.16(b)), the resultant image is free of pixelation artifact and has much better contrast compared to Figure 8.16(c).

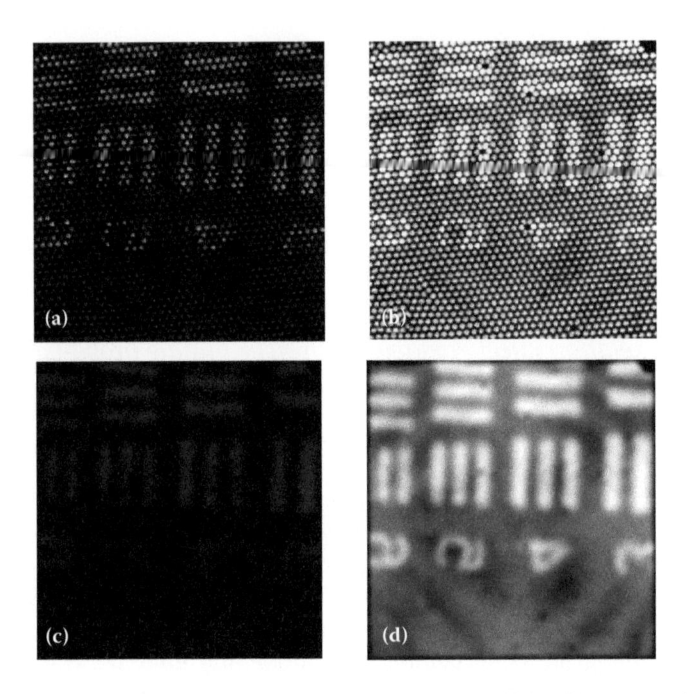

Figure 8.16 (a) Original image with fiber pixelation effect; (b) image after histogram equalization; (c) Gaussian smoothing filtered result with the original image shown in Figure 8.11(a); (d) Gaussian smoothing filtered result with a pre-histogram equalized image.

8.3.4 *Gene transfection efficacy assessment of human cervical cancer cells using dual-mode fluorescence microendoscopy*

Gene therapy is a new therapeutic approach for cancer and has been shown to be effective not only with cancer but also with many other types of diseases (Rubanyi 2001). In gene therapy, it is important to measure how well various gene vectors are delivered and influence the target cells. The most common optical method for gene transfection efficacy assessment is to detect tagged fluorescence signals from transfected cells using endogenous fluorescent probes. By comparing bright field illumination and fluorescence images from a bench-top fluorescence microscope, gene transfection efficiency can be calculated as the ratio of fluorescence-expressing cells to total cells; however, such an assessment is usually performed in vitro due to the limitation of accessibility in standard bench-top fluorescence microscopes. Flexible endoscopes that can perform subcellular imaging in vivo based on high-resolution fiber bundles have been developed. A dual-modality microendoscope system based on the fiber bundle, as shown in Figure 8.17, can simultaneously register both fluorescing and entire cells and can, therefore, be used to assess intracellular gene

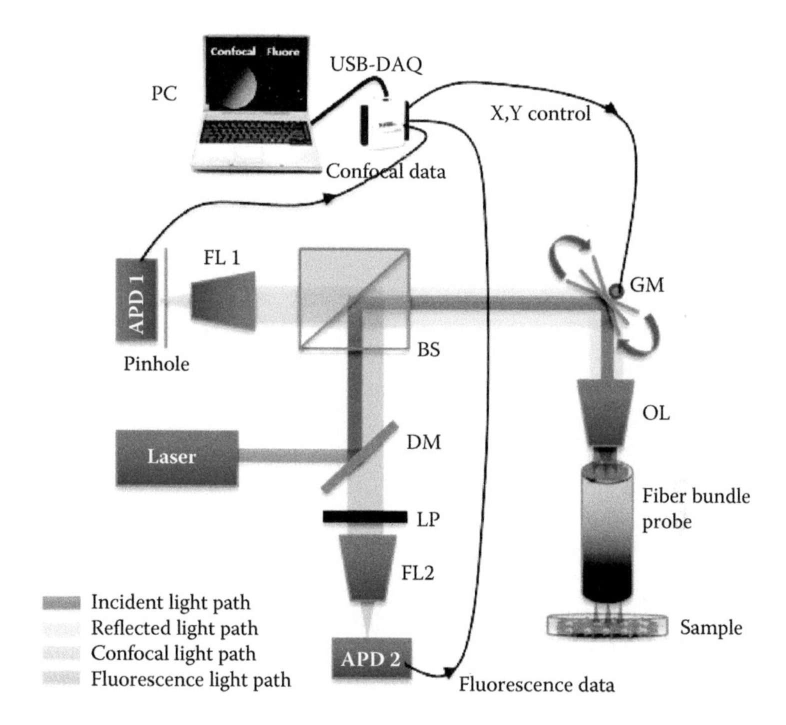

Figure 8.17 **(See color insert.)** System configuration (DM, dichroic mirror; BS, 50:50 beam splitter; GM, galvo mirror; OL, objective lens; FL 1&2, focusing lens; LP, longpass filter; APD 1&2, avalanche photodetector; DAQ, digital-to-analog and analog-to-digital).

delivery efficacy (Cha et al. 2013). The laser-scanning confocal microscope is connected to a coherent fiber bundle imaging probe. Incident light blue laser is reflected by a dichroic, a 50:50 beam splitter, and an X-Y galvo in series and coupled into a multicore fiber bundle by a microscope objective lens. Returning light from the specimen is divided into two pathways by the beam splitter. One pathway is used for the reflectance imaging (blue color in Figure 8.17) which is directed through the beam splitter and spatially filtered using a focusing lens and a 100-μm pinhole to reject background and out-of-focus light. A highly sensitive Si avalanche photodiode, with an active area diameter of 1 mm^2, is used to detect the reflectance imaging light. The second pathway is used for the fluorescence imaging (green color in Figure 8.17), which gets reflected by the beam splitter, is transmitted through the dichroic mirror, and is then filtered by a longpass filter to obtain only the fluorescence signal. Similar to the reflectance imaging path, the longpass filtered beam is spatially filtered using a focusing lens 2 and a 100-μm pinhole. The resulting fluorescence signals are detected by a UV-enhanced Si avalanche photodiode.

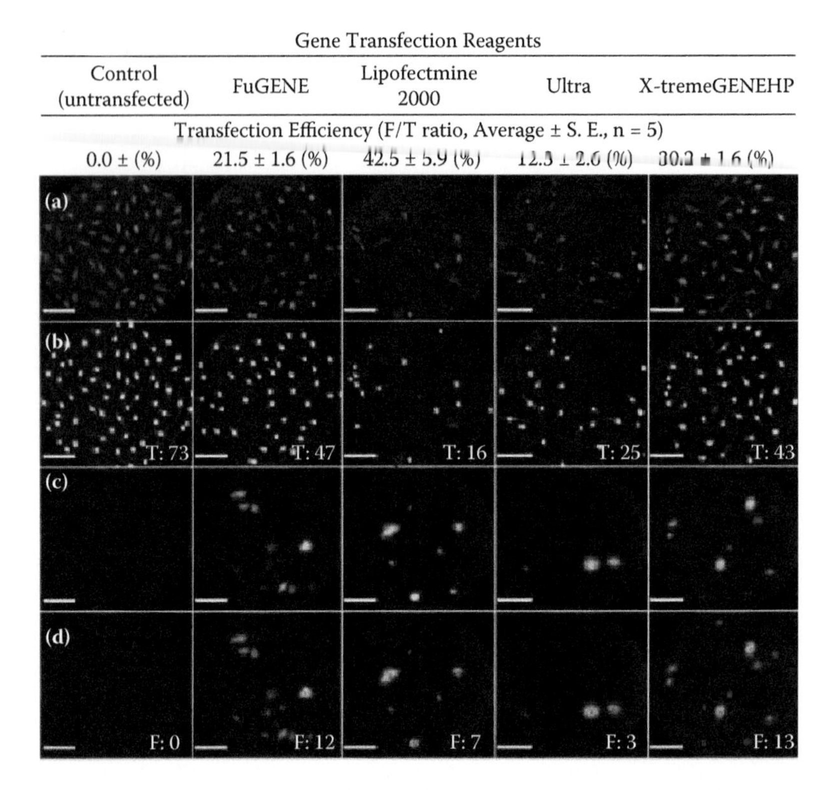

Figure 8.18 Representative dual-modality microendoscope sample imaging results in four different reagent groups: (a) reflectance image, (b) total cell counting from the reflectance image, (c) fluorescence image, (d) fluorescent cell counting from the fluorescence image. (All white bars represent 100 µm, T, total cell counting number; F, fluorescent cell counting number; pseudo-color applied on the fluorescence images.)

To assess gene transfection efficacy, the human cervical cancer cells (HeLa cells) were transfected with four different transfection reagents (FuGENE 6, Lipofectamine 2000, Ultra, and X-tremeGENE HP). Representative images of HeLa cells obtained with the dual-modality microendoscope are shown in Figure 8.18.

An automated image-processing algorithm can be used to analyze the images obtained and calculate the ratio between the total and the number of fluorescence-expressing cells that can be used to determine how many cells were transfected by each of the vector systems. The gene transfection efficiency measured by the dual-modality microendoscope is compared with the measurement obtained using a commercial bench-top microscope in Figure 8.19. The bench-top microscope and the dual-modality microendoscope results were found to be clearly correlated, with the p-value being less than 0.05 by using the two-tailed Student's t test.

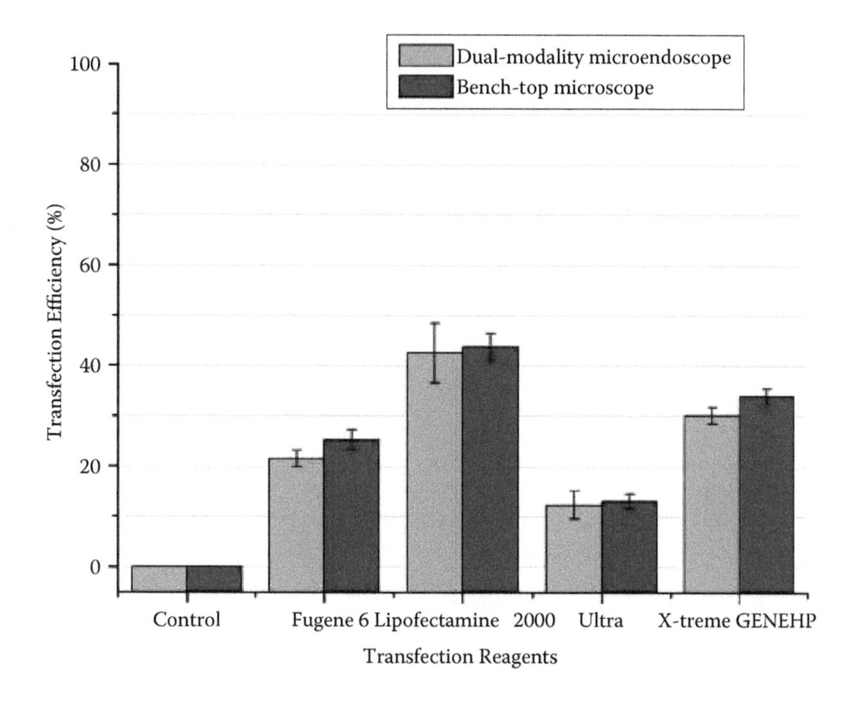

Figure 8.19 Comparison of transfection efficiency in the four reagent groups.

References

Agrawal, G. P. (2010). *Optical Fibers, in Fiber-Optic Communication Systems.* New York: John Wiley & Sons.

Azadeh, M. (2009). *Fiber Optics Engineering.* New York: Springer.

Bird, D. and M. Gu (2003). "Two-photon fluorescence endoscopy with a micro-optic scanning head." *Optics Letters* **28**(17): 1552–1554.

Born, M. and E. Wolf (1999). *Principles of Optics: Electromagnetic Theory of Propagation, Interference and Diffraction Of Light.* Cambridge, UK: Cambridge University Press Archive.

Cha, J. et al. (2013). "Gene transfection efficacy assessment of human cervical cancer cells using dual-mode fluorescence microendoscopy." *Biomedical Optics Express* **4**(1): 151.

Chen, X. et al. (2008). "Experimental and theoretical analysis of core-to-core coupling on fiber bundle imaging." *Optics Express* **16**(26): 21598–21607.

Delaney, P. M. et al. (1994). "Fiber-optic laser scanning confocal microscope suitable for fluorescence imaging." *Applied Optics* **33**(4): 573–577.

Flusberg, B. A. et al. (2005). "Fiber-optic fluorescence imaging." *Nature Methods* **2**(12): 941–950.

Flusberg, B. A. et al. (2008). "High-speed, miniaturized fluorescence microscopy in freely moving mice." *Nature Methods* **5**(11): 935–938.

Fu, L. and M. Gu (2007). "Fibre-optic nonlinear optical microscopy and endoscopy." *Journal of Microscopy* **226**(3): 195–206.

Gmitro, A. F. and D. Aziz (1993). "Confocal microscopy through a fiber-optic imaging bundle." *Optics Letters* **18**(8): 565–567.

Han, J.-H. et al. (2010). "Pixelation effect removal from fiber bundle probe based optical coherence tomography imaging." *Optics Express* **18**(7): 7427.

Han, J.-H. et al. (2009). "Common path optical coherence tomography with fibre bundle probe." *Electronics Letters* **45**(22): 1110–1112.

Han, J.-H. and S. M. Yoon (2011). "Depixelation of coherent fiber bundle endoscopy based on learning patterns of image prior." *Optics Letters* **36**(16): 3212–3214.

Izatt, J. A. et al. (1994). "Optical coherence microscopy in scattering media." *Optics Letters* **19**(8): 590–592.

Jonkman, J. and E. Stelzer (2002). "Resolution and contrast in confocal and two-photon microscopy." In *Confocal and Two-Photon Microscopy: Foundations, Applications, and Advances*, edited by A. Diaspro, 101-125. New York: Wiley-Liss.

Kim, D.-H. (2007). *Fiber Optic Confocal Microscope for Bio-Medical Applications*, Doctoral dissertation, Johns Hopkins University, Baltimore.

Kimura, S. and T. Wilson (1991). "Confocal scanning optical microscope using single-mode fiber for signal detection." *Applied Optics* **30**(16): 2143–2150.

Liang, C. et al. (2001). "Fiber confocal reflectance microscope (FCRM) for in-vivo imaging." *Optics Express* **9**(13): 821–830.

Liu, X. et al. (2011). "Dark-field illuminated reflectance fiber bundle endoscopic microscope." *Journal of Biomedical Optics* **16**(4): 046003–046007.

Muldoon, T. J. et al. (2007). "Subcellular-resolution molecular imaging within living tissue by fiber microendoscopy." *Optics Express* **15**(25): 16413.

Oh, W. et al. (2006). "Spectrally-modulated full-field optical coherence microscopy for ultrahigh-resolution endoscopic imaging." *Optics Express* **14**(19): 8675.

Rubanyi, G. M. (2001). "The future of human gene therapy." *Molecular Aspects of Medicine* **22**(3): 113–142.

Sun, J. et al. (2010). "Needle-compatible single fiber bundle image guide reflectance endoscope." *Journal of Biomedical Optics* **15**(4): 040502–040503.

Tearney, G. J. et al. (1998). "Spectrally encoded confocal microscopy." *Optics Letters* **23**(15): 1152–1154.

Villiger, M., E. Pache, and T. Lasser (2011). Dark Field Optical Coherence Microscopy, US Patent 2,011,083,420, filed Jan. 4, 2011.

Vo-Dinh, T., ed. (2003). *Biomedical Photonics Handbook*. Boca Raton, FL: CRC Press.

Wang, L. V. and H.-I. Wu (2012). *Biomedical Optics: Principles and Imaging*. New York: Wiley.

Xie, T. et al. (2005). "Fiber-optic-bundle-based optical coherence tomography." *Optics Letters* **30**(14): 1803–1805.

Yariv, A. (1997). *Optical Electronics in Modern Communications*. New York: Oxford University Press.

Problems

1. Given a fiber with a core diameter of 10 microns, a core index of 1.43, and a cladding index of 1.42, if the operating wavelength is 500 nm, calculate the NA of the fiber and the total number of modes supported by the fiber.
2. If the fiber described in Problem 1 is used to build a fiber optic microscope without any lens at the distal end of the fiber, calculate both axial and lateral resolution of the fiber optic microsocpe.
3. Explain the principle of confocal gating for a fiber optic microscope based on single-mode fiber (i.e., how does an optical fiber act as a confocal gating.)
4. Explain why coherent light sources are preferred for single-mode fiber-based microscopes whereas low coherent light sources are preferred for full-field fiber-bundle-based microscopes?
5. Spatial resolution of optical microscopy scales with wavelength. Consider the same fiber optic microscope based on the fiber in Problem 1 is illuminated by 0.8-μm and 1.3-μm lights. Calculate axial and lateral resolutions for these wavelengths.
6. Assume a fiber optic confocal microscope with an objective with NA = 0.2, light source $\lambda = 1.3$ μm is used in air where refractive index $n = 1$. Calculate the axial and lateral resolution of this confocal microscope. Repeat the calculation assuming NA = 0.8.
7. Consider a fiber optic microscope based on a fiber bundle, as shown in Figure 8.14.
   (1) The fiber bundle uses an individual fiber core to discretely sample the optical field from the specimen. Assume the average interval between adjacent fiber cores with the fiber bundle equals *C* and assume the lens system projects the fiber bundle surface to the sensor plane of the camera with a magnification of **M**. Estimate f_{max}, the highest spatial frequency of the image obtained from the fiber bundle.
   (2) According to Nyquist sampling theorem, a signal has to be sampled at at least twice its highest frequency. Calculate the minimum sampling rate R_{min} to resolve a fiber bundle image with highest spatial frequency f_{max}. R_{min} is essentially the reciprocal of the pixel width (**W**) of the array detector assuming a square shape for pixels. Calculate the maximum pixel size $\mathbf{W}_{\min}$ of the array detector.

chapter nine

Scanning ion conductance microscopy

Zhicong Fei, Hui Shi, Yanjun Zhang, and Yuchun Gu

Contents

9.1 Background and principle of operation

The scanning ion conductance microscope (SICM) is another member of the scanning probe microscope family, which is specially designed for scanning soft nonconductive materials that are bathed in electrolyte solution. It was originally developed by Hansma (Hansma et al. 1989) and based on an electrolyte-filled micropipette used as a local probe for insulating samples immersed in an electrolytic solution, as schematically illustrated in Figure 9.1.

A three-dimensional (3D) piezo scanner is actuated to translate the sample relative to the micropipette tip in the x-, y-, and z-directions. As the micropipette is made to approach the sample, the flow of ions through the opening of the pipette is reduced at small probe-to-sample separations, resulting in a decrease of the ion conductance between an electrode inside the pipette and an electrode in the electrolyte reservoir. The distance-dependence of ion conductance provides the feedback control to perform noncontact surface profiling and is thus less likely to damage the sample surface, which is the advantage of the SICM in comparison with

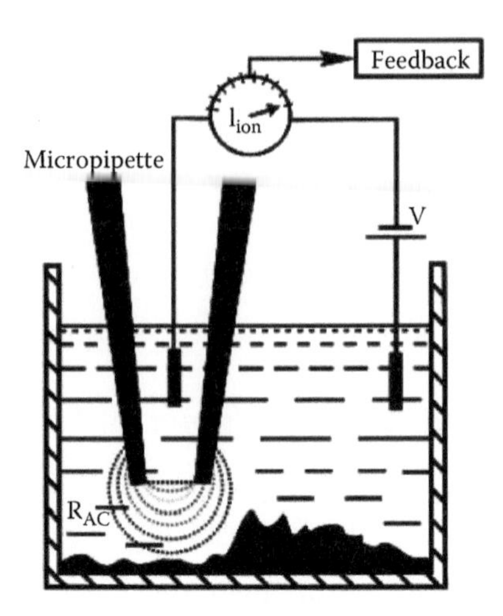

Figure 9.1 Schematic of the working principle of the scanning ion conductance microscope.

Atomic Force Microscopy (AFM) techniques. To put it briefly, the decrease in the ion conductance reduces the ion current (I_{ion}), which is monitored by a feedback control unit (CU), and is also used as a feedback input signal (V_p) on the z-direction of the 3D piezo-stage to keep the probe tip and sample separate during the scanning procedure. Sequentially recording the positions of z-direction piezo at every point of a given x-y plane results in a topographical image of the surface. Positioning of the probe is usually achieved by means of piezoelectric elements or step motors.

Since the ion current flow into the pipette is sensitive to the distance between the tip and the sample surface due to partial blockage of ion flow, in principle, the probe does not come into contact with the surface. However, since the dependence of the ionic current on the distance between the tip and the sample is often nonlinear, which makes the feedback control more difficult, the tip could crash into a rough surface during routine imaging. It is even worse when performing a high-resolution imaging or during high-speed scans; the fragile micropipettes of the SICM often break on contact with the insulator sample, thus limiting its usefulness. As a practical remedy, the operating distance from the sample surface has been purposely kept large, which in turn leads to a considerable degradation of the sensitivity and the resolution. Microfabrication of more robust silicon probes with a small aperture at the tip apex improved matters to some extent, and allowed high scan speeds. In order to achieve higher contrast and less apparent sample damage, a "tapping mode" SICM has been developed (Proksch et al. 1996), where a bent micropipette is used both as a cantilever and as an ion conductance sensitive probe.

However, for a long time the SICM technique was limited to imaging of flat synthetic membranes and not biological samples or living cells.

9.2 *Improvements in scanning ion conductance microscopy*

The SICM technique was first applied to the imaging of living cells by the Korchev group (Korchev et al. 1997). Since then, the SICM has been used for investigating biological samples, such as cells and tissues in vitro (Korchev 2000; Happel et al. 2003; Korchev et al. 1997; Mann et al. 2002).

In order to achieve 3D and high-precision movement of the probe over the sample, the SICM micropipette is mounted on a 3D piezo-translation stage (Tritor 100, Piezosystem Jena, Germany) with a range of 100 μm, thus the imaging of relatively large areas of biological samples is possible. In addition, to make sure a long-range coarse 3D adjustment and the approaching procedure more flexible, this piezo stage is further mounted on mechanical micromanipulators with 8-mm working distance (OptoSigma Corporation, USA). To make things convenient for cell recognition as well as environmental control during scanning biological samples, and the combination with other microscopy techniques, the scanning head of SICM is placed on an inverted optical microscope Diaphot 200 (Nikon Corporation, Tokyo, Japan).

By using a distance-modulated feedback control protocol (Pastre et al. 2001; Shevchuk et al. 2001), and an extra special high-resonant frequency modulation piezo actuator, novel SICM has achieved sufficient stability to permit real-time monitoring of membrane dynamics in living cells with nanometer resolution (Gorelik 2004). In this approach, we have modulated the distance between the pipette and sample to create an AC component in the ionic current. A lock-in amplifier is used to detect the changes in the AC current amplitude, which is then used as a feedback control signal for the SICM. It has been shown that this method of control has several advantages because it makes scanning more sensitive to the signal than those previously employed based on DC current (Hansma et al. 1989; Korchev et al. 1997). Owing to this improved sensitivity, the tip can be operated at a much closer distance from the sample surface, even for a relatively rough topography of biological specimen. Moreover, it also provides a reliable distance control mechanism even over contracting heart cells (Shevchuk et al. 2001). This is not affected by changes in the value of the ion current drift during the long periods of scanning, the osmotic strength changes of the solution, or even partial pipette blockage by contaminants in biological media. This control method has enabled us to reliably image the living renal epithelial cell surface for more than 24 hours (Gorelik 2004).

9.2.1 Hopping scanning mode

In 2009, Novak and colleagues developed an innovative scanning mode on the SICM (hopping probe ion conductance microscopy, HPICM), which achieved the capability to scan cells with complex structure, such as neurons. Instead of continuous negative feedback scanning mode, the SICM system applies a hopping scanning mode, which meshes the targeted area into smaller grids of the same size. Each grid is scanned separately.

Prior to scanning a grid, the SICM system will conduct a roughness test (i.e., the height change in a single grid) by measuring the altitudes of its four corners first. And then the system will perform high- or low-resolution scanning depending on the outcome of its measured roughness. A high roughness grid corresponds to high-resolution scanning and a low roughness gird corresponds to low-resolution scanning. In addition, the SICM system will revise the hopping range of its pipette according to its measured roughness test (i.e., extend the hopping range on a high roughness grid or vice versa). Thus, by conducting a roughness test on each grid, it drastically improves the system's performance in terms of scanning speed and image resolution (Zhang 2008).

9.2.2 The relationship between pipette resistance and scanning

The theory of SICM scanning is based on a curve that expresses the relationship between ionic current and distance from the pipette tip to the sample surface. The slope of the curve is the key factor that decides the pipette's sensitivity on the z-axis and resolution in scanning. The slope of the curve depends on resistances on a pipette. Figure 9.2 shows that the resistance of pipette R_t consists of R_h, R_p, and R_a.

1. R_h is the resistance generated from the oxide on the surface of the electrode.
2. R_p is the resistance generated on the tip of the pipette.
3. R_a is the resistance between the sample surface and the pipette tip.

$$I(z) = \frac{U}{R_t} \tag{9.1}$$

$$R_t = R_p + R_a + R_h \tag{9.2}$$

Because R_h is much smaller (possesses a lesser impact on Rt) compared to the other two resistances R_a and R_p, it can be ignored. The equation developed as (9.3)

$$R_t = R_p + R_a \tag{9.3}$$

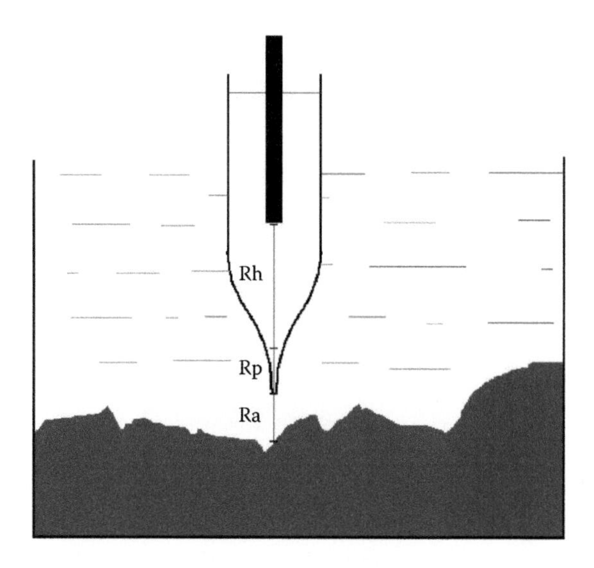

Figure 9.2 The structure of resistances in the pipette.

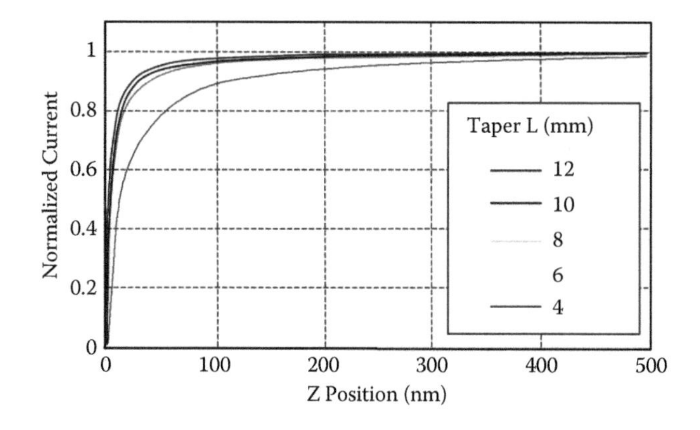

Figure 9.3 Distance-current curve under different tip heights (r_i = 25 nm; r_o = 60 nm).

Figure 9.3 demonstrates the relationship between tip length and its resistance given the outer and inner diameter of the pipette tip are designated as 60 nm and 25 nm, respectively.

9.2.3 Pressure: The factor affecting the shape of the pipette tip

The geometry changes of the pipette tips can be obtained by adjusting the pressure on the pipette puller. As illustrated from the following photos taken from an electron microscope, the shape and dimension are different

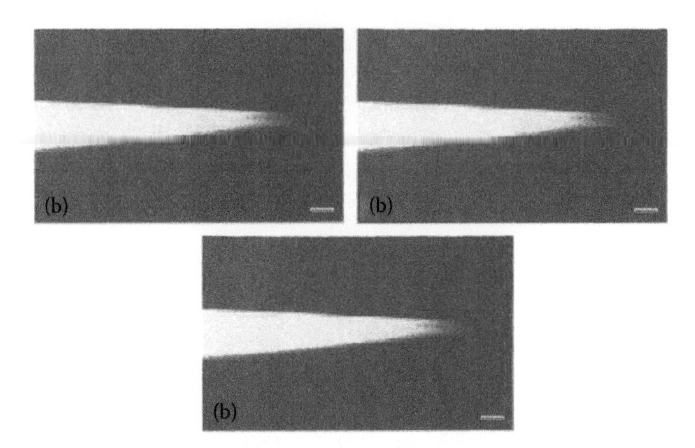

Figure 9.4 Pipette tip images under an electron microscope.

for each pipette tip, which produced under 350, 400, and 500 PSI air pressure, respectively (Figure 9.4).

Figure 9.5 demonstrates the patchy fracture surface on the tip of the pipette. On a closer look, Figure 9.6 reveals the extent of the fracture and its overall unevenness.

It was found that the higher pressure set in the pulling process (Sutter P100) can cause more roughness on the fracture surface of the pipette tip (Figure 9.5). This roughness and its surrounding resulted in an increased interference of leakage current. The conclusion can be drawn, therefore, that the higher the pressure in the pulling process, the less accurate the pipette is for measuring the sample's altitude.

Figure 9.5 Pipette tip fracture surface.

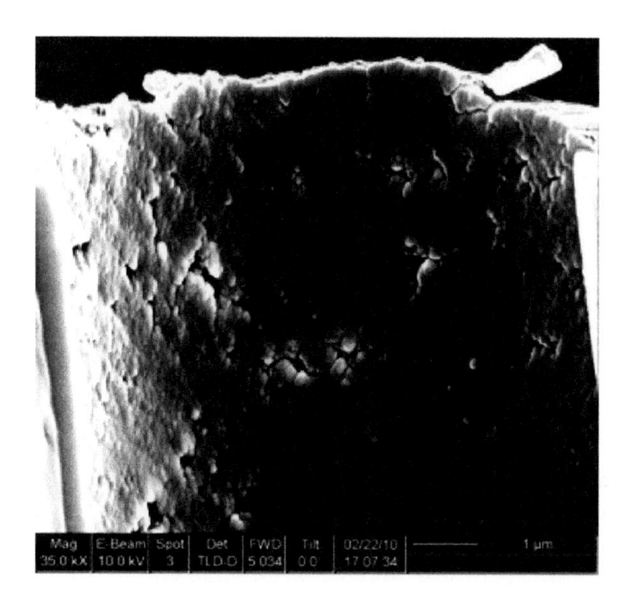

Figure 9.6 Pipette tip surface.

Moreover, the study shows that the saturated current and pipette resistance are negatively correlated to each other exclusively. Therefore, the resistance of a pipette can be regarded as an important aspect to assess the quality of pipette produced. And if the saturated current is detected during a scanning process, generally it is the result of a damaged pipette tip.

Figure 9.7(a) shows that pipettes produced under different pressure had different saturated currents. The higher the pressure applied in the pipette pulling process, the stronger the saturated current a pipette possessed. Given the same magnitude between the tip of the pipette and the sample surface, the pipettes with various tip lengths detect different

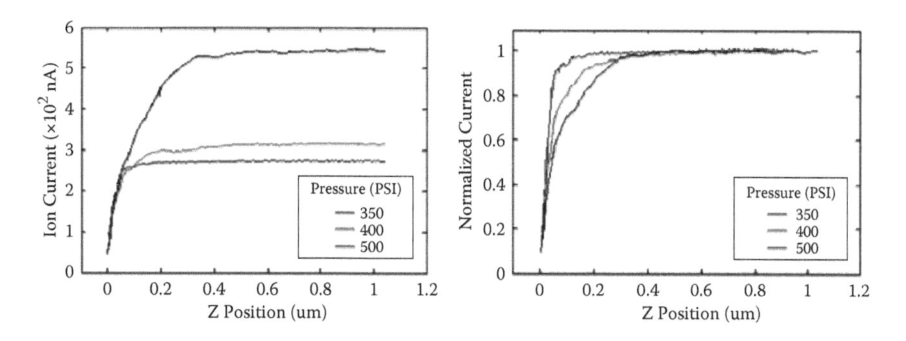

Figure 9.7 (a) The curve of current and distance generated under fixed voltage. (b) The curve of current and distance generated under fixed saturated current.

current. (The pipette with a 4-mm tip length detects twice as much current as a 10-mm tip length). In order to draw a comparison between the different shapes of pipettes, we redefined the maximum value of saturated current as 1 nA in the experiment. The shorter pipette tip has a larger effective working distance as shown in Figure 9.7(b).

9.3 Scanning ion conductance microscopy and "smart" patch-clamp

The patch-clamp technique is an extremely powerful and versatile method for studying electrophysiological properties of biological membranes. Soon after its development by Erwin Neher and Bert Sakmann, it was adopted by numerous laboratories and subsequently caused revolutionary advances in many areas of research in both cellular and molecular biology. Subsequent key improvements in this methodology have refined its use and applicability to virtually all biological preparations. In recent years, even more technical and experimental variations of the patch-clamp method have emerged and further expanded its power to address previously unapproachable questions in cell biology.

Today, the patch clamp is the method of choice when it comes to investigating cellular and molecular aspects of electrophysiology, and has shown that ion channels frequently associate with specific sub-cellular structures and are not uniformly distributed on the cell surface, which is important for their function (Korchev 2000; Gu 2002; Gorelik 2002). However, the conventional patch-clamp technique does not allow the precise selection of a region of interest on the cell to be investigated and thus provides limited information on the important question of ion channel localization (Marrion and Tavalin 1998; Almers, Stanfield, and Stuhmer 1983). For example, it is hard to record from fine structures such as microvilli or to patch opaque samples that are cultured on membrane filters, because of the difficulty of optically controlling patch pipettes in their approach to such samples. Typically, in the patch-clamp method an electrode is positioned using manual adjustments while focusing between the pipette and sample under a light microscope. Therefore, during the approach, the sample and pipette electrode are not in the same focal plane, and under these circumstances it is easy to damage the pipette tip and the fine structure of living cells (Gu 2002; Gorelik 2002).

The smart patch-clamp method was first reported by Gorelik (2002) and Gu (2002), in which the same micropipette can be used first in SICM protocols to image the cell surface and identify membrane structures of interest, and then as a patch pipette for electrophysiological recording, as shown schematically in Figure 9.8.

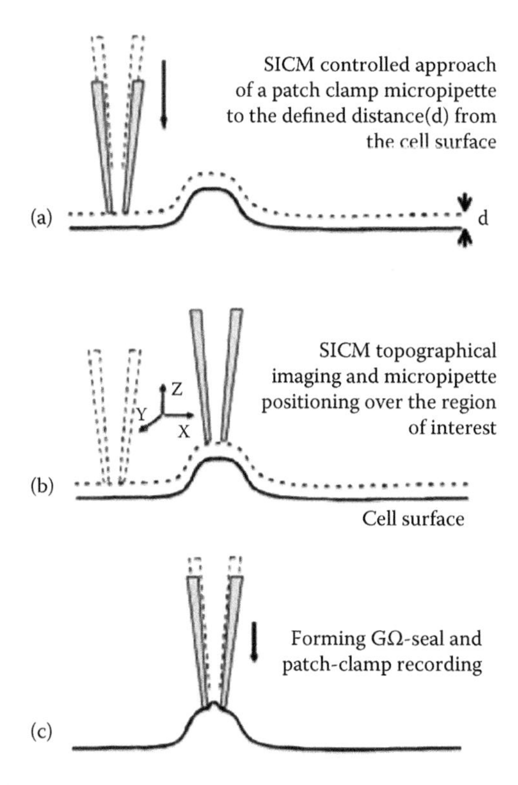

Figure 9.8 Scanning patch-clamp principle of operation. (a) A micropipette approaches the cell surface and reaches a defined separation distance *d*, whereupon the distance is kept constant by SICM feedback control. (b) The SICM scans this micropipette over the cell surface and positions it at a place of interest for patch-clamp recording. (c) The micropipette is lowered to form the giga-ohm seal for patch-clamp recording from the selected structure.

In this method the scanning micropipette is arranged vertically and manipulated by SICM computer control. A feedback control system operates when the pipette approaches the cell surface. As soon as the pipette reaches a distance of about 50 nm from the surface, the SICM feedback control maintains this constant tip-sample separation. This procedure makes the approach straightforward and safe, because the patch pipette is prevented from touching the cell membrane until it is desired to do so in order to form a seal. Once the SICM protocol has obtained a topographic image of the cell surface, it can be used to position the patch pipette precisely over a place of interest for patch recording (Figure 9.9).

Finally, the feedback control is switched off, the pipette is lowered and suction applied, resulting in the formation of a GΩ-seal. Ion channel recording is then performed by conventional methods.

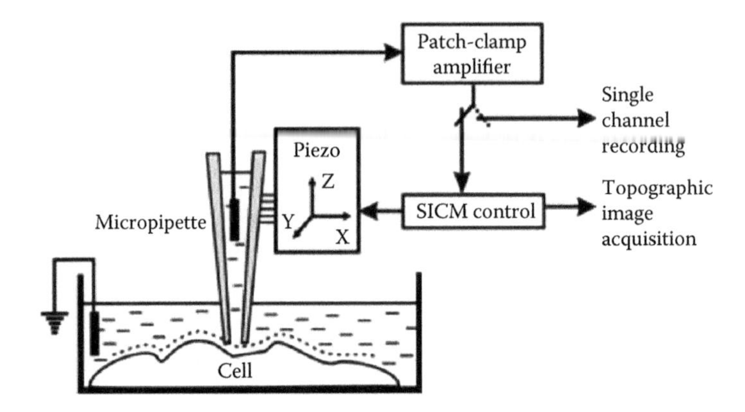

Figure 9.9 Schematic diagram of the scanning patch-clamp setup. The micropipette is mounted on a three-axis piezo actuator controlled by a computer. The ion current that flows through the pipette is measured by a patch-clamp amplifier, and it is used for feedback control to keep a constant distance between the micropipette and the sample during scanning. Upon completion of the scanning procedure, computer control is used to position the micropipette at a place of interest based on the topographic image acquired, and finally the same patch-clamp amplifier is used for electrophysiological recording.

9.4 *Applications of scanning ion conductance microscopy to living cells*

SICM has already been reported as capable of imaging the topography of living cardiac myocytes with lateral resolution of 50 nm and vertical resolution of 10 nm (Korchev et al. 1997), and measuring changes in cell volume (Figure 9.10; Marrion and Tavalin 1998; Almers, Stanfield, and Stuhmer 1983). In previous studies we have also shown that the SICM is capable of scanning microvilli of renal epithelial cells at high resolution (Korchev et al. 2000; Gu 2002; Zhang 2005).

Typically Korchev's lab uses 100- to 150-nm inner diameter pipettes, controlled 50 to 75 nm from the cell surface so that the pipette "rolls" over the cell surface without touching it. Lateral topographic resolution is given by the diameter of the pipette and is typically 100 to 150 nm. The SICM has been combined with confocal microscopy to simultaneously measure the contraction of cardiac myocytes and the intracellular calcium (Gorelik 2002; Shevchuk et al. 2001). The pipette has been used as a local light source for scanning near-field optical microscopy to obtain images of cardiac myocyte and sperm (Figure 9.11; Korchev 2000; Gu 2002; Rothery et al. 2003).

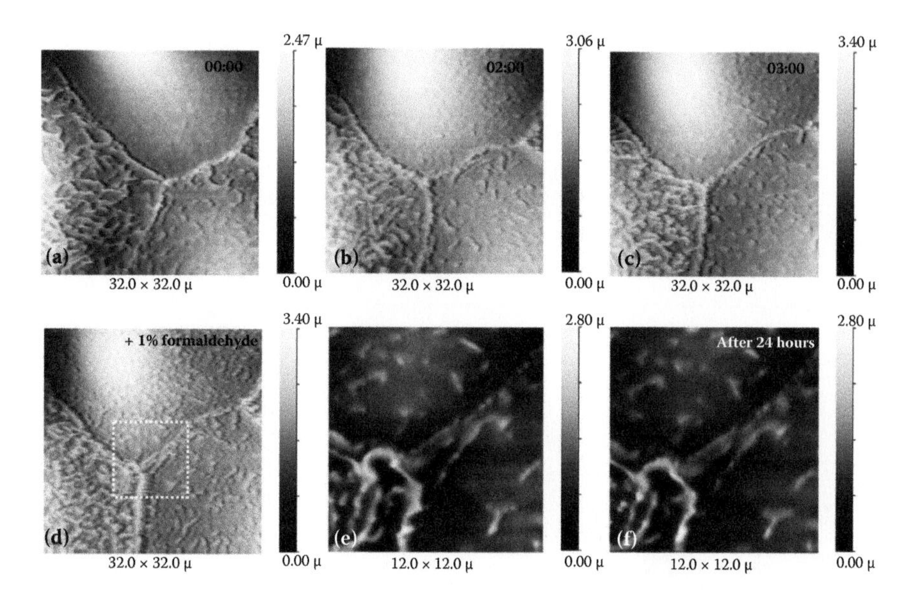

Figure 9.10 SICM images of a specific region of A6 cells. Parts (a–c) are images from the same area of live A6 cells in this monolayer imaged over a 4-h period. This monolayer was then fixed with formaldehyde and the same region imaged, shown in (d). Parts (e) and (f) show a zoom of the dotted region marked in (d).

The SICM method also combines the capabilities of patch-clamping with the advanced features of SICM. We have used the pipette to locally deliver ions for the mapping of ion channels in cardiac myocytes (Korchev 2000) and extended this work to "smart" patch clamps (Figure 9.12; Gu 2002; Gorelik 2002).

This method is employed to perform ion channels patch-clamp recording on subcellular features such as microvilli of epithelial cells (Figure 9.13), head of sperm, living renal cells, and neuronal dendrites (Gu 2002; Gorelik 2002, 2004).

SICM is often used in drug testing and tissue necrosis research. In 2008, Corlik and colleagues studied isolated ventricular myocytes from patients with ischemic heart disease, idiopathic dilated cardiomyopathy, and hypertrophic obstructive cardiomyopathy and determined T-tubule structure with SICM and fluorescence microscopy.

9.5 Conclusion

The scanning ion conductance microscope is an important member of the scanning probe microscope family. It is able to apply noncontact scanning on the surface of a sample by measuring the ion current detected from

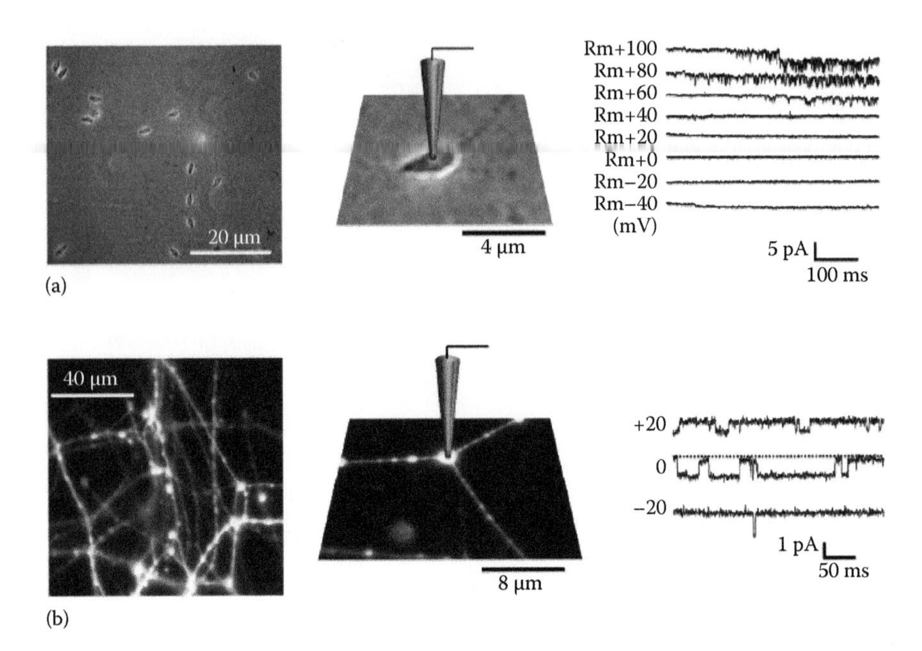

Figure 9.11 Ion channel recordings from sea urchin spermatozoa and from fine neuronal processes. (a) Smart patch-clamp recording from sea urchin spermatozoa: optical image of sea urchin sperm (left), a schematic of the pipette positioned over a sperm (middle), and a cell-attached current recording showing Ba^{2+} currents through Ca^{2+} channels (right). (b) Smart patch-clamp recording from a neurite varicosity structure: optical image of SCG neurites loaded with a lipid-binding dye DiI (left), the pipette placed above the junction between neurites (middle), and a cell-attached recording of Ba^{2+} currents through Ca^{2+} channels (right).

the micropipette, thereby constructing a visual image of the sample. The SICM has the capability of high-resolution imaging, easy pipette production, and ensuring that the sample surface remains undisturbed during the scanning process. It has emerged as a valuable tool in studies of living cells under physiological conditions, and is complementary to the scanning chemical microscope and atomic force microscope in biological research.

Over the years because of the development in hardware and software in the SICM, there has been significant improvement in the speed of capturing an image as well as quality and accuracy of the image constructed, and the SICM technique is compatible with other technologies such as fluorescence, patch-clamp, and SECM. The SICM has become the new benchmark for nanometer-scale fabrication and is widely used in other areas in the life sciences.

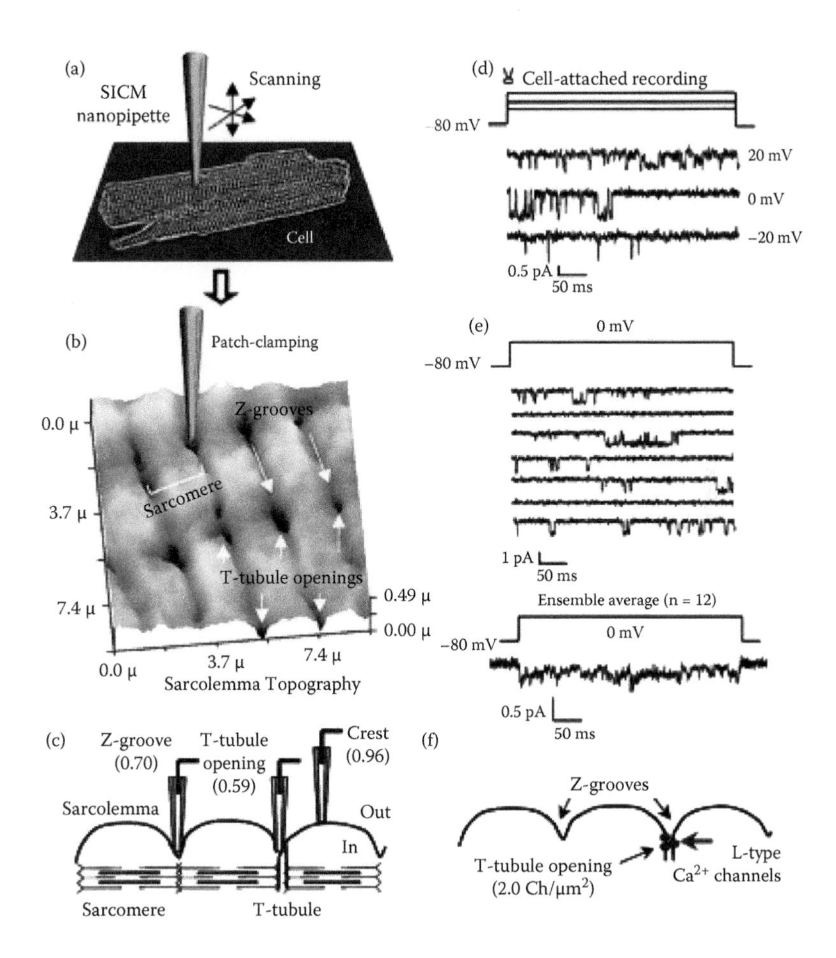

Figure 9.12 L-type Ca^{2+} channel distribution in the cardiac myocyte sarcolemma: mapping of ion channels by the high-resolution scanning patch-clamp technique. (a) To perform patch-clamp recording from different regions on the cardiac myocyte sarcolemma, we selected an area of interest (white dotted square). The patch-clamp nanopipette with a backfill solution for investigating Ca^{2+} channels is used to image the cell surface topography controlled by SICM. (b) Experimental topographic image of a representative rat cardiomyocyte membrane. Z-grooves, T-tubule opening, and characteristic sarcomere units are marked. (c) Functional schematic of sarcomere units showing the position of the probed region (Z-groove, T-tubule opening, and scallop crest). Probabilities of forming a GΩ seal as a function of surface position are shown in parenthesis. (d) Cell-attached Ba^{2+} current transients at voltages of +20, 0, –20 mV. (e) Several current transients elicited at 0 mV from one patch and ensemble average of 12 transients showing typical L-type inactivation kinetics. (f) Statistical distribution of L-type Ca^{2+} channels with the highest density near the T-tubule opening.

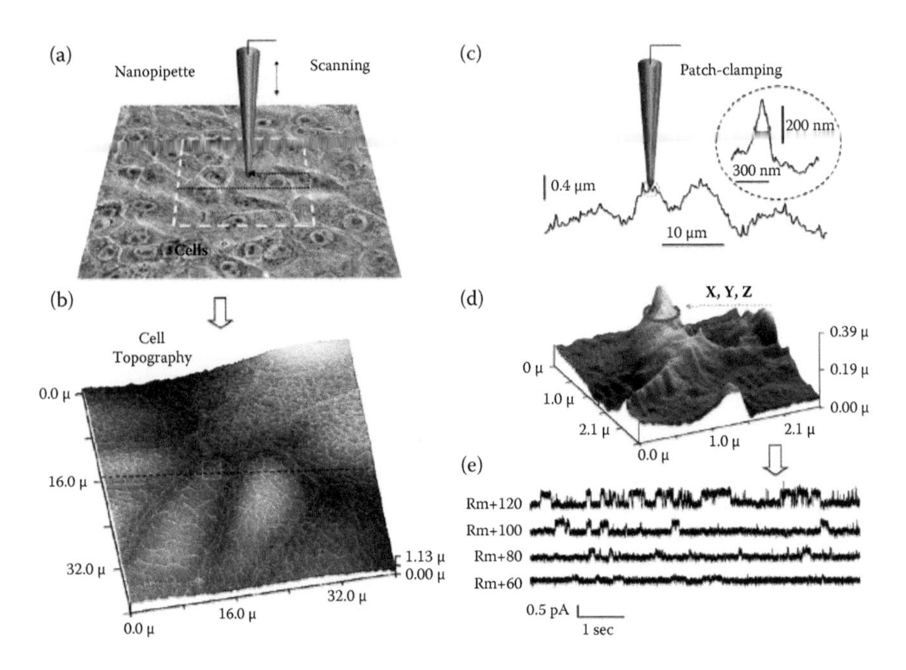

Figure 9.13 High-resolution scanning patch-clamp technique: ion channel recording on the tip of a single microvillus.

References

Almers, W., P. R. Stanfield, and W. Stuhmer. 1983. Lateral distribution of sodium and potassium channels in frog skeletal muscle: measurements with a patch-clamp technique. *J Physiol* 336:261–84.

Gorelik, J. 2002. Ion channels in small cells and subcellular structures can be studied with a smart patch-clamp system. *Biophys J* 83 (6):8.

Gorelik, J. 2004. The use of scanning ion conductance microscopy to image A6 cells. *Mol Cell Endocrinol* 31:8.

Gorelik, J. 2008. Loss of T-tubules and other changes to surface topography in ventricular myocytes from failing human and rat heart. PNAS, 106:6854–6859.

Gu, Y. 2002. High-resolution scanning patch-clamp: new insights into cell function. *FASEB J* 16:3.

Hansma, P. K., B. Drake, O. Marti, S. A. Gould, and C. B. Prater. 1989. The scanning ion-conductance microscope. *Science* 243 (4891):641–3.

Happel, P., G. Hoffmann, S. A. Mann, and I. D. Dietzel. 2003. Monitoring cell movements and volume changes with pulse-mode scanning ion conductance microscopy. *J Microsc* 212 (Pt 2):144–51.

Korchev, Y. E. 2000. Cell volume measurement using scanning ion conductance microscopy. *Biophys J* 78:7.

Korchev, Y. E., C. L. Bashford, M. Milovanovic, I. Vodyanoy, and M. J. Lab. 1997. Scanning ion conductance microscopy of living cells. *Biophys J* 73 (2):653–8.

Korchev, Y. E., Y. A. Negulyaev, C. R. Edwards, I. Vodyanoy, and M. J. Lab. 2000. Functional localization of single active ion channels on the surface of a living cell. *Nature Cell Biol* 2 (9):616–9.

Mann, S. A., G. Hoffmann, A. Hengstenberg, W. Schuhmann, and I. D. Dietzel. 2002. Pulse-mode scanning ion conductance microscopy: a method to investigate cultured hippocampal cells. *J Neurosci Methods* 116 (2):113–7.

Marrion, N. V., and S. J. Tavalin. 1998. Selective activation of Ca^{2+}-activated K^+ channels by co-localized Ca^{2+} channels in hippocampal neurons. *Nature* 395 (6705):900–5.

Novak, P., C. Li et al. (2009). Nanoscale live-cell imaging using hopping probe ion conductance microscopy. *Nat Methods* 6 (4):279–281.

Pastre, D., H. Iwamoto, J. Liu, G. Szabo, and Z. Shao. 2001. Characterization of AC mode scanning ion-conductance microscopy. *Ultramicroscopy* 90 (1):13–9.

Proksch, R., R. Lal, P. K. Hansma, D. Morse, and G. Stucky. 1996. Imaging the internal and external pore structure of membranes in fluid: tapping Mode scanning ion conductance microscopy. *Biophys J* 71 (4):2155–7.

Rothery, A. M., J. Gorelik, A. Bruckbauer, W. Yu, Y. E. Korchev, and D. Klenerman. 2003. A novel light source for SICM-SNOM of living cells. *J Microsc* 209 (Pt 2):94–101.

Shevchuk, A. I., J. Gorelik, S. E. Harding, M. J. Lab, D. Klenerman, and Y. E. Korchev. 2001. Simultaneous measurement of Ca^{2+} and cellular dynamics: combined scanning ion conductance and optical microscopy to study contracting cardiac myocytes. *Biophys J* 81 (3):1759–64.

Zhang, Y. 2005. Scanning ion conductance microscopy reveals how a functional renal epithelial monolayer maintains its integrity. *Kidney Int* 68:8.

Zhang, Y. 2008. Scanning ion conductance microscopy and its applications in nanobiology. *Acta Biophys Sin* 28 (10):644.

Problems

1. How does SICM detect the distance between the pipette tip and the sample surface?
2. What kind of samples or specimens are suitable for application of the SICM?
3. What is the most important advantage of SICM? How does it help your study?
4. What kind of sample information/data can you obtain by applying the SICM technique?
5. How does pipette tip length and tip opening size affect the scanning?
6. What is the operational principle for SICM?
7. What determines the resolution of the SICM image?

chapter ten

Advanced photoacoustic microscopy

Yichen Ding, Qiushi Ren, and Changhui Li

Contents

10.1 Introduction

After more than 400 years of development, optical microscopy is not only a key instrument, but also important in promoting developments in biomedicine (Pluta 1989; Slayter and Slayter 1992; Abramowitz 1993; Herma and Lemasters 1993; Pawley 1995; Inoué and Spring 1997; Sheppard and Shotton 1997; Bradbury and Bracegirdle 1998; Herman 1998). However, two obstacles remained for centuries in optical microscopy. One is the diffraction limit, which shrank the optical focal point to half of the light wavelength at most, but this was broken during the last decade (Fernández-Suárez and Ting 2008; Hell 2009; Huang et al. 2010); the other one is the diffusion limit, which results from the scattering characteristic of photons in turbid medium, and leads to a very shallow imaging depth for high-resolution optical microscopic methods in biological tissues (L.V. Wang and Wu 2007; Ntziachristos 2010). Moreover, although the optical absorption property provides essential information on the biomolecule, it does not play an important role in the traditional optical microscopic methods.

Primarily since the 1990s, a novel biomedical imaging modality called photoacoustic tomography (PAT) has gained significant progress. PAT is based on the photoacoustic effect (Bell 1880; Oraevsky and Karabutov 2003; Xu and Wang 2006; C. Li and Wang 2009a), which refers to the generation of sound waves after the item absorbing intensity-varying electromagnetic (EM) waves. The photoacoustic effect is highly sensitive to the optical absorption properties of tissues. As one of the major categories of PAT, photoacoustic microscopy (PAM) breaks the diffusion barrier and opens a new window for optical microscopy. Different from traditional pure optical microscopic methods, where the multiscattering mechanism can significantly impact the imaging quality, PAM can take advantage of those diffused photons to achieve deeper imaging, because PAM detects signals from much less scattered ultrasound waves instead of light. In general, the scattering coefficient for ultrasound in soft tissue is two to three orders less than light (L.V. Wang 2009). Thus, the soft tissue is nearly "transparent" to ultrasound. By detecting much less scattered ultrasound waves, PAM can therefore conserve high-resolution imaging of optically absorbing targets in deep tissue. PAM uniquely combines the optical contrast with ultrasonic detection, emerging as a novel hybrid biomedical imaging method. During the past decade, PAM has successfully imaged multiscale tissues, from submolecular organelles to subcutaneous cancer tissues in vivo (L.V. Wang 2008, 2009; C. Li and Wang 2009a; L.V. Wang and Hu 2012; J. Yao and Wang 2013).

Based on different mechanisms in determining the lateral resolution, PAM has two major types: acoustic-resolution PAM (AR-PAM) and optical-resolution PAM (OR-PAM). Several books and book chapters, as well as review journal articles, have provided comprehensive coverage of PAM, focusing on instrumentation (L.V. Wang and Wu 2007; Ntziachristos 2010), contrast agents (C. Li and Wang 2009a; L.V. Wang and Hu 2012), or biomedical applications (Guittet et al. 1999; Maslov et al. 2005, 2008; H.F. Zhang et al. 2006c, 2007b; J. Yao and Wang 2013). In this chapter, we first describe the basic mechanism of PAM, then discuss two major types of PAM, and finally review the studies in contrast agents for PAM. In the final section, we discuss the prospects of PAM.

10.2 PAM mechanism

10.2.1 Photon scattering and absorption in tissue

Scattering and absorption are two major interactions between EM waves and biological tissue. Since the refractive index mismatch exists not only between various biological components but also between tissues and the environmental medium, photons undergo multiple scattering soon after they enter living tissue. A typical scattering coefficient is 100 cm^{-1}, that is, the mean free path of a photon in tissue is about 100 μm. However, multiple scattering will significantly alter the direction of the photon, which is the major reason that all traditional optical microscopic methods can only image targets within a very shallow region into the turbid tissue, typically ~100 μm to 1 mm.

Unlike light scattering, which is closely related to the shape and size of the scatter, optical absorption is a characteristic of matter that is highly related to the molecular structure. When photons are absorbed by molecules, the EM energy can convert into heat, chemical energy, or re-emit as other photons, such as fluorescence. Among these, the heating effect is the most common phenomenon. According to thermodynamics, tissue generally undergoes thermal expansion after it was heated, which will generate sound waves propagating outward. This phenomenon is called the photoacoustic effect. Unlike scattering, whether the photons are scattered or not, they can all be absorbed and contribute to the photoacoustic signal. That is the reason why PAM can image deeper in a turbid medium compared to traditional optical microscopic methods.

Moreover, the optical absorption characteristic is a "fingerprint" for various molecules. Figure 10.1 presents typical absorption spectra for several biomolecules. Relying on the absorption contrast, PAM has been successfully used to image various targets, such as cell nuclei (D.K. Yao et al. 2010), hemoglobin (H.F. Zhang et al. 2007a), melanin (Staley et al. 2010), lipid (H.W. Wang et al. 2011), and water (Z. Xu et al. 2010). In addition, Figure 10.1 also shows that the absorption values can be highly dependent on optical wavelengths. Thus, multi-wavelength PAM has the potential to differentiate molecular constitutions.

10.2.2 Generation of photoacoustic signal

Production of the photoacoustic effect takes place in several steps: (1) EM pulses illumination, (2) photon absorption, (3) temperature and thermal expansion. In the following, the mathematical formulations for the generation and propagation of the ultrasound waves are presented.

The photoacoustic effect relies on the heating phenomenon. For optical illumination, the heating function H can be explicitly written as (C. Li et al. 2008):

$$H(\mathrm{r},t) = \eta\mu_a(\mathrm{r}) \cdot \Phi(\mathrm{r},t), \tag{10.1}$$

where $H(r, t)$ refers to the thermal energy converted at spatial position r and time t by the EM radiation per unit volume per unit time; η represents

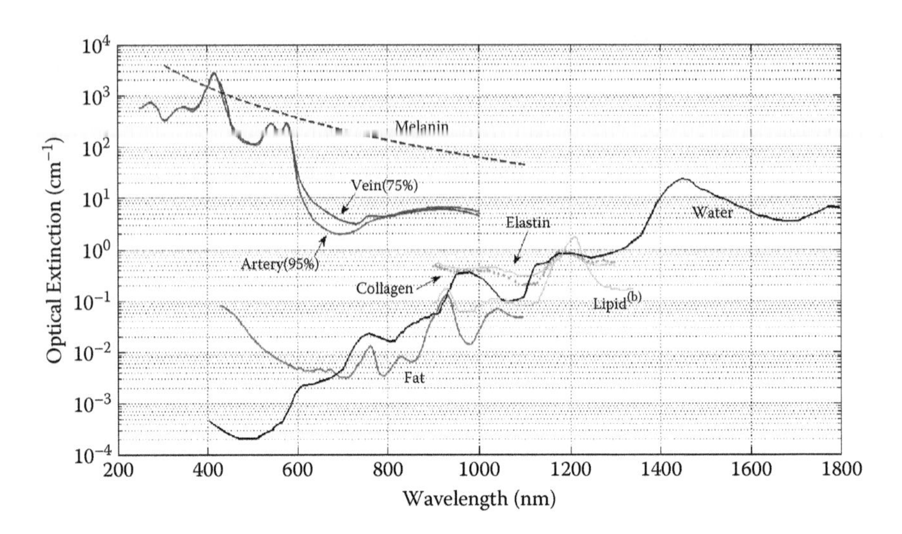

Figure 10.1 **(See color insert.)** Absorption coefficients of variable components in tissue at different wavelengths. Data for artery (HbO_2 95%, 150 gl^{-1}, red) and vein (HbO_2 75%, 150 gl^{-1}, blue) from http://omlc.ogi.edu/spectra/hemoglobin/summary.html; water (black) from Hale and Querry 1973; melanin (green dashed) from http://oomlc.ogi.edu/spetra/melanin/mua.html; lipid(b) (purple), elastin (yellow), and collagen (blue dashed) from Tsai et al. 2001); and fat (orange) from http://omlc.ogi.edu/spectra/fat/fat.txt.

the ratio of how much absorbed energy can be converted into heat. It demonstrates that generating heat in tissue depends on both absorption coefficient $\mu_a(\mathrm{r})$ and optical radiation fluence rate $\Phi(\mathrm{r},t)$. In PAM, short laser pulses are generally used for the excitation source, and the pulse duration τ satisfies the thermal confinement condition, which is

$$\tau << \tau_{th} = \frac{d_c^2}{D_T}, \tag{10.2}$$

where d_c is the characteristic dimension size of the target, and D_T is thermal diffusivity (~0.14 mm²/s for soft tissue), τ_{th} is called the thermal confinement threshold, which reflects the time of thermal diffusion.

Under this condition, the photoacoustic pressure p generated in an acoustically homogeneous, nonviscous, and nondispersive medium can be described by the following wave function (Morse and Ingard 1986):

$$\nabla^2 p(\mathrm{r},t) - \frac{1}{v_s^2}\frac{\partial^2}{\partial t^2} p(\mathrm{r},t) = -\frac{\beta}{C_p}\frac{\partial}{\partial t} H(\mathrm{r},t), \tag{10.3}$$

where v_s is the acoustic velocity, β is the isobaric volume expansion coefficient in K^{-1}, and C_p is the isobaric specific heat in J/K·kg. Under the thermal confinement condition, the solution of Equation (10.3) by using a Green function approach in an infinite homogeneous medium is

$$\begin{aligned} p(\mathrm{r},t) &= \int dt \int d\mathrm{r}' G(\mathrm{r},t;\mathrm{r}',t') \frac{\beta}{C_p} \frac{\partial H(\mathrm{r}',t')}{\partial t'} \\ &= \frac{\beta}{4\pi C_p} \int d\mathrm{r}' \frac{1}{|\mathrm{r}-\mathrm{r}'|} \frac{\partial H\left(\mathrm{r}',t-|\mathrm{r}-\mathrm{r}'|/v_s\right)}{\partial t} \\ &= \frac{\beta}{4\pi C_p} \frac{\partial}{\partial t} \int d\mathrm{r}' \frac{1}{|\mathrm{r}-\mathrm{r}'|} H\left(\mathrm{r}',t-|\mathrm{r}-\mathrm{r}'|/v_s\right), \end{aligned} \tag{10.4}$$

where $G(r, t; r', t')$ is the Green function and is solved as:

$$G(\mathrm{r},t;\mathrm{r}',t') = \frac{\delta\left(t-t'-|r-\mathrm{r}'|/v_s\right)}{4\pi|r-\mathrm{r}'|}. \tag{10.5}$$

In some cases, compared with the system spatial resolution d_c, the radiation pulse duration τ can be so short that the volume expansion of absorbed contrast agents during the heating process in tissue can be negligible. It is named the acoustic stress confinement. The mathematical description of the confinement is

$$\tau << \tau_{st} = \frac{d_c}{v_s}, \tag{10.6}$$

where τ_{st} is the acoustic stress confinement threshold.

If both confinement conditions can be satisfied, the EM radiation can be well approximated to be a temporal delta pulse. Thus, the heating process can be simplified as the absorbed energy density $A(r)$ at location r multiplied by a delta function $\delta(t)$:

$$H(r,t) \approx A(r) \cdot \delta(t). \tag{10.7}$$

Then, the process is treated as a Dirac delta heating function. One can derive the initial buildup pressure by an absorber as:

$$p_0(\mathrm{r}) = \Gamma \cdot A(\mathrm{r}), \tag{10.8}$$

where Γ denotes the Grueneisen parameter, a dimensionless parameter, defined as:

$$\Gamma = \frac{v_s^2 \beta}{C_p}. \tag{10.9}$$

Equation (10.8) can be used to estimate the initial pressure. Under the delta heating condition, Equation (10.4) can be simplified to:

$$p(\mathrm{r}, t) = \frac{\beta}{4\pi C_p} \frac{\partial}{\partial t} \int d\mathrm{r}' \frac{1}{|\mathrm{r} - \mathrm{r}'|} A(\mathrm{r}')\delta\left(t - |\mathrm{r} - \mathrm{r}'|/v_s\right) \tag{10.10}$$

$$= \frac{v_s^2}{4\pi} \frac{\partial}{\partial t} \frac{1}{v_s t} \iint_{|\mathrm{r}-\mathrm{r}'|=v_s t} p_0(\mathrm{r}')ds'.$$

Equation (10.10) indicates the photoacoustic pressure at r and the time t comes from initial photoacoustic sources over a spherical shell with a radius of $v_s t$. Notably, during the propagation of the photoacoustic wave, the pressure value is proportional to the size of the absorber, but inversely proportional to the distance from the source. In addition, the temporal width of the photoacoustic signal relates to the size of the absorber: the smaller the absorber, the narrower the signal width that contains more higher-frequency components.

However, although the scattering of the acoustic waves can be ignored in soft tissue, the attenuation of the acoustic wave is not generally negligible, especially at high frequencies or in highly absorptive acoustic media. Thus, with the increase of ultrasound frequency, tissue works as a low-pass filter, which reduces spatial resolution at depth. That is the reason why high-frequency acoustic detector cannot be applied in deep tissue imaging, that is, both the amplitude and the temporal profile of the photoacoustic signal are changed by ultrasound attenuation. These characteristics are useful for analyzing the size and location of the source target during the observation of the photoacoustic signal.

Although PAM detects much less scattered ultrasound waves, the signal generation requires the presence of a photon absorption. Therefore, the imaging depth of the PAM also highly depends on the light propagation in tissue. Light propagation in tissue is governed by several parameters, the absorption coefficient μ_a, the scattering coefficient μ_s, and the anisotropy factor g. In tissue, μ_s is on the scale of ~100 cm^{-1}, and g is approximately 0.9. However, the absorption coefficient is highly dependent on tissue composition, which can range from 0.01 cm^{-1} to 200 cm^{-1}. Mathematically, the light transportation is determined by the radioactive transfer equation, which can be solved by Monte Carlo simulation or other approximate analytical methods, such as the diffusion approximation. This issue has been addressed in many books and articles and we will not discuss it in detail (C. Li and Wang 2009). In general, the mean free path of a photon in tissue is about 0.1 mm, which is the primary factor that limits the imaging depth of high-resolution pure optical microscopy. For instance, the fluence rate of a planar illumination source in the diffusion regime can be calculated as:

$$\Phi(z) \approx \Phi_0 \exp(-\mu_{eff} \cdot z), \tag{10.11}$$

where Φ_0 is the illumination fluence, z is the penetration depth, and the effective attenuation coefficient $\mu_{eff} = \sqrt{3\mu_a(\mu_a+\mu_s(1\text{-}g))}$. Both absorption and scattering contribute to the transportation of light. Due to the safety limitation, maximum illumination strength in PAM is strictly constrained.

10.3 PAM techniques

The ultrasound transducer used in PAM is generally a focused transducer. Based on different methods to determine the lateral resolution, PAM can be classified into acoustic-resolution PAM (AR-PAM) and optical-resolution PAM (OR-PAM).

10.3.1 Acoustic-resolution photoacoustic microscopy

AR-PAM aims to break through the diffusion limit in traditional high-resolution optical microscopy. It takes advantage of the diffused photons and weakly scattered ultrasound in tissue, and achieves high lateral resolution by acoustic focusing. Since not only ballistic but also diffused photons can be used to generate photoacoustic signals, the imaging depth of AR-PAM is primarily determined by the transportation depth of diffused photons, which can be up to several centimeters in living tissue by using near infrared light.

Based on acoustic focusing, the lateral resolution of AR-PAM is similar to optical microscopy. The acoustic focus spot R_L is calculated as:

$$R_L = 0.71\frac{\lambda_A}{NA_A}, \tag{10.12}$$

where λ_A is the wavelength of the acoustic wave, and NA_A denotes the numerical aperture of the ultrasonic transducer. The constant 0.71 reflects the difference between acoustic and optical detection. In acoustic detection, it measures the amplitude of the signal, whereas in optical detection it measures intensity. With the use of a focused transducer, the lateral resolution is only well maintained within a limited range centered at the focal point. Focal zone Z_A generally represents the region. Similar to optical focus, it is two times the Rayleigh range. The Rayleigh range corresponds with the place of lateral resolution degraded $\sqrt{2}$ times due to diffraction along the focusing axis. Therefore, in AR-PAM, it is:

$$Z_A = 2.4\frac{\lambda_A}{NA_A^2}. \tag{10.13}$$

This indicates the focal zone is related to NA_A, that is, the higher the value of NA_A, the shorter the focal zone.

In terms of axial resolution, the photoacoustic signal is also very different from conventional optical microscopy. In optical microscopy, the axial resolution is the full width at half maximum (FWHM) of the focal depth along the optical axis. However, PAM detects the time-resolved ultrasound, so the axial resolution is determined by the bandwidth of the system, as in Equation (10.14).

$$R_A = 0.88\frac{v_A}{\Delta f_A}, \tag{10.14}$$

where v_A and Δf_A are the acoustic speed and system bandwidth, respectively. Since the bandwidth is generally proportional to the transducer's central frequency, increasing central frequency leads to higher axial resolution, but at the expense of penetration depth, which is due to the acoustic attenuation.

Figure 10.2(a) demonstrates one of the most successful AR-PAM systems: the dark-field AR-PAM (Maslov et al. 2005). The dark-field design minimizes the interference caused by strong photoacoustic signals from superficial structures. Unlike in pure optical confocal microscopy where confocal setup helps reach high resolution, in PAM the light can only be weakly focused into the diffusive tissue, helping to increase the signal-to-noise ratio (SNR). As shown in Figure 10.2(a), a conical lens is used to reshape the illumination light from a beam to a ring, which forms a donut-shaped illumination pattern after passing a condenser, and slightly focuses in tissue to overlap ultrasonic focus coaxially. The subcutaneous vasculature of mouse hind limb is shown in Figure 10.2(b) with this setup (Ye et al. 2012a). Figure 10.2(c) presents the different detected signals of mice from one to three dimensions (H.F. Zhang et al. 2006a). Based on the dark-field design, signals from the epidermis were eliminated manifestly, which can be seen from the A-scan and B-scan results. When applying multi-wavelengths for different absorbers in tissue, this AR-PAM can

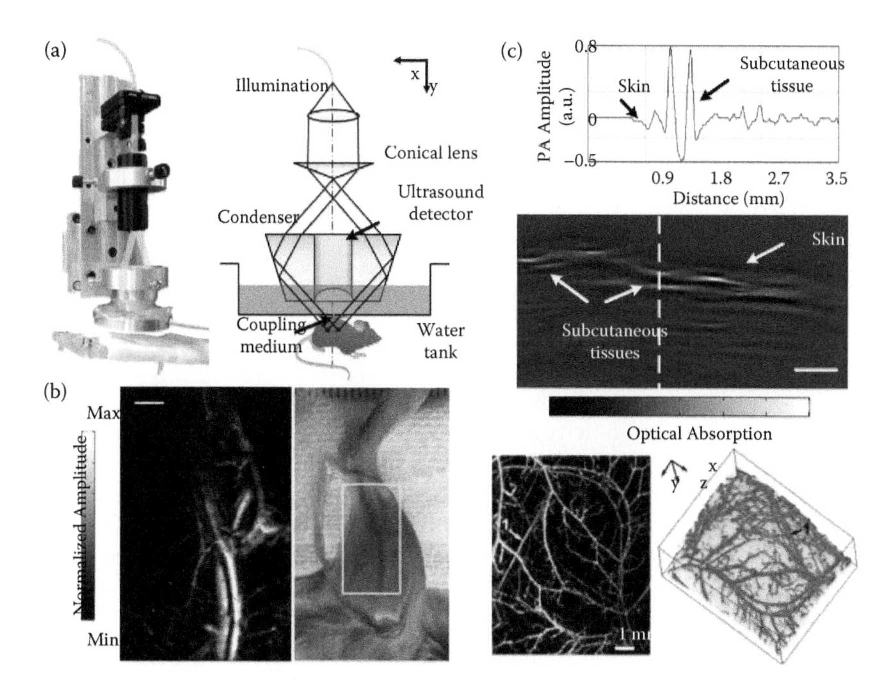

Figure 10.2 **(See color insert.)** AR-PAM system and imaging results. (a) System setup. (b) The result of subcutaneous vasculature of the mouse hind limb with 20 MHz central frequency, scale bar: 1 mm.(c) The A-scan, B-scan, maximum amplitude projection, and three-dimensional vasculature results of mouse, scale bar: 1 mm. (From Ye et al. 2012a (panel b); H.F. Zhang et al. 2006a (panel c); H.F. Zhang (panel c). Reproduced with permission.)

offer functional imaging (H.F. Zhang et al. 2006). Here, with a 50-MHz (70%) bandwidth and a transducer with NA_A of 0.44, its lateral resolution is 45 μm and axial resolution is 15 μm, while the imaging depth is 3 mm (Maslov et al. 2005, H.F. Zhang et al. 2006c). Detailed procedures of operating dark-field AR-PAM can be found in H.F. Zhang et al. (2007b).The resolution of AR-PAM is scalable by changing transducers with different central frequencies. For instance, if the transducer is changed to 20 MHz with an acoustic lens having $NA_A = 0.46$, the lateral resolution and axial resolutions are 100 μm and 105 μm, respectively (Ye et al. 2012a); if the transducer is changed to a 5 MHz or lower one, typically lateral resolution is 560 μm and axial resolution is 144 μm, with maximum penetration depth of 38 mm in chicken breast (K.H. Song and Wang 2007; K.H. Song et al. 2008). How to choose the transducer depends on the imaging target and imaging depth. As the central frequency decreases, for example, to 5 MHz, the photoacoustic modality is also called photoacoustic macroscopy (PAMac) since its imaging depth can be of the order of centimeters with millimeter resolution.

To improve the out-of-focus lateral resolution of AR-PAM with a large NA_A transducer, a synthetic-aperture focusing technique or virtual-detector combined with coherence weighting can be useful in AR-PAM (Liao et al. 2004, M.L. Li et al. 2006, C. Li and Wang 2009b). Recently, a novel bright-field AR-PAM demonstrated functional imaging at video rate, based on voice-coil scanning (L. Wang et al. 2012), also proposing a solution for higher imaging speed in AR-PAM.

10.3.2 Optical-resolution photoacoustic microscopy

In AR-PAM, the diffused photons are used to achieve imaging at greater depth, with the resolution determined by acoustic detection. However, PAM can also use ballistic photons in the shallow region to achieve optical microscopic lateral resolution. Distinct from AR-PAM, OR-PAM relies on optical focusing for high lateral resolution, at the expense of imaging depth, in the same region as pure optical microscopy. Although OR-PAM loses advantage in the depth of imaging it allows, its inherent optical absorption contrast is significantly different from other pure optical microscopy methods.

According to the basic mechanism of the photoacoustic effect, only the region that absorbs illuminated light can generate photoacoustic signals. The ballistic photons can be focused to a much smaller size compared with the ultrasound wavelength in OR-PAM. Therefore, the lateral resolution of OR-PAM is dependent on optics, not ultrasound. Based on classic optics, the FWHM of the focused spot, that is, the lateral resolution, with the optical wavelength λ_o and numerical aperture NA_o, is calculated by:

$$R_o = 0.51\frac{\lambda_o}{NA_o}. \tag{10.15}$$

In this equation, each parameter is related to optics instead of ultrasound. Compared with Equation (10.12), Equation (10.15) has a smaller coefficient of 0.51, because the photoacoustic signal in OR-PAM is proportional to the light intensity, not the amplitude. The axial resolution of OR-PAM is the same as AR-PAM, both relying on the temporal resolution of the acoustic signal in Equation (10.14).

The focal zone of OR-PAM also relies on optical focusing, defined by the axial Rayleigh range of the focus. Similar to lateral resolution in OR-PAM, the intensity of the amplitude is used instead. Thus, the constant in Equation (10.13) should decrease and other parameters should take optical focus into account:

$$Z_o = 1.8\frac{\lambda_o}{NA_o^2}. \tag{10.16}$$

For instance, for λ_o = 532 nm and NA_o = 0.1, the depth of focus is about 100 μm. Definitely, the imaging depth is constricted by the diffusion limit, that is, nearly 1 mm.

The first high-resolution OR-PAM system emerged in 2008 (Maslov et al. 2008). Many more systems based on this idea were proposed in the following few years, making significant progress in both resolution and imaging speed. The basic design of the first working OR-PAM has two right-angle prisms, which sandwiched a thin layer of silicon oil as the optical coupling medium. Illumination light can transmit while ultrasound is reflected by the layer. The optical and acoustic foci need to be aligned confocally to achieve maximum SNR. Figure 10.3 illustrates several other designs of OR-PAM. An upgraded system was built in 2011, as shown in Figure 10.3a. Compared with earlier ones, the system in Figure 10.3(a) significantly improved the sensitivity by 18.4 dB (Hu et al. 2011), which is achieved by taking account of ultrasound mode transform on the liquid-solid interface. As a departure from mechanical scanning, researchers also applied laser scanning in reflection-mode OR-PAM, as shown in Figure 10.3(b) (Xie et al. 2009). The scanning light and stationary ultrasonic transducer made this device capable of integrating with other optical imaging modalities. Besides reflection-mode OR-PAM, several transmission-mode systems have also been developed. The transmission-mode OR-PAM is very suitable for imaging thin biological objects, such as mouse ear, tissue samples, and small animal models. Figure 10.3(c) to (d) demonstrates the imaging result of cells and zebrafish larvae. In addition, owing to much less working distance, the transmission-mode systems have been proposed with better resolution by using high NA_o objective lens (C. Zhang et al. 2010; Ye et al. 2012b). With an immersion objective lens, the OR-PAM can now achieve sub-wavelength lateral resolution (C. Zhang et al. 2010). In order to achieve greater imaging depth, double illumination OR-PAM is demonstrated (Yao et al. 2012b). It illuminates the thin sample from both sides with reflection and transmission modes, and can improve the focal zone up to 260 μm, with penetration depth of 2 mm and lateral resolution of 4 μm.

Imaging speed is a key factor in the application of OR-PAM. Unlike AR-PAM, which requires higher laser energy per pulse to deliver enough light into deeper tissue, the pulse energy used in OR-PAM can be thousands of times less than in AR-PAM. Therefore, the laser repetition rate can be so high that it is no longer a limitation for imaging speed, and the scanning method determines the imaging speed in OR-PAM. Traditional OR-PAM uses mechanical scanning driven by stepper or servo motors, which makes it challenging for real-time in vivo experiments. Much effort has gone into improving imaging speed. A high-speed voice-coil stage has been developd recently, which almost achieves real-time OR-PAM

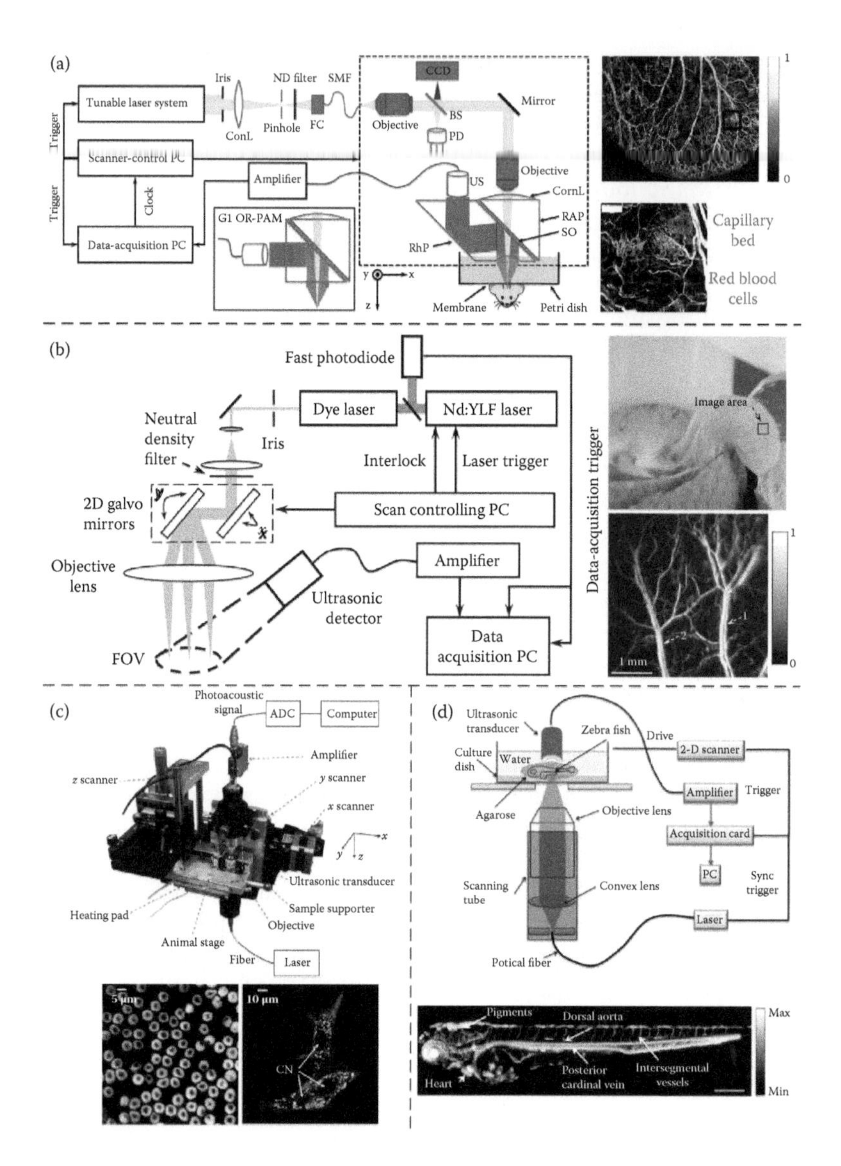

Figure 10.3 **(See color insert.)** OR-PAM implementation and imaging results. (a) One type of reflection-mode system and total hemoglobin concentration of the living mouse ear, with 50 MHz central frequency. Scale bar: 150 μm. (b) Laser scanning reflection-mode system and vasculature in a mouse ear in vivo, with 50 MHz central frequency. Scale bar: 1 mm. (c) Transmission-mode OR-PAM and ex vivo images of red blood cells (left, scale bar: 5 μm) and melanoma cells (right, scale bar: 10 μm), with 40 MHz central frequency. (d) Transmission-mode system and in vivo image of zebra fish larva, with 40 MHz central frequency. Scale bar: 250 μm. (From Hu et al. 2011 (panel a); Xie et al. 2009 (panel b); C. Zhang et al. 2010 (panel c); Ye et al. 2012b (panel d). Reproduced with permission.)

imaging (L. Wang et al. 2011). For instance, the voice-coil scanning system applied in OR-PAM produces a frame rate up to 40 Hz over 1 mm, and dynamic processes can be observed in real time (L. Wang et al. 2011). A single red blood cell flowing in a mouse ear capillary is reported to have been functionally detected in vivo. Although mechanical scanning can conserve the confocal alignment, the imaging speed is still not enough to monitor fast biological processes. In the pure optical microscopy field, optical scanning has been widely used successfully. OR-PAM also borrows this method to achieve faster imaging speed. In optical scanning OR-PAM, the light is focused in the effective field of view (FOV) of the ultrasonic transducer, while the transducer is weakly focused or even flat. Two-dimensional (2D) galvanometers are essential components in traditional laser scanning optical systems. The 2D B-scan scanning rate can be increased up to hundreds to thousands of hertz by galvanometers. One type of galvanometer-based OR-PAM employs raster scanning with unfocused ultrasonic receivers (Jiao et al. 2009, 2010; Xie et al. 2009). It maintains large FOV, generally a few millimeters, during scanning but relatively weak SNR because it uses an unfocused transducer. The other type uses a focused detector to improve SNR by nearly 40 dB (Hajireza et al. 2011; Rao et al. 2011). But because of its static ultrasound detector, the FOV is limited to a scale of hundreds of micrometers. Thus, it is a tradeoff between the FOV and sensitivity. Besides galvanometers, a microlens array for optical illumination and an ultrasound array for photoacoustic detection with cylinder lens has been demonstrated (L. Song et al. 2011). This multifocal system provided three to four times faster mechanical scanning in practice, with 38 dB contrast-to-noise ratio and a millimeter-scale imaging area. But the more focal points that are being scanned in parallel, the more energy is needed. Recently, a water-immersible microelectro-mechanical system (MEMS) OR-PAM (J. Yao et al. 2012a) has been presented. In this setup, both illumination and detection share the same scanning system submerged in the water, with the B-scan rate up to 400 Hz over 3 mm, and the detecting sensitivity and FOV are both maintained. This development presents a novel perspective in studying highly dynamic biological phenomena.

In addition to using higher NA_o objective lens to improve the lateral resolution (C. Zhang et al. 2010; C. Zhang et al. 2012b), shorter illumination wavelength was also applied for imaging specific biological targets. By employing 266-nm wavelength ultraviolet laser, OR-PAM can image label-free DNA and RNA in cell nuclei (D.K. Yao et al. 2010). In this modality, lateral and axial resolutions can reach 0.7 and 28.5 μm, respectively. Because of its good absorption of ultraviolet light in the nuclei, it has a great potential for label-free in vivo cell nuclei noninvasive imaging.

10.3.3 Multimodal PAM system

Two major photoacoustic microscopic imaging methods, AR-PAM and OR-PAM, have been discussed in this chapter. PAM intrinsically combines ultrasound and optics, so it is very suitable to integrate current imaging methods to carry out multimodal imaging approaches. Multimodality imaging can provide comprehensive information, which is essential for biological and clinical research.

The most inherent dual-modality PAM is combined with pulse-echo ultrasound. One is absorption contrast and the other is mechanical contrast (Niederhauser et al. 2005). In most cases, ultrasound imaging provides the structural information and PAM provides functional or molecular imaging. PAM with ultrasound dual-modality imaging has great potential to study tumor physiology and drug/contrast agent delivery (Niederhauser et al. 2005; Harrison et al. 2009; Kong et al. 2009). A combination of photoacoustic and ultrasound systems at high frequencies (>20 MHz) was proposed in 2009 (Harrison et al. 2009), in order to provide a structural context for PAM by ultrasound. In the meantime, it can improve frame rate due to the voice-coil scanning system.

Besides integrating ultrasound, various optical imaging methods, including optical coherence tomography (OCT), confocal microscopy, and fluorescence microscopy, have been combined with OR-PAM. Compared to PAM, OCT provides backscattering information on a structure with comparable lateral resolution and imaging depth (Jiao et al. 2009; L. Li et al. 2009; E.Z. Zhang et al. 2011). The integration of PAM and a confocal microscopy imaging system has also been proposed (H.F. Zhang et al. 2010; Y. Wang et al. 2011). When back-scattered photons of illumination are collected from confocal microscopy, information on the structure is provided (H.F. Zhang et al. 2010). In addition to structural imaging, pure optical microscopy can use fluorescence labeling to achieve other functional information, such as the oxygen partial pressure (Y. Wang et al. 2011), to complement the results of OR-PAM. Besides photons absorbed to generate heat, some are produced to excite autofluorescence in tissue. A dual-modality PAM is employed to image distinctive molecular contrasts provided by the same wavelength (X. Zhang et al. 2010), but with an acoustic transducer and an avalanched photodiode, respectively. This system is useful in the diagnosis of ocular diseases, and it reveals potential relationships among different molecules.

10.4 Contrast agents in PAM

The unique characteristic of PAM is its absorption contrast (see Figure 10.4). That is why although OR-PAM cannot penetrate beyond the diffusion limit, it has still garnered much interest in the field of optical microscopy.

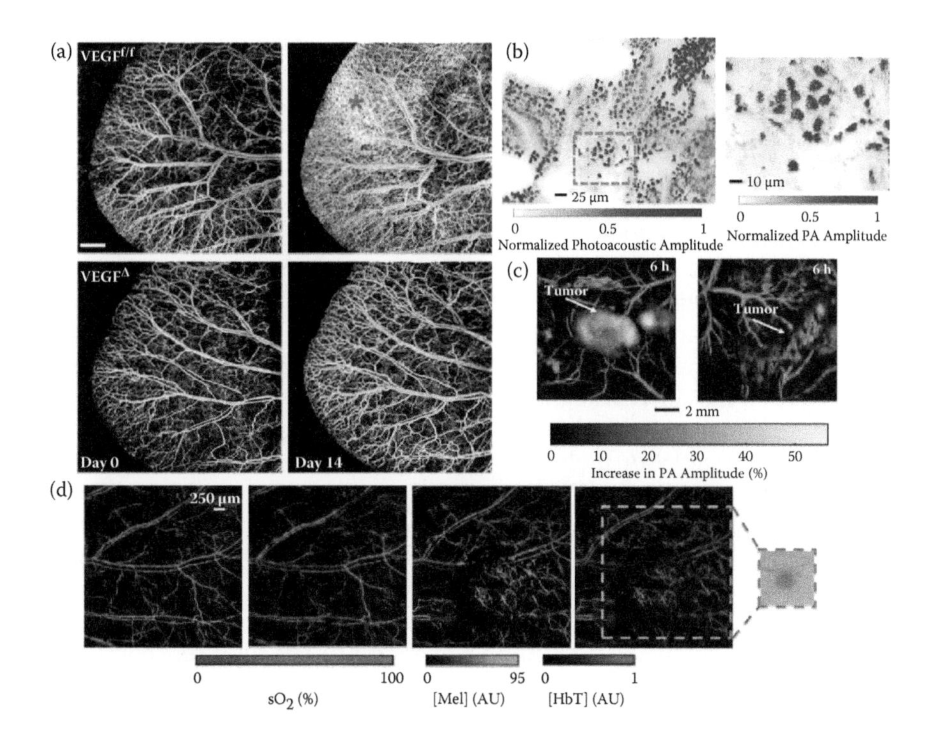

Figure 10.4 **(See color insert.)** (a) In vivo label-free study of neovasculature, scale bar: 500 µm. (b) Label-free cell nuclei images. (c) Noninvasive photoacoustic results of melanomas labeled by gold nanocages. (d) Tyrosinase reporter gene in vivo PAM study. (From Oladipupo et al. 2011 (panel a); Yao et al. 2010 (panel b); Kim et al. 2010 (panel c); Krumholz et al. 2011 (panel d). Reproduced with permission.)

Different molecules have different absorption spectra, from ultraviolet to infrared. Thus, any component with absorption behaviors distinct enough from the surrounding medium can be imaged by PAM at appropriate illumination wavelengths. In the application of PAM, optical absorbers can be classified into two groups: one based on endogenous molecules, and the other on exogenous agents. In the following, these two types of contrast agent and their applications are discussed in detail.

10.4.1 Endogenous contrast agent

Endogenous contrast agents offer PAM great label-free imaging capability. In the ultraviolet region, molecules in cell nuclei (D.K. Yao et al. 2012), especially DNA or RNA (D.K. Yao et al. 2010), have strong absorption (Yao et al. 2010, 2012). Unlike confocal microscopy and multiphoton microscopy, for the first time OR-PAM can provide specific, positive, and high-image contrast of nuclear material using optical microscopy (D.K. Yao et al. 2012). In the study of unstained cell nuclei, confocal microscopy depends on different refractive indices between the nuclei and surrounding medium (Dwyer et al. 2006, Nehal et al. 2000), but this is generally not suitable for in vivo experiments due to the lack of contrast. For multiphoton microscopy (Li et al. 2010; B.G. Wang et al. 2010), nicotinamide adenine dinucleotide (NADH) is generally the biomarker in the study of cell nuclei; however, it introduces negative contrast because NADH associates with mitochondria, which always locate around the nuclei instead of inside them. Compared to the gold-standard hematoxylin and eosin (H&E) staining method, PAM is able to provide in vivo label-free images equivalent to the conventional method.

One of the most widely used endogenous PA contrast agents is in blood, the hemoglobin molecule. There are two forms of hemoglobin, oxygenated hemoglobin (HbO_2) and deoxygenated hemoglobin (HbR), respectively. Whether the hemoglobin molecule combines with oxygen has a significant impact on the optical absorption behavior. Therefore, PAM can reveal a crucial biological parameter quantitatively: the oxygen saturation of hemoglobin (sO_2) by spectroscopic imaging (H.F. Zhang et al. 2007a; Hu et al. 2011; Krumholz et al. 2012). The total concentration of hemoglobin (C_{Hb}) can also be acquired at absorption isometric wavelengths, such as 498 nm, 532 nm, and 568 nm (J. Yao et al. 2011), if the local fluence is known. For most cancer research, hypoxia is a manifest indicator. Therefore, PAM provides an opportunity to study tissue metabolic activity by monitoring sO_2 level. PAM can also be employed to measure blood flow based on Doppler frequency shift (Fang et al. 2007a, 2007b; J. Yao and Wang 2010), or based on thermal diffusion (Sheinfeld and Eyal 2012). Furthermore, depending on the functional parameters C_{Hb} and sO_2 and the fluid-dynamic factor acquired by Doppler imaging, PAM can reveal the metabolic rate of oxygen (MRO_2), which directly indicates the oxygen consumption rate of tissue. This factor is critical to various diseases, especially in oncology (Hanahan and Weinberg 2011), and PAM is the only noninvasive label-free imaging modality for quantification of MRO_2 (Liu et al. 2011; Yao et al. 2011). Because of it's advantage of being robust and label-free, PAM has great potential to be applied in the early diagnosis of diseases associated with metabolism.

Melanin is another marvelous endogenous contrast owing to its superior high optical absorption coefficient (H.F. Zhang et al. 2006; Jiao et al. 2010; Staley et al. 2010; Y. Zhang et al. 2010; W. Song et al. 2012). It is a main constituent in the melanoma, and so it offers a novel way to noninvasively image this kind of malignant tumor in vivo by PAM (H.F. Zhang et al. 2006c; Staley et al. 2010; Y. Zhang et al. 2010). Besides melanoma, the retinal pigment epithelium (RPE) layer in eyes also contains abundant melanin. PAM has the potential to aid in both fundamental investigation and clinical diagnosis of eye diseases (Jiao et al. 2010; W. Song et al. 2012). Melanin, cytochrome, and myoglobin combined were used in PAM imaging of myocardial sheet architecture in unfixed, unstained, and unsliced tissue (C. Zhang et al. 2012a).

In the region of near-infrared to infrared spectra, water becomes a major absorber (Xu et al. 2010). In addition, lipid (Yakovlev et al. 2010; H.W. Wang et al. 2011) and glucose (MacKenzie et al. 1999; Kottmann et al. 2012) also present distinct optical absorption at certain wavelengths, and they are under investigation for potential use in PAM. By concentrations and locations of water, lipid, and glucose, PAM can image these endogenous contrasts with high resolution. It can reveal more comprehensive information for diagnosing corresponding diseases in situ (Yakovlev et al. 2010; Wang et al. 2011; Kottmann et al. 2012).

10.4.2 Exogenous contrast

Although a few endogenous contrast agents, such as blood and melanin, have been successfully used in PAM, there is a much wider selection of exogenous contrast agents. Based on different composition, exogenous contrast agents include dyes (X. Wang et al. 2004; J. Yao et al. 2009; Erpelding et al. 2010), biological proteins (Razansky et al. 2009; Filonov et al. 2012), and nanoparticles (Yang et al. 2007; Galanzha et al. 2009; Chen et al. 2010; Kim et al. 2010; Cai et al. 2011; Cobley et al. 2011; Moon et al. 2011; Y. Zhang et al. 2011; Avti et al. 2012; Cho et al. 2013). The absorption spectra of exogenous agents cover almost the entire optical spectrum. For instance, various agents with high near-infrared absorption, such as gold nanorods, are very critical for deep tissue PAM imaging. With the aid of exogenous contrast agents, PAM can now image tissue that does not have enough natural optical absorption contrast with the environmental medium, such as lymphatic vessels (K.H. Song et al. 2008; Kim et al. 2009; Manojit et al. 2009; Erpelding et al. 2010; Cai et al. 2011). Moreover, due to the enhanced permeability and retention (EPR) effect at the tumor site, various exogenous contrast agents, particularly nanoparticles, have been used for tumor imaging by PAM (Kim et al. 2010).

Another key role of exogenous contrast agents in PAM is for molecular imaging. Typically, the contrast agents with high optical absorption conjugate with probing molecules, such as antibodies. Once the agents accumulate in the location through biochemical conjugation, PAM can provide molecular imaging with high spatial resolution, even in the optical diffusion region. Up to now, many exogenous contrast agents have been designed to target tumors, like the result shown in Figure 10.4(c). Due to the enhanced optical absorption by targeted contrast agents, the tumor can be seen by PAM in its early stage with high resolution, which creates a new method for in vivo tumor study besides the traditional fluorescence imaging.

One of the major concerns of the exogenous contrast agents is biosafety. Although novel particles, including biodegradable nanoparticles (Kohl et al. 2011; Zha et al. 2013), are made and implemented for photoacoustic effect, clinical approval of contrast agents is still very limited. Recently, another promising method is to apply genetic coding techniques to let tissue generate optical absorbers, such as inducing tyrosinase to produce pigment eumelanin (Krumholz et al. 2011; Paproski et al. 2011). Unlike the aforementioned contrast agents, these genes can be expressed during growth and have less toxicity. They serve as endogenous contrast agents with better specificity, instead of fluorescent dyes and proteins. The main obstacle in this work is transferring genes to encode as the contrast.

In addition to targeting tumors or other diseases, exogenous contrast agents can be treated as biosensors for a biochemical environment in tissue. Aided by a pH-sensitive dye, PAM quantified this parameter with less than 2% error at a depth to 2 mm (Chatni et al. 2011). More accurate measurement required the solution of photon attenuation, which is also dependent on spectra in tissue. Another parameter, partial oxygen pressure (pO_2), is also measured with the help of an oxygen-sensitive dye for PAM (Ray et al. 2012). It eliminates the influences of optical properties and light fluence at depth, at the expense of sensitivity because of low fluorescence quantum yield. Obviously, PAM provides diverse possibilities for biomedical applications by exogenous contrast, but much work is still required.

10.5 Discussion and conclusions

Since the first successful dark-field AR-PAM was presented in 2005, significant progress has been made in PAM in both theory and instrument development. PAM has now become one of the fastest-developing biomedical imaging methods.

Several challenges still exist for PAM. The axial resolution of PAM is determined by the frequency bandwidth of the acoustic detection system. However, traditional ultrasonic detection based on the piezoelectric effect has a limited bandwidth, which is not enough for PAM to image multiscale structures. Recently, the emergence of novel ultrasound detection techniques has a great potential to significantly improve the bandwidth for PAM. The pure optical detection methods have seen significant progress in wideband ultrasonic detection. In 2011, a PAM system with a pure optical wideband microring resonator was developed (Xie et al. 2011). Having a superior high Q value, the axial resolution was improved to be 8 μm, close to the lateral resolution of 5 μm. At the same time, the sensitivity was also enhanced greatly.

To reduce the cost of PAM, a low-cost power-modulated continuous-wave (CW) laser for PAM is also under investigation (Maslov and Wang 2008). Even with the lock-in detection method, PAM based on a CW laser has not obtained comparable sensitivity to pulse laser systems. In addition to the sensitivity, it is also challenging to improve the axial resolution for CW-based PAM.

Besides instrument development, novel optical absorption mechanisms are also under investigation, including multiphoton absorption (Yamaoka et al. 2007; Yamaoka and Takamatsu 2009), stimulated Raman scattering (Yakovlev et al. 2010), polarization-sensitive absorption (Hu et al. 2012) and Förster resonance energy transfer (Y. Wang and Wang 2012). Most of these studies relied on ballistic photons, so they generally use OR-PAM. In the future, as new optical-absorption-related interactions are discovered, PAM can find more important roles in basic biological science.

PAM is also moving forward into clinical use. Due to its higher imaging depth than traditional optical microscopy, AR-PAM has great advantages in imaging subcutaneous diseases, such as skin tumors or burn wounds (H.F. Zhang et al. 2006b). For OR-PAM, its high sensitivity to blood circulation brings great opportunities to study microcirculation systems in shallow or optical clearing tissue. For instance, a human finger cuticle has been imaged in vivo with superior contrast (Hu and Wang 2013). Another very important application is ocular imaging, which has been successfully demonstrated in animal studies (Jiao et al. 2009, 2010; de la Zerda et al. 2010). PAM systems are becoming more compact and stable, and have great potential to be a powerful clinical imaging instrument.

In conclusion, PAM is a revolutionary modality for optical microscopy. PAM has two unique characteristics: breaking through the optical diffusion limit, and sensitive optical absorption contrast. Many biological molecules have unique optical absorption spectra, which all can be potential imaging targets for PAM. Even for those molecules that have no distinctive absorption, exogenous contrast agents are essential to enhance imaging contrast. Moreover, exogenous contrast agents play important roles in molecular PAM. Owing to its unique characteristics and with the aid of exogenous contrast agents, PAM has successfully imaged tissues, including in such applications as oncology, angiogenesis, neurology, and the circulatory system. Moreover, PAM can also feasibly combine with other current imaging techniques, such as ultrasound, OCT, DOT, and other optical microscopes. Overall, both fundamental life science studies and clinical applications will benefit from the development of PAM.

Acknowledgments

This work was supported by National Natural Science of China (Grant No. 61078073), and the National Basic Research Program of China (973 Program, 2011CB707502).

References

Abramowitz, M. (1993). *Fluorescence Microscopy: The Essentials*. New York: Olympus America Inc.

Avti, P. K. et al. (2012). "Detection, mapping, and quantification of single walled carbon nanotubes in histological specimens with photoacoustic microscopy." *Plos One* **7**(4): e35064.

Bell, A. G. (1880). "On the production of sound by light." *Am. J. Sci.* **20**: 305–324.

Bradbury, S. and B. Bracegirdle (1998). *Introduction to Light Microscopy*. Oxford, UK: BIOS Scientific Publishers Ltd.

Cai, X. et al. (2011). "In vivo quantitative evaluation of the transport kinetics of gold nanocages in a lymphatic system by noninvasive photoacoustic tomography." *ACS Nano* **5**(12): 9658–9667.

Chatni, M. R. et al. (2011). "Functional photoacoustic microscopy of pH." *J. Biomed. Opt.* **16**(10): 100503.

Chen, J. et al. (2010). "Gold nanocages: a novel class of multifunctional nanomaterials for theranostic applications." *Adv. Funct. Mater.* **20**(21): 3684–3694.

Cho, E. C. et al. (2013). "Quantitative analysis of the fate of gold nanocages in vitro and in vivo after uptake by U87-MG tumor cells." *Angew. Chem.* **125**(4): 1190–1193.

Cobley, C. M. et al. (2011). "Gold nanostructures: a class of multifunctional materials for biomedical applications." *Chem. Soc. Rev.* **40**(1): 44–56.

de la Zerda, A. et al. (2010). "Photoacoustic ocular imaging." *Opt. Lett.* **35**(3): 270–272.

Dwyer, P. J. et al. (2006). "Confocal reflectance theta line scanning microscope for imaging human skin in vivo." *Opt. Lett.* **31**(7): 942–944.

Erpelding, T. N. et al. (2010). "Sentinel lymph nodes in the rat: noninvasive photoacoustic and US imaging with a clinical US system1." *Radiology* **256**(1): 102–110.

Fang, H. et al. (2007a). "Photoacoustic Doppler effect from flowing small light-absorbing particles." *Phys. Rev. Lett.* **99**(18): 184501.

Fang, H., et al. (2007b). "Photoacoustic Doppler flow measurement in optically scattering media." *Appl. Phys. Lett.* **91**(26): 264103.

Fernández-Suárez, M. and A. Y. Ting (2008). "Fluorescent probes for super-resolution imaging in living cells." *Nat. Rev. Mol. Cell Biol.* **9**: 929–943.

Filonov, G. S. et al. (2012). "Deep tissue photoacoustic tomography of a genetically encoded near infrared fluorescent probe." *Angew. Chem.* **124**(6): 1477–1480.

Galanzha, E. I. et al. (2009). "In vivo magnetic enrichment and multiplex photoacoustic detection of circulating tumour cells." *Nat. Nanotechnol.* **4**(12): 855–860.

Guittet, C. et al. (1999). "In vivo high-frequency ultrasonic characterization of human dermis." *IEEE Trans. Biomed. Eng.* **46**(6): 740–746.

Hajireza, P. et al. (2011). "Label-free in vivo fiber-based optical-resolution photoacoustic microscopy." *Opt. Lett.* **36**(20): 4107–4109.
Hale, G. M. and M. R. Querry (1973). "Optical-constants of water in 200 nm to 200 mm wavelength." *Appl. Opt.* **12**: 555–563.
Hanahan, D. and R. A. Weinberg (2011). "Hallmarks of cancer: the next generation." *Cell* **144**(5): 646–674.
Harrison, T. et al. (2009). "Combined photoacoustic and ultrasound biomicroscopy." *Opt. Express* **17**(24): 22041–22046.
Hell, S. W. (2009). "Microscopy and its focal switch." *Nat. Methods* **6**(1): 24–32.
Herma, B. and J. J. Lemasters (1993). *Optical Microscopy: Emerging Methods and Applications*. New York: Academic Press.
Herman, B. (1998). *Fluorescence Microscopy*. Oxford, UK: BIOS Scientific Publishers Ltd.
Hu, S. et al. (2011). "Second-generation optical-resolution photoacoustic microscopy with improved sensitivity and speed." *Opt. Lett.* **36**(7): 1134–1136.
Hu, S. et al. (2012). *Dichroism Optical-Resolution Photoacoustic Microscopy*. Photons Plus Ultrasound: Imaging and Sensing, International Society for Optics and Photonics.
Hu, S. and L. V. Wang (2013). "Optical-resolution photoacoustic microscopy: auscultation of biological systems at the cellular level." *Biophys. J.* **105**(4): 841–847.
Huang, B. et al. (2010). "Breaking the diffraction barrier: super-resolution imaging of cells." *Cell* **143** (7): 1047–1058.
Inoué, S. and K. R. Spring (1997). *Video Microscopy: The Fundamentals*. New York: Plenum Press.
Jiao, S. et al. (2010). "Photoacoustic ophthalmoscopy for in vivo retinal imaging." *Opt. Express* **18**(4): 3967–3972.
Jiao, S. et al. (2009). "Simultaneous multimodal imaging with integrated photoacoustic microscopy and optical coherence tomography." *Opt. Lett.* **34**(19): 2961–2963.
Kim, C. et al. (2010). "In vivo molecular photoacoustic tomography of melanomas targeted by bioconjugated gold nanocages." *ACS Nano* **4**(8): 4559–4564.
Kim, J. W. et al. (2009). "Golden carbon nanotubes as multimodal photoacoustic and photothermal high-contrast molecular agents." *Nat. Nanotechnol.* **4**(10): 688–694.
Kohl, Y. et al. (2011). "Preparation and biological evaluation of multifunctional PLGA-nanoparticles designed for photoacoustic imaging." *Nanomed. Nanotechnol. Biology Med.* **7**(2): 228–237.
Kong, F. et al. (2009). "High-resolution photoacoustic imaging with focused laser and ultrasonic beams." *Appl. Phys. Lett.* **94**(3): 3.
Kottmann, J. et al. (2012). "Glucose sensing in human epidermis using mid-infrared photoacoustic detection." *Biomed. Opt. Express* **3**(4): 667–680.
Krumholz, A. et al. (2011). "Photoacoustic microscopy of tyrosinase reporter gene *in vivo*." *J. Biomed. Opt.* **16**(8): 080503.
Krumholz, A. et al. (2012). "Functional photoacoustic microscopy of diabetic vasculature." *J. Biomed. Opt.* **17**(6): 060502.
Li, C. et al. (2010). "Multiphoton microscopy of live tissues with ultraviolet autofluorescence." *IEEE J. Sel. Top. Quant.* **16**(3): 516–523.
Li, C. and L. V. Wang (2009a). "Photoacoustic tomography and sensing in biomedicine." *Phys. Med. Biol.* **54**(19): R59–R97.

Li, C. and L. V. Wang (2009b). "Photoacoustic tomography of the mouse cerebral cortex with a high-numerical-aperture-based virtual point detector." *J. Biomed. Opt.* **14**(2): 024047.

Li, C. H. et al. (2008). "Image distortion in thermoacoustic tomography caused by microwave diffraction." *Phys. Rev. E.* **77**: 031923.

Li, L. et al. (2009). "Three-dimensional combined photoacoustic and optical coherence microscopy for in vivo microcirculation studies." *Opt. Express* **17**(19): 16450–16455.

Li, M. L. et al. (2006). "Improved in vivo photoacoustic microscopy based on a virtual-detector concept." *Opt. Lett.* **31**(4): 474–476.

Liao, C. K. et al. (2004). "Optoacoustic imaging with synthetic aperture focusing and coherence weighting." *Opt. Lett.* **29**(21): 2506–2508.

Liu, T. et al. (2011). "Combined photoacoustic microscopy and optical coherence tomography can measure metabolic rate of oxygen." *Biomed. Opt. Express* **2**(5): 1359–1365.

MacKenzie, H. A. et al. (1999). "Advances in photoacoustic noninvasive glucose testing." *Clin. Chem.* **45**(9): 1587–1595.

Manojit, P. et al. (2009). "In vivo carbon nanotube-enhanced non-invasive photoacoustic mapping of the sentinel lymph node." *Phys. Med. Biol.* **54**(11): 3291.

Maslov, K. et al. (2005). "In vivo dark-field reflection-mode photoacoustic microscopy." *Opt. Lett.* **30**(6): 625–627.

Maslov, K. and L. V. Wang (2008). "Photoacoustic imaging of biological tissue with intensity-modulated continuous-wave laser." *J. Biomed. Opt.* **13**(2): 024006.

Maslov, K. et al. (2008). "Optical-resolution photoacoustic microscopy for in vivo imaging of single capillaries." *Opt. Lett.* **33**(9): 929–931.

Moon, G. D. et al. (2011). "A new theranostic system based on gold nanocages and phase-change materials with unique features for photoacoustic imaging and controlled release." *J. Am. Chem. Soc.* **133**(13): 4762–4765.

Morse, P. M. and K. U. Ingard (1986). *Theoretical Acoustics*. Princeton, NJ: Princeton University Press.

Nehal, K. S. et al. (2008). *Skin Imaging With Reflectance Confocal Microscopy*. Seminars in Cutaneous Medicine and Surgery. Philadelphia: W.B. Saunders.

Niederhauser, J. J. et al. (2005). "Combined ultrasound and optoacoustic system for real-time high-contrast vascular imaging in vivo." *IEEE Trans. Med. Imaging* **24**(4): 436–440.

Ntziachristos, V. (2010). "Going deeper than microscopy: the optical imaging frontier in biology." *Nat. Methods* **7**(8): 603–614.

Oladipupo, S. et al. (2011). "VEGF is essential for hypoxia-inducible factor-mediated neovascularization but dispensable for endothelial sprouting." *Proc. Natl. Acad. Sci. USA* **108**(32): 13264–13269.

Oraevsky, A. A. and A. A. Karabutov (2003). "Optoacoustic tomography." In *Biomedical Photonics Handbook,* edited by T. Vo-Dinh. Boca Raton, FL: CRC Press.

Paproski, R. et al. (2011). "Tyrosinase as a dual reporter gene for both photoacoustic and magnetic resonance imaging." *Biomed. Opt. Express* **2**(4): 771–780.

Pawley, J. B. (1995). *Handbook of Biological Confocal Microscopy.* New York: Plenum Press.

Pluta, M. (1989). *Advanced Light Microscopy*. New York: Elsevier.

Rao, B. et al. (2011). "Real-time four-dimensional optical-resolution photoacoustic microscopy with Au nanoparticle-assisted subdiffraction-limit resolution." *Opt. Lett.* **36**(7): 1137–1139.

Ray, A. et al. (2012). "Lifetime-based photoacoustic oxygen sensing *in vivo*." *J. Biomed. Opt.* **17**(5): 057004.

Razansky, D. et al. (2009). "Multispectral opto-acoustic tomography of deep-seated fluorescent proteins in vivo." *Nat. Photonics* **3**(7): 412–417.

Sheinfeld, A. and A. Eyal (2012). "Photoacoustic thermal diffusion flowmetry." *Biomed. Opt. Express* **3**(4): 800–813.

Sheppard, C. J. R. and D. M. Shotton (1997). *Confocal Laser Scanning Microscopy*. Oxford, UK: BIOS Scientific Publishers Ltd.

Slayter, E. M. and H. S. Slayter (1992). *Light and Electron Microscopy*. Cambridge, UK: Cambridge University Press.

Song, K. H. et al. (2008). "Noninvasive photoacoustic identification of sentinel lymph nodes containing methylene blue in vivo in a rat model." *J. Biomed. Opt.* **13**(5): 054033.

Song, K. H. and L. V. Wang (2007). "Deep reflection-mode photoacoustic imaging of biological tissue." *J. Biomed. Opt.* **12**(6): 060503.

Song, L., et al. (2011). "Multifocal optical-resolution photoacoustic microscopy in vivo." *Opt. Lett.* **36**(7): 1236–1238.

Song, W. et al. (2012). "Integrating photoacoustic ophthalmoscopy with scanning laser ophthalmoscopy, optical coherence tomography, and fluorescein angiography for a multimodal retinal imaging platform." *J. Biomed. Opt.* **17**(6): 061206.

Staley, J. et al. (2010). "Growth of melanoma brain tumors monitored by photoacoustic microscopy." *J. Biomed. Opt.* **15**(4): 040510.

Tsai, C. L. et al. (2001). "Near-infrared absorption property of biological soft tissue constituents." *J. Med. Biol. Eng.* **21**: 7–14.

Wang, B. G. et al. (2010). "Two-photon microscopy of deep intravital tissues and its merits in clinical research." *J. Microscopy* **238**(1): 1–20.

Wang, H. W. et al. (2011). "Label-free bond-selective imaging by listening to vibrationally excited molecules." *Phys. Rev. Lett.* **106**(23): 238106.

Wang, L. et al. (2012). "Video-rate functional photoacoustic microscopy at depths." *J. Biomed. Opt.* **17**(10): 106007.

Wang, L. et al. (2011). "Fast voice-coil scanning optical-resolution photoacoustic microscopy." *Opt. Lett.* **36**(2): 139–141.

Wang, L. V. (2008). "Prospects of photoacoustic tomography." *Med. Phys.* **35**(12): 5758–5767.

Wang, L. V. (2009). "Multiscale photoacoustic microscopy and computed tomography." *Nat. Photonics* **3**(9): 503–509.

Wang, L. V. and S. Hu (2012). "Photoacoustic tomography: in vivo imaging from organelles to organs." *Science* **335**(6075): 1458–1462.

Wang, L. V. and H. I. Wu (2007). *Biomedical Optics: Principles and Imaging*. New York: Wiley Interscience.

Wang, X. et al. (2004). "Noninvasive photoacoustic angiography of animal brains in vivo with near-infrared light and an optical contrast agent." *Opt. Lett.* **29**(7): 730–732.

Wang, Y. et al. (2011). "*In vivo* integrated photoacoustic and confocal microscopy of hemoglobin oxygen saturation and oxygen partial pressure." *Opt. Lett.* **36**(7): 1029–1031.

Wang, Y. and L. V. Wang (2012). "Förster resonance energy transfer photoacoustic microscopy." *J. Biomed. Opt.* **17**(8): 086007.
Xie, Z. et al. (2011). "Pure optical photoacoustic microscopy." *Opt. Express* **19**(10): 9027–9034.
Xie, Z. et al. (2009). "Laser-scanning optical-resolution photoacoustic microscopy." *Opt. Lett.* **34**(12): 1771–1773.
Xu, M. and L. V. Wang (2006). "Photoacoustic imaging in biomedicine." *Rev. Sci. Instrum.* **77**(4): 041101–041122.
Xu, Z. et al. (2010). "Photoacoustic tomography of water in phantoms and tissue." *J. Biomed. Opt.* **15**(3): 036019.
Yakovlev, V. V. et al. (2010). "Stimulated Raman photoacoustic imaging." *Proc. Natl. Acad. Sci. USA* **107**(47): 20335–20339.
Yamaoka, Y. et al. (2007). "Improvement of depth resolution on photoacoustic imaging using multiphoton absorption." Presented at European Conference on Biomedical Optics, International Society for Optics and Photonics.
Yamaoka, Y. and T. Takamatsu (2009). "Enhancement of multiphoton excitation-induced photoacoustic signals by using gold nanoparticles surrounded by fluorescent dyes." Presented at SPIE BiOS: Biomedical Optics, International Society for Optics and Photonics.
Yang, X. et al. (2007). "Photoacoustic tomography of a rat cerebral cortex in vivo with Au nanocages as an optical contrast agent." *Nano Lett.* **7**(12): 3798–3802.
Yao, D. K. et al. (2012). "Optimal ultraviolet wavelength for in vivo photoacoustic imaging of cell nuclei." *J. Biomed. Opt.* **17**(5): 056004.
Yao, D. K. et al. (2010). "In vivo label-free photoacoustic microscopy of cell nuclei by excitation of DNA and RNA." *Opt. Lett.* **35**(24): 4139–4141.
Yao, J. et al. (2012a). "Wide-field fast-scanning photoacoustic microscopy based on a water-immersible MEMS scanning mirror." *J. Biomed. Opt.* **17**(8): 080505.
Yao, J. et al. (2009). "Evans blue dye-enhanced capillary-resolution photoacoustic microscopy in vivo." *J. Biomed. Opt.* **14**(5): 054049.
Yao, J. et al. (2012b). "Double-illumination photoacoustic microscopy." *Opt. Lett.* **37**(4): 659–661.
Yao, J. et al. (2011). "Label-free oxygen-metabolic photoacoustic microscopy *in vivo*." *J. Biomed. Opt.* **16**(7): 076003.
Yao, J. and L. V. Wang (2010). "Transverse flow imaging based on photoacoustic Doppler bandwidth broadening." *J. Biomed. Opt.* **15**(2): 021304.
Yao, J. and L. V. Wang (2012). "Photoacoustic microscopy." *Laser Photon. Rev.* **7**(5): 758–778.
Ye, S. et al. (2012a). "Studying murine hindlimb ischemia by photoacoustic microscopy." *Chin. Opt. Lett.* **10**(12): 121701.
Ye, S. et al. (2012b). "Label-free imaging of zebrafish larvae in vivo by photoacoustic microscopy." *Biomed. Opt. Express* **3**(2): 360–365.
Zha, Z. et al. (2013). "Biocompatible polypyrrole nanoparticles as a novel organic photoacoustic contrast agent for deep tissue imaging." *Nanoscale* **5**(10): 4462–4467.
Zhang, C., et al. (2012a). "Label-free photoacoustic microscopy of myocardial sheet architecture." *J. Biomed. Opt.* **17**(6): 060506.
Zhang, C. et al. (2012b). "Reflection-mode submicron-resolution in vivo photoacoustic microscopy." *J. Biomed. Opt.* **17**(2): 020501–020504.
Zhang, C. et al. (2010). "Subwavelength-resolution label-free photoacoustic microscopy of optical absorption in vivo." *Opt. Lett.* **35**(19): 3195–3197.

Zhang, E. Z. et al. (2011). "Multimodal photoacoustic and optical coherence tomography scanner using an all optical detection scheme for 3D morphological skin imaging." *Biomed. Opt. Express* **2**(8): 2202–2215.
Zhang, H. F. et al. (2006a). "In vivo volumetric imaging of subcutaneous microvasculature by photoacoustic microscopy." *Opt. Express* **14**(20): 9317–9323.
Zhang, H. F. et al. (2007a). "Imaging of hemoglobin oxygen saturation variations in single vessels in vivo using photoacoustic microscopy." *Appl. Phys. Lett.* **90**(5): 053901–053903.
Zhang, H. F. et al. (2006b). "Imaging acute thermal burns by photoacoustic microscopy." *J. Biomed. Opt.* **11**(5): 054033.
Zhang, H. F. et al. (2006c). "Functional photoacoustic microscopy for high-resolution and noninvasive in vivo imaging." *Nat. Biotechnol.* **24**(7): 848–851.
Zhang, H. F., et al. (2007b). "In vivo imaging of subcutaneous structures using functional photoacoustic microscopy." *Nat. Protoc.* **2**(4): 797–804.
Zhang, H. F. et al. (2010). "Collecting back-reflected photons in photoacoustic microscopy." *Opt. Express* **18**(2): 1278–1282.
Zhang, X. et al. (2010). "Simultaneous dual molecular contrasts provided by the absorbed photons in photoacoustic microscopy." *Opt. Lett.* **35**(23): 4018–4020.
Zhang, Y. et al. (2010). "Chronic label-free volumetric photoacoustic microscopy of melanoma cells in three-dimensional porous scaffolds." *Biomaterials* **31**(33): 8651–8658.
Zhang, Y. et al. (2011). "Noninvasive photoacoustic microscopy of living cells in two and three dimensions through enhancement by a metabolite dye." *Angew. Chem.* **123**(32): 7497–7501.

Problems

1. Compared with other traditional pure optical microscopy methods, what are the major advantages of AR-PAM?
2. According to Equation 10.10 in this chapter, derive the photoacoustic pressure as a function of time observed outside a uniformly spherical absorber excited by a temporal delta laser pulse.
3. Which parameters primarily affect the intensity of a photoacoustic signal?
4. What are the critical factors to determine the spatial resolution and penetration depth in AR-PAM and OR-PAM, respectively?
5. Describe the role the laser pulse width and ultrasound bandwidth play in PAM.
6. What is the difference between AR-PAM and OR-PAM?
7. List three endogenous contrast agents and three exogenous contrast agents.
8. List three limitations of PAM.
9. Explain how to perform functional PAM. Give an example.
10. Design a molecular probe for imaging a specific biomarker on tumor cells by OR-PAM.

Index

Q

R

S